AF361021

Advances in Modelling of Rainfall Fields

Advances in Modelling of Rainfall Fields

Editors

Davide Luciano De Luca
Andrea Petroselli

MDPI • Basel • Beijing • Wuhan • Barcelona • Belgrade • Manchester • Tokyo • Cluj • Tianjin

Editors
Davide Luciano De Luca   Andrea Petroselli
University of Calabria   Tuscia University
Italy   Italy

Editorial Office
MDPI
St. Alban-Anlage 66
4052 Basel, Switzerland

This is a reprint of articles from the Special Issue published online in the open access journal *Hydrology* (ISSN 2306-5338) (available at: https://www.mdpi.com/journal/hydrology/special_issues/MR).

For citation purposes, cite each article independently as indicated on the article page online and as indicated below:

LastName, A.A.; LastName, B.B.; LastName, C.C. Article Title. *Journal Name* **Year**, *Volume Number*, Page Range.

ISBN 978-3-0365-5747-2 (Hbk)
ISBN 978-3-0365-5748-9 (PDF)

Contents

About the Editors

Davide Luciano De Luca

Davide Luciano De Luca is an Assistant Professor in Hydrology and Hydraulics Structures c/o University of Calabria (Italy), Department of Informatics, Modeling, Electronics and System Engineering, from 28/12/2012 to date. His research activities can be divided into three main areas: 1) rainfall field modeling; 2) analysis of hydrological phenomena at slope scale; 3) uncertainty evaluation in rainfall–runoff modeling. His scientific production, updated in September 2022, includes: a) 1 book; b) 35 papers in journals which are indexed on Scopus and/or WoS databases; c) 46 other papers and abstracts.

Andrea Petroselli

Andrea Petroselli is Associate Professor at Tuscia University (Italy), Department of Economics, Engineering, Society, Business Organization. He is an expert in modeling and monitoring hydrological processes. Recent research topics span from infiltration modeling to rainfall–runoff modeling. He is a member of GISTAR - GIS Terrain Analysis Research Group (www.gistar.org), a web portal for researchers and professionals involved in the investigation, development, and application of GIS-based terrain analysis tools for hydrologic and geomorphic models, and a member of MechHydroLab - Mechanical Engineering for Hydrology and Water Science (www.mechydrolab. org/), a multidisciplinary laboratory composed of mechanical engineers, hydrologists, and water scientists with the goal of combining mechanical engineering technologies and hydrological sciences toward the development of novel experimental systems for advanced environmental monitoring.

hydrology

MDPI

Editorial

Advances in Modelling of Rainfall Fields

Davide Luciano De Luca [1,*] **and Andrea Petroselli** [2,*]

1   Department of Informatics, Modelling, Electronics and System Engineering, University of Calabria, Arcavacata, 87036 Rende, Italy
2   Department of Economics, Engineering, Society and Business Organization (DEIM), Tuscia University, 01100 Viterbo, Italy
*   Correspondence: davide.deluca@unical.it (D.L.D.L.); petro@unitus.it (A.P.)

Citation: De Luca, D.L.; Petroselli, A. Advances in Modelling of Rainfall Fields. *Hydrology* **2022**, *9*, 142. https://doi.org/10.3390/hydrology9080142

Received: 25 July 2022
Accepted: 8 August 2022
Published: 10 August 2022

Rainfall is the main input for all hydrological models, such as rainfall–runoff models and the forecasting of landslides triggered by precipitation, with its comprehension being clearly essential for effective water resource management as well. The need to improve the modelling of rainfall fields constitutes a key aspect both for efficiently realizing early warning systems and for carrying out analyses of future scenarios related to occurrences and magnitudes for all induced phenomena.

The aim of this Special Issue was to provide a collection of innovative contributions for rainfall modelling, focusing on hydrological scales and a context of climate changes. The first group of papers regarded the study of global precipitation products and their downscaled versions [1], the estimation of peak discharges in rainfall–runoff modeling under different rainfall depth–duration–frequency formulations [2], stormwater infiltration practices in rapidly urbanizing cities with the aim of designing resilient urban environments [3], and a novel temporal stochastic rainfall simulator [4] aiming to generate long and high-resolution rainfall time series, with the advantage of being strongly user friendly and parsimonious in terms of employed input parameters. Moreover, other works focused on determining the quantities of runoff by knowing the amount of rainfall in order to calculate the required quantities of water storage in reservoirs and to determine the likelihood of flooding [5], some analyzed intrastorm pattern recognition through fuzzy clustering [6], and others investigated the use and combination of pluviograph and daily records to assess rain behavior in urban areas, selecting a suitable method that would provide the best results of IDF relationships [7]. Finally, a sensitivity analysis of the rainfall–runoff modeling parameters in data-scarce urban catchment areas was performed aiming to improve the rainfall–runoff model calibration process [8], satellite-based rainfall estimations were compared with ground data [9], machine learning and process-based models for rainfall–runoff simulations were applied [10], and deep convective systems associated with extreme rainfall storms were examined in tropical regions [11].

We believe that the contribution from the latest research outcomes presented in this Special Issue can shed novel insights on the comprehension of the hydrological cycle and all the phenomena that are a direct consequence of rainfall.

Moreover, all these proposed papers can clearly constitute a valid base of knowledge for improving specific key aspects of rainfall modelling, mainly concerning climate change and how it induces modifications in properties such as magnitude, frequency, duration, and the spatial extension of different types of rainfall fields. The goal should also consider providing useful tools to practitioners for quantifying important design metrics in transient hydrological contexts (quantiles of assigned frequency, hazard functions, intensity–duration–frequency curves, etc.).

Author Contributions: Writing—original draft preparation, D.L.D.L. and A.P.; writing—review and editing, D.L.D.L. and A.P. All authors have read and agreed to the published version of the manuscript.

Funding: This research received no external funding.

Conflicts of Interest: The authors declare no conflict of interest.

References

1. Tadesse, K.E.; Melesse, A.M.; Abebe, A.; Lakew, H.B.; Paron, P. Evaluation of Global Precipitation Products over Wabi Shebelle River Basin, Ethiopia. *Hydrology* **2022**, *9*, 66. [CrossRef]
2. Gioia, A.; Lioi, B.; Totaro, V.; Molfetta, M.G.; Apollonio, C.; Bisantino, T.; Iacobellis, V. Estimation of Peak Discharges under Different Rainfall Depth–Duration–Frequency Formulations. *Hydrology* **2021**, *8*, 150. [CrossRef]
3. Bastia, J.; Mishra, B.K.; Kumar, P. Integrative Assessment of Stormwater Infiltration Practices in Rapidly Urbanizing Cities: A Case of Lucknow City, India. *Hydrology* **2021**, *8*, 93. [CrossRef]
4. De Luca, D.L.; Petroselli, A. STORAGE (STOchastic RAinfall GEnerator): A User-Friendly Software for Generating Long and High-Resolution Rainfall Time Series. *Hydrology* **2021**, *8*, 76. [CrossRef]
5. Hamdan, A.N.A.; Almuktar, S.; Scholz, M. Rainfall-Runoff Modeling Using the HEC-HMS Model for the Al-Adhaim River Catchment, Northern Iraq. *Hydrology* **2021**, *8*, 58. [CrossRef]
6. Vantas, K.; Sidiropoulos, E. Intra-Storm Pattern Recognition through Fuzzy Clustering. *Hydrology* **2021**, *8*, 57. [CrossRef]
7. Gámez-Balmaceda, E.; López-Ramos, A.; Martínez-Acosta, L.; Medrano-Barboza, J.P.; Remolina López, J.F.; Seingier, G.; Daesslé, L.W.; López-Lambraño, A.A. Rainfall Intensity-Duration-Frequency Relationship. Case Study: Depth-Duration Ratio in a Semi-Arid Zone in Mexico. *Hydrology* **2020**, *7*, 78. [CrossRef]
8. Ballinas-González, H.A.; Alcocer-Yamanaka, V.H.; Canto-Rios, J.J.; Simuta-Champo, R. Sensitivity Analysis of the Rainfall–Runoff Modeling Parameters in Data-Scarce Urban Catchment. *Hydrology* **2020**, *7*, 73. [CrossRef]
9. Hamal, K.; Sharma, S.; Khadka, N.; Baniya, B.; Ali, M.; Shrestha, M.S.; Xu, T.; Shrestha, D.; Dawadi, B. Evaluation of MERRA-2 Precipitation Products Using Gauge Observation in Nepal. *Hydrology* **2020**, *7*, 40. [CrossRef]
10. Bhusal, A.; Parajuli, U.; Regmi, S.; Kalra, A. Application of Machine Learning and Process-Based Models for Rainfall-Runoff Simulation in DuPage River Basin, Illinois. *Hydrology* **2022**, *9*, 117. [CrossRef]
11. Velásquez, N. Assessment of Deep Convective Systems in the Colombian Andean Region. *Hydrology* **2022**, *9*, 119. [CrossRef]

Article

Evaluation of MERRA-2 Precipitation Products Using Gauge Observation in Nepal

Kalpana Hamal [1,2], **Shankar Sharma** [3], **Nitesh Khadka** [2,4], **Binod Baniya** [5], **Munawar Ali** [2,6], **Mandira Singh Shrestha** [7], **Tianli Xu** [8], **Dibas Shrestha** [3,*] and **Binod Dawadi** [3,8,*]

1   International Center for Climate and Environment Sciences, Institute of Atmospheric Physics, Chinese Academy of Sciences, P.O. Box 9804, Beijing 100029, China; kalpana@mail.iap.ac.cn
2   University of Chinese Academy of Sciences, Beijing 100049, China; nkhadka@imde.ac.cn (N.K.); munawarali092@itpcas.ac.cn (M.A.)
3   Central Department of Hydrology and Meteorology, Tribhuvan University, Kirtipur, Kathmandu 44613, Nepal; sharmash@itpcas.ac.cn
4   Institute of Mountain Hazards and Environment, Chinese Academy of Sciences, Chengdu 610041, China
5   Department of Environmental Science, Patan Multiple Campus, Tribhuvan University, Lalitpur 44700, Nepal; binod.baniya@pmc.tu.edu.np
6   Institute of Tibetan Plateau Research, Chinese Academy of Sciences, Beijing 100101, China
7   International Center for Integrated Mountain Development (ICIMOD), Kathmandu 44700, Nepal; mandira.shrestha@icimod.org
8   Kathmandu Centre for Research and Education, Chinese Academy of Sciences-Tribhuvan University, Kirtipur, Kathmandu 44613, Nepal; xutianli@itpcas.ac.cn
*   Correspondence: st.dibas@yahoo.com (D.S.); dawadibinod@gmail.com (B.D.)

Received: 11 June 2020; Accepted: 9 July 2020; Published: 13 July 2020

Abstract: Precipitation is the most important variable in the climate system and the dominant driver of land surface hydrologic conditions. Rain gauge measurement provides precipitation estimates on the ground surface; however, these measurements are sparse, especially in the high-elevation areas of Nepal. Reanalysis datasets are the potential alternative for precipitation measurement, although it must be evaluated and validated before use. This study evaluates the performance of second-generation Modern-ERA Retrospective analysis for Research and Applications (MERRA-2) datasets with the 141-gauge observations from Nepal between 2000 and 2018 on monthly, seasonal, and annual timescales. Different statistical measures based on the Correlation Coefficient (R), Mean Bias (MB), Root-Mean-Square Error (RMSE), and Nash–Sutcliffe efficiency (NSE) were adopted to determine the performance of both MERRA-2 datasets. The results revealed that gauge calibrated (MERRA-C) underestimated, whereas model-only (MERRA-NC) overestimated the observed seasonal cycle of precipitation. However, both datasets were able to reproduce seasonal precipitation cycle with a high correlation (R ≥ 0.95), as revealed by observation. MERRA-C datasets showed a more consistent spatial performance (higher R-value) to the observed datasets than MERRA-NC, while MERRA-NC is more reasonable to estimate precipitation amount (lower MB) across the country. Both MERRA-2 datasets performed better in winter, post-monsoon, and pre-monsoon than in summer monsoon. Moreover, MERRA-NC overestimated the observed precipitation in mid and high-elevation areas, whereas MERRA-C severely underestimated at most of the stations throughout all seasons. Among both datasets, MERRA-C was only able to reproduce the observed elevation dependency pattern. Furthermore, uncertainties in MERRA-2 precipitation products mentioned above are still worthy of attention by data developers and users.

Keywords: MERRA-2; Nepal; precipitation; rain-gauge; reanalysis datasets

1. Introduction

Precipitation is an important variable in atmospheric circulation for weather and climatic studies and is the dominant driver of land surface hydrologic conditions [1–4]. However, precipitation is a complex variable to predict and estimate as it varies highly in space and time due to large-scale atmospheric circulation patterns and the geographic and topographic factors of the region [2,5]. Recent climate change has impacted precipitation distribution and triggered its extremes, such as drought, floods, and soil erosion around the globe [6,7].

The precipitation is one of the crucial factors for causing major environmental changes and disasters, such as drought, floods and landslides across Nepal [8–11]. Thus, understanding the precipitation patterns and other changes in the hydrometeorological cycle requires high-resolution long-term precipitation records. Rain gauge-based measurement typically provides precise measurement of precipitation on the earth's surface [12,13]. However, the spatial extent and temporal resolution of rain gauge-based precipitation measurements in the country are inadequate to support the creation of regional precipitation datasets. This is mainly true for high-elevation areas due to the complex geography and remote location [9,14]. Meanwhile, the discontinuity and missing values in meteorological data records even worsened the results and interpretation of precipitation [15]. Consequently, the newly developed climate and weather prediction model (reanalysis) and satellite-based precipitation products (SBPPs) are the potential alternatives to precipitation measurement for the station sparse region [16,17]. However, the existing satellite-based products have limited (short period) period of precipitation records. For example, recent GPM IMERG and TRMM are only available after 1998, and GSMaP product is no exception [18–21]. PERSIANN products are available for a longer period, although most of the satellite products only provide precipitation data [22]. In contrary, reanalysis product provides comprehensive insights of weather and climate conditions at the regular grid for long periods [23]. These reanalysis datasets assimilate the in situ and remote sensing observations into Numerical Weather Prediction (NWP) model to provide global estimates of land surface, oceanic and atmospheric conditions at different timescale [3]. Therefore, evaluation and validation of reanalysis precipitation products are necessary for the improvement of product quality and hydrometeorological research.

Reanalysis precipitation datasets have been evaluated for different applications around the globe [24–29]. The performance of reanalysis precipitation datasets was relatively more reliable and accurate in the flatter regions than that of complex terrains [30]. Different climatic models used in reanalysis datasets may not resolve precipitation in areas of complex topographic relief with drastic elevation changes (e.g., Nepal). The moisture convergence, in such areas, is locally determined and local convective precipitation is strongly dependent upon local thermal forcing of the terrain. The reanalysis datasets (ERA-Interim) overestimated the precipitation over the northern slope of the Himalaya [31]. Whereas, Modern-ERA Retrospective analysis for Research and Applications (MERRA) datasets show more consistent performance to observed precipitation than ERA-Interim and CFSR in a complex topography region of Central Asia [5]. Similarly, MERRA precipitation datasets reproduced observed spatial patterns and their extremes over the United States [32]. The MERRA datasets were more reasonable to capture inter-annual variation and long-term precipitation trends than other reanalysis datasets over the East Asian domain [33].

A comprehensive evaluation and validation of global reanalysis datasets are very limited in Nepal. For example, Barros and Lang [34] evaluated the performance of the National Center for Environmental Prediction (NCPE-NCAR) reanalysis and found that precipitation was consistently underestimated in Marshyangdi river basin (mid-elevation areas of central region) using a limited number of stations. The authors also mentioned that these reanalysis datasets were able to present precipitation trends and spatial variability. Meanwhile, Ichiyanagi et al. [35] found that NCEP datasets showed a positive correlation between monthly, annual and seasonal precipitation with All India-Rainfall (AIR) in western Nepal, while precipitation in eastern Nepal was negatively correlated. However, both studies did not validate or compare their performance concerning observed datasets over the country. Moreover, systematic evaluation of new generation reanalysis product and inter-comparison (spatio-temporal performance) with observed datasets, to date, has not yet been performed in Nepal. Therefore, this study aims to address this research gap by evaluating the performance of second-generation MERRA (MERRA-2) with 141-gauge observations from Nepal. Among the MERRA-2 product, model-only datasets are selected to evaluate the accuracy of MERRA simulation, while gauge calibrated to quantify the performance improvement after the gauge adjustment. The evaluation is based on monthly, seasonal, and annual timescale, spanning the period of 19 years from 2000 to 2018. Further, the evaluation of MERRA-2 precipitation product over unique topographic and climatic regions helps to select these datasets for different hydrometeorological application and identify its tendency and discrepancies, specific weaknesses, and strengths of the product under different circumstances.

2. Materials and Methods

2.1. Study Area

The study area includes the southern slopes of central Himalaya, Nepal. The lowlands (~60 m) in the south to the high Himalayas (up to 8848 m) in the north represent varying landscapes, topography, weather, and vegetation within a small width (average ~192 km) of the country (Figure 1). Nepal is broadly divided into three Ecological zones, named as Terai (Lowland areas), hills, and mountains. The wettest (Lumle) and driest (Mustang and Manang) areas of Nepal are located in the Gandaki river basin. These areas explain how the topography mediates the precipitation distribution, i.e., being located in the windward (Lumle) and leeward (Manang and Mustang) side of the Annapurna mountain ranges [9,36]. The Eastern, Central and Western regions are divided based on three major river basins, namely, Koshi, Gandaki, and Karnali, respectively. Furthermore, the South Asian monsoon and Western disturbances determine the climatology and distribution of the precipitation in Nepal. South Asian monsoon brings widespread precipitation from June to September through the Bay of Bengal, with higher amounts in the Central region than Eastern and Western regions [8,35]. Western disturbances bring precipitation from December to February, which are eastward-moving air currents in mid-latitudes that enter the Western region of Nepal passing through Iran, Afghanistan, Pakistan, and northwest India [37,38]. The average annual precipitation during 1982–2015 in Nepal is 1428 mm [39].

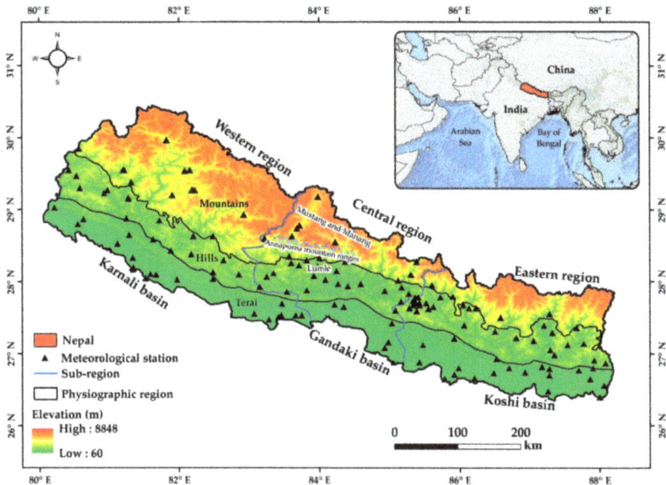

Figure 1. The spatial distribution of 141 rain gauge network, and the elevation pattern over the study area. The blue lines divide the three subregions (Western, Central, and Eastern) of the country. The base map in the inset is from ESRI.

2.2. Datasets

The meteorological station network of Nepal is maintained by the Department of Hydrology and Meteorology (DHM). This network of gauge stations is irregularly distributed [4]; denser on the southern lowlands (below 2500 m elevation) and sparse, especially in the northern mountainous region (above 2500 m elevation), where the terrain is very complex (Figure 1). Such an irregular distribution of network creates an information gap, which ultimately hinders precipitation-related studies over the study areas. Most of the gauge-based datasets are manually collected and are subjected to personnel and instrumental errors [40]. Additional errors for the gauges that are located in high-elevation regions come from the wind effect. Initially, data from 400 stations were collected from DHM (https://www.dhm.gov.np/contents/resources), and quality control was performed for each station; according to WMO standard [41], a year having missing data more than 15% is excluded from the analysis. After quality control, daily data from 141 DHM stations (with ~1044 km^2 coverage by single station) were selected from January 2000 to December 2018 for this study.

MERRA is an emerging reanalysis product of the National Aeronautics and Space Administration (NASA) that provides long-term atmospheric and surface records globally [3,42]. MERRA is a modern reanalysis system, which applies advanced numerical models and assimilation schemes to combine gauge observations from multiple sources [42]. Based on the grid to point interpolation, MERRA precipitation product applies three-dimension variational (3D-Var) data assimilation. The second version of the MERRA reanalysis system (MERRA-2) is the latest atmospheric reanalysis of the modern era, produced by NASA's Global Modeling and Assimilation Office (GMAO) [43]. To provide an advanced product for weather and climate applications, MERRA-2 uses the latest V5 Goddard Earth Observing System Model (GOES) data assimilation with an updated grid to point statistical interpolation system [44,45]. MERRA-2 includes two precipitation datasets, namely PRECTOT (MERRA-NC) and PRECTOTCOR (MERRA-C). MERRA-NC is the model generated precipitation data, while MERRA-C is corrected with CPC Unified Gauge-Based Analysis of Global Daily Precipitation (CPCU) product and the CPC Merged Analysis of Precipitation (CMAP) based precipitation product. With such an advanced and updated data assimilation system, MERRA-2 can be a good alternative to monitor precipitation and hydrological application in the unique physiographic region, where the station is very sparse. This study applies to mean monthly total precipitation (M2TMNXFLX) MERRA-2 data, with a spatial resolution of 0.50° × 0.625°, which was downloaded from NASA's website (https://disc.gsfc.nasa.gov/datasets/M2TMNXFLX).

2.3. Methodology

Observed precipitation datasets are in point scale (station), while both MERRA-2 datasets provide precipitation at a 0.5 km grid box. The complex topographic gradients and heterogeneous distribution of rain gauge station within the study region restrain the accurate rainfall interpolation [46]. Thus, the point-to-pixel method was adopted to compare rain gauge observation with grid-based MERRA-2 datasets [47–50]. Grid-based precipitation datasets were extracted to rain-gauge station location using the original resolution of MERRA-2, instead of interpolating the gauge observations to avoid accumulating additional errors by gridding the observed data [51,52]. Meanwhile, stations falling under the same grid were averaged for better representation of the pixel precipitation with station-based datasets.

Observed datasets are in daily timescale, while MERRA-2 in the monthly timescale. To make consistent timescale at each station, monthly data are computed for the station with the availability of more than 25 days of precipitation data in a month; else, the precipitation in a particular month is considered a missing value. If the corresponding monthly data was missing from the observed datasets, then the monthly precipitation data from MERRA-2 were also considered as a missing value for consistency. The number of rain gauge with complete data series in each year during the study period are presented in Figure S1.

Mean, annual and seasonal precipitation (pre-monsoon, summer monsoon, post-monsoon, and winter season) of the observed and both MERRA-2 datasets were calculated for each station. Further, spatial consistency was performed by comparing the spatial distribution of mean precipitation at different seasons, while for temporal consistency, monthly time-series, and annual cycle of precipitation in all datasets are analyzed, respectively. Similarly, both of these datasets are compared for different elevation bins to quantify the performance from low-elevation to high-elevation areas. Additionally, two percentile-based precipitation indices (95th and 5th percentile threshold values of observed datasets at each station) were calculated for different seasons. The total frequency of very wet events (R95p) and very dry (R5p) events [53,54] is relevant for floods and agriculture management (drought) over the region, respectively. The bias in the frequency of these two percentile indices (R95p and R5p) for MERRA-2 product were calculated for each station.

Four different statistical metrics were calculated to quantify the performance of MERRA-2 datasets (Equations (1)–(4)). The Root Mean Square Error (RMSE) measures the average magnitude of the deviation of MERRA-2 dataset from the observed data. Bias (difference) measures any persistent tendency of a dataset to either overestimate or underestimate; correlation coefficient (R) reflects the strength and direction of the linear association between datasets; Nash–Sutcliffe efficiency (NSE) determines the relative magnitude of the residual variance (noise) compared to the observed data variance. RMSE and Bias Equations (1) and (2) indicate a perfect match between observed and predicted values when it equals to 0, with increasing RMSE and mean bias (MB) values indicating an increasingly poor consistency. The NSE Equation (3) ranges between $-\infty$ and 1 and 1 being a perfect score. The R-value in Equation (4) ranges between 0 and 1, with higher values indicating less error variance.

$$\text{RMSE} = \sqrt{\frac{\sum_{i=1}^{n}(E_i - O_i)^2}{n}} \tag{1}$$

$$\text{MB} = \frac{\left(\overline{E} - \overline{O}\right)}{n} \tag{2}$$

$$\text{NSE} = 1 - \frac{\sum_{i=1}^{n}\left(O_i - \overline{O}\right)^2 - \sum_{i=1}^{n}(E_i - O_i)^2}{\sum_{i=1}^{n}\left(O_i - \overline{O}\right)^2} \tag{3}$$

$$R = \frac{\sum_{i=1}^{n}\left(O_i - \overline{O}\right)\left(E_i - \overline{E}\right)}{\sqrt{\sum_{i=1}^{n}\left(O_i - \overline{O}\right)^2}\sqrt{\sum_{i=1}^{n}\left(E_i - \overline{E}\right)^2}} \tag{4}$$

where, O is the observed data, E is the estimated precipitation by both MERRA-2 datasets, and n is the sample size.

3. Results

3.1. Seasonal Pattern of Precipitation

The temporal evaluation of precipitation is critical for many different hydrometeorological applications. The monthly precipitation time series (mm) of observed and MERRA-2 product average over Nepal during the study period is shown in Figure 2. Among the MERRA-2 datasets, MERRA-NC overestimated, while MERRA-C underestimated the observed precipitation throughout the study period. The observed data revealed that precipitation peaks during the monsoon (June-September) season. Meanwhile, both the datasets can capture high peaks of summer precipitation during June-September. Further, the estimation of the pecipitation peaks was higher and lower for MERRA-NC and MERRA-C during the monsoon season as compared to another season, respectively. MERRA-C largely underestimated the observed precipitation between 2000 and 2003, while MERRA-NC overestimated mainly after 2015. Meanwhile, MERRA-NC showed higher precipitation with more than

800 mm/month in 2016, 2017, and 2018. MERRA-NC appeared more consistent with observation during 2000–2006, whereas MERRA-C during 2012–2018 (Figure 2). Large errors in MERRA-C during the early 2000s could be attributed to considerable changes in the gauge network over the reanalysis period.

The annual precipitation during the study period was in continuous fluctuation from 2000 to 2018, and the overall pattern of both datasets was consistent with the observation (Figure A1). However, some deviations were found for specific years: the observed rainfall reached a maximum and minimum value in 2007 and 2005, respectively. Whereas, both MERRA-2 datasets were not in agreement: MERRA-C and MERRA-NC showed maximum values in 2013 and 2017, respectively (Figure A1). Meanwhile, changes in the performance of the MERRA-2 datasets may be related to improvements in algorithms and additional data in recent years. The result indicates that rainfall simulation ability in MERRA-2 still needs further enhancement.

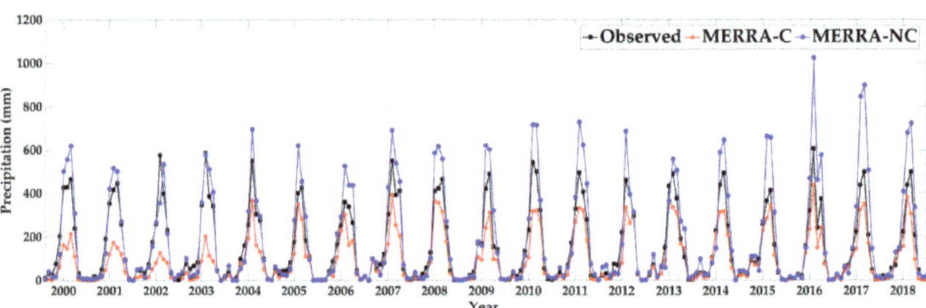

Figure 2. Monthly precipitation time-series (mm) from observed, MERRA-C, MERRA-NC average over Nepal during 2000–2018.

Additionally, boxplots of the annual performance metrics (R, RMSE, MB) for MERRA-2 product and observation are presented in Figure 3. These metrics were generated by comparing point data for interannual timescale. MERRA-C datasets show the median R, RRMSE and MB of 0.64, 190.16 mm/year and −152.44 mm/year, respectively. As similar to Figure A1, the model-only MERRA-NC overestimated the annual precipitation amount with a larger error as indicated by higher RMSE values than MERRA-C. Overall, MERRA-C depicts a slightly high median of R-value; however, the difference is nominal.

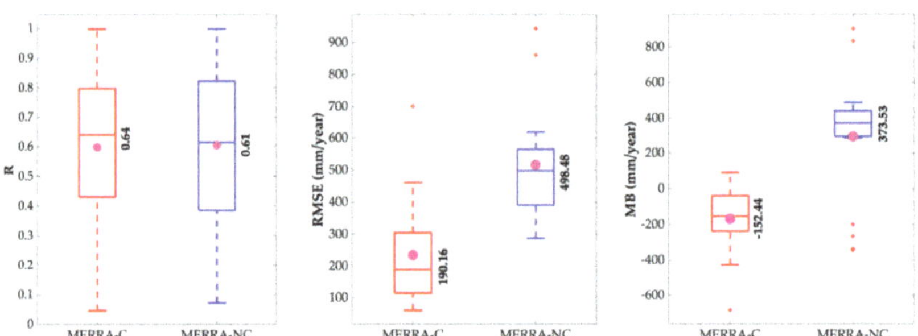

Figure 3. Boxplots for performance metrics (R, Root-Mean-Square Error (RMSE), mean bias (MB)) of MERRA-2 (MERRA-C and MERRA-NC) with observed datasets for annual timescale. The bold values and magenta dot in the boxplot represent the median and mean values of the statistical metrics, respectively.

The monthly cycle of the precipitation average over 2000–2018, shows the maximum precipitation from June to September (summer, 80% of annual precipitation) in the observed dataset, followed by pre-monsoon (13%), post-monsoon (4%), and winter (3%) (Figure 4). Precipitation initially increases in June, reaching a peak during July, and decreasing in August and September. The highest precipitation of ~480 mm was observed in July and the lowest precipitation of ~30 mm during November and December. A similar pattern is also shown by both MERRA-C and MERRA-NC (Figure 4). MERRA-C and MERRA-NC showed the highest precipitation of ~300 mm and 650 mm in July, respectively. It is worth noting that both datasets showed a significant error during the monsoon season as compared to other seasons. Statistics metrics were also calculated by averaging all stations for monthly mean precipitation value (Table 1). Among both datasets, MERRA-NC outperforms MERRA-C in estimating the annual precipitation cycle with lower MB (39.35 mm/month) and RMSE (75.10 mm/month). However, both datasets can capture monthly precipitation variation with a high correlation (R ≥ 0.95) across the country.

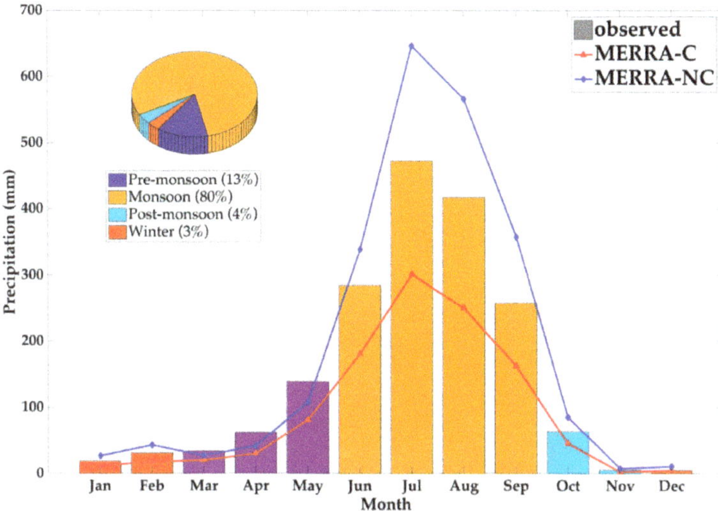

Figure 4. Monthly precipitation cycle (mm/month) derived from observations, MERRA-C, and MERRA-NC for study region from January 2000 to December 2018.

Table 1. The RMSE, MB, R, and Nash–Sutcliffe efficiency (NSE) between MERRA-2 and observed datasets based on the monthly mean value from 2000 to 2018 over the study area. SD represents the standard deviation of respective datasets.

Datasets	Mean (mm/Month)	RMSE (mm/Month)	MB (mm/Month)	R	NSE	SD
MERRA-C	92.73	82.72	−56.72	1	0.73	104.48
MERRA-NC	188.80	75.10	39.35	0.99	0.78	229.42

3.2. Spatial Distribution of Precipitation

The mean annual precipitation is calculated (Figure 5) to study the spatially distributed precipitation in observed and both MERRA-2 datasets. In general, the topographical characteristics establish the large spatial variability of precipitation in the country. Precipitation tends to decrease from east to west, with maximum annual precipitation (>3000 mm/year) was observed in mid-elevation areas of the central region (Lumle areas), whereas the minimum precipitation (<500 mm/year) appeared in the high-elevation areas of central and western region (Figure 5a). The presence of the high mountain range in the central region (Figure 1) acts as a south-north (uplifting) moisture barrier and creates an orographic ascent. This phenomenon generates a "rain shadow" to the northern slope (high elevation areas of the central and western region) and considerably increase precipitation in the southern slope of the mountainous region (Lumle). Among the MERRA-2 datasets, only MERRA-C demonstrated a spatial pattern similar to that of the observation, with a noticeable decreasing pattern from the east to west (Figure 5b); and maximum precipitation (~1800 mm/year) in lower reaches of the eastern region, followed by low-elevation areas of the central region (~1500 mm/year). Meanwhile, MERRA-NC showed the annual precipitation with the range of 2500 to 4500 mm/year in mid-elevation areas of the country, with the highest precipitation (~4500 mm/year) in the high-elevation of the central region (Figure 5c). In general, MERRA-C underestimated, and MERRA-NC overestimated mean annual precipitation across the country. However, the spatial patterns depicted by MERRA-C resemble the known precipitation regime over the country, although differences exist in both datasets.

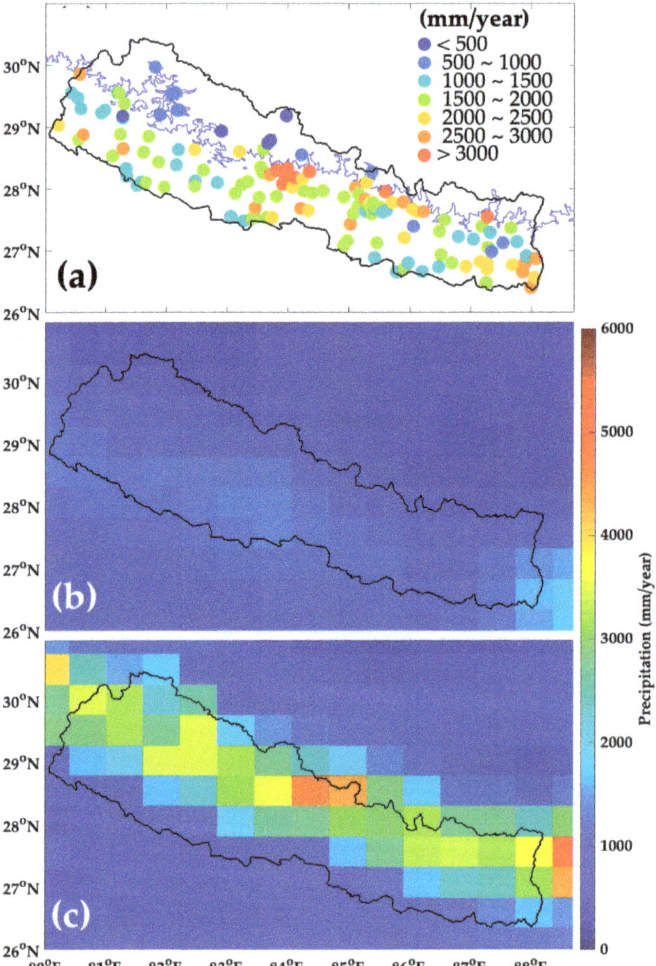

Figure 5. Spatial distribution of mean annual precipitation (mm/year) in (**a**) Observation, (**b**) MERRA-C, and (**c**) MERRA-NC during 2000–2018.

Figure 6 shows the spatial distribution of MB at each station for different seasons calculated by averaging monthly precipitation values of each station. The spatial distribution of MB varies for different stations and seasons over the country. The large MB was observed in the monsoon season as compared to other seasons. During the pre-monsoon season, MERRA-C overestimated, and MERRA-NC underestimated at most of the stations in low-elevation areas, while the latter overestimated in high-elevation areas (Figure 6a,b). Both the datasets showed monsoon distribution as similar to pre-monsoon, but with larger MB (Figure 6c,d). Meanwhile, in post-monsoon and winter season, both datasets showed a very smaller MB and mostly underestimated by MERRA-C, while overestimated by MERRA-NC at most of the stations across the country (Figure 6e–h). Model-only datasets (MERRA-NC) overestimated the observed precipitation, especially in mid and high-elevation areas, whereas, gauge adjusted MERRA-C severely underestimated the observed precipitation, except for few stations. Among the different seasons, both datasets were more consistent with observed datasets in the winter season followed by a post-monsoon, pre-monsoon, and summer monsoon. This might be related to the amount of precipitation in respective seasons (less consistent for a higher amount of precipitation). Overall, the result indicates that the gauge calibration in MERRA-2 effectively reduces positive bias.

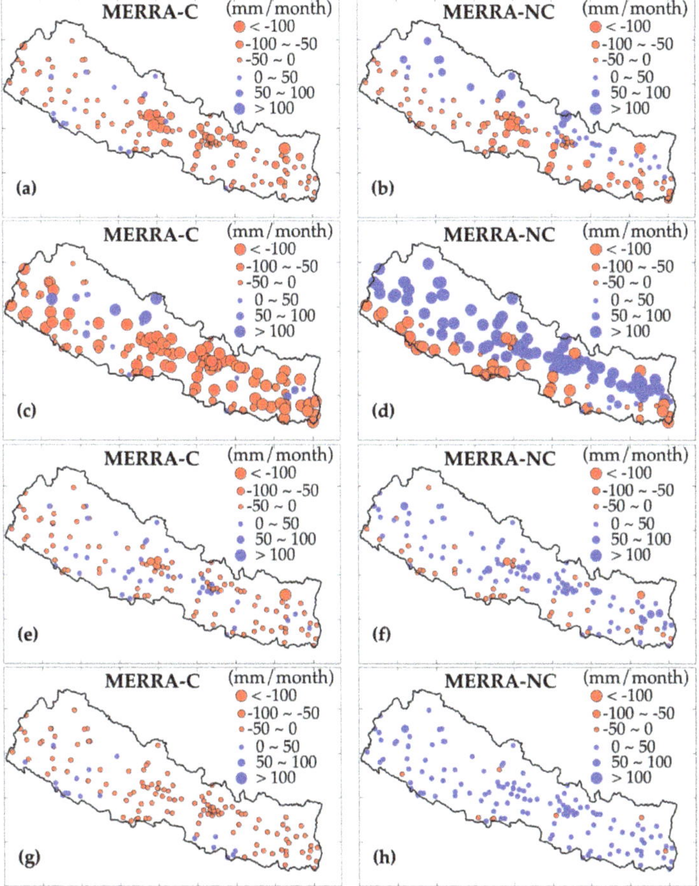

Figure 6. Spatial distribution of mean bias (MB) at each station in MERRA-C and MERRA-NC for a different season, (**a**,**b**) pre-monsoon, (**c**,**d**) summer monsoon, (**e**,**f**) post-monsoon, and (**g**,**h**) winter, respectively. Red and blue circles represent the underestimated and overestimated of precipitation amount across the country, respectively.

Statistical metrics were calculated to quantify the overall performance of the MERRA-2 product, by averaging monthly datasets at each season and presented in Table 2. Performance metrics showed noticeable differences between both datasets for different seasons. MERRA-C underestimated the mean precipitation in all seasons, while MERRA-NC only underestimated during the pre-monsoon season. RMSE in both datasets was very similar, indicating that these two datasets have a similar magnitude of error distribution; however, slightly smaller RMSE and higher R in MERRA-C revealed more spatial consistency with observed datasets. Gauge corrected MERRA-C performed reasonably well with the lowest RMSE, and higher R in all seasons, although estimated precipitation (lower MB) by MERRA-NC was more reliable during pre-monsoon. For the annual performance, MERRA-C archived best overall performance with lower RMSE (86.46 mm/month), higher R (0.34), and NSE of −0.87, although estimated precipitation amount by MERRA-NC was more reliable (lower MB) than MERRA-C. In general, MERRA-C datasets showed more consistent spatial performance (lower RMSE, higher R, and NSE closer to 1) with observed datasets, indicating that the gauge calibrated MERRA-C is more reasonable to reproduce the spatial pattern of observed precipitation over the study area.

Table 2. Spatial performance of both MERRA-2 datasets for the different seasons during 2000–2018. SD represents the standard deviation of respective datasets. The blue font shows the best performance in the respective metrics.

Season	Datasets	Mean (mm/Month)	RMSE (mm/Month)	MB (mm/Month)	R	NSE	SD	Mean OBS (mm/Month)
Pre-monsoon	MERRA-C	44.61	50.75	−33.82	0.44	−0.38	13.01	78.43
	MERRA-NC	62.47	49.27	−15.95	0.37	−0.39	41.18	
Summer monsoon	MERRA-C	222.64	205	−130.02	0.3	−0.53	48.75	353.46
	MERRA-NC	484.11	267.82	130.64	0.1	−1.62	183.28	
Post-monsoon	MERRA-C	24.75	22.37	−10.51	0.5	0.10	8.55	35.26
	MERRA-NC	49.21	26.96	13.95	0.46	−0.42	22.07	
Winter	MERRA-C	10.48	7.35	−4.7	0.65	0.12	4.84	15.17
	MERRA-NC	23.9	12.53	8.72	0.77	−1.85	13.39	
Annual	MERRA-C	94.23	86.48	−57.04	0.34	-0.87	20.64	151.27
	MERRA-NC	196.9	103.97	45.62	0.18	−1.27	77.08	

Figure 7 shows the spatial distribution of correlation coefficient (R) and RMSE in summer monsoon for both MERRA-2 datasets at each station during the study period (2000–2018). The previous result shows that the large error occurs during the monsoon season (Table 2 and Figure 6) at most of the stations across the country. Both of these MERRA-2 datasets showed a very similar correlation at most of the station, although MERRA-C achieves a slightly higher correlation value at mid-elevation areas of the central region (Figure 7a). MERRA-C showed the larger RMSE (>400 mm/month) at most of the station located at high precipitation areas (central region), while MERRA-NC showed large RMSE at most of the station over mid- and high elevation areas of central and western region (Figure 7b). It is possible that the systematic error in MERRA-C was reduced from MERRA-NC due to bias correction, but the random error might be the same in both the datasets. Hence, the results of comparison over these mountainous regions are uncertain, indicating complex terrain lead to higher uncertainty in rainfall estimates over these regions.

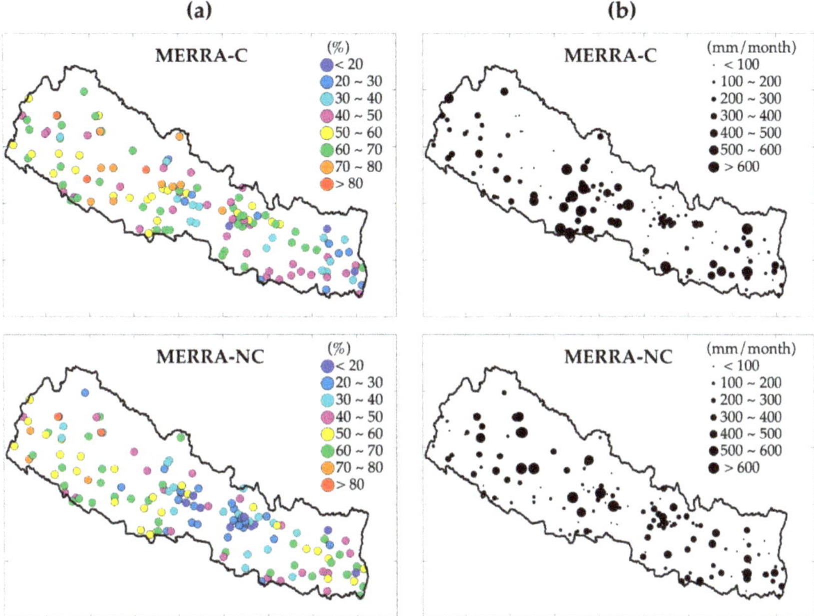

Figure 7. Spatial distribution of (**a**) correlation coefficient (R) and (**b**) RMSE (mm/month) in MERRA-2 datasets for summer monsoon (JJAS) at each station during the study period.

3.3. Extreme Precipitation Events

The bias distribution of very wet (R95p) and very dry (R5p) precipitation events in both MERRA-2 datasets was calculated for different seasons. Figure 8 shows the spatial distribution of bias in the total frequency of R95p at each station across the country. Both datasets show the large positive bias (>10 events) at high-elevation areas of the country during the pre-monsoon (Figure 8a,b). In the monsoon season, MERRA-NC largely overestimated (>20 events) at most of the stations in mid-and high-elevation areas, whereas MERRA-C shows more consistent performance with smaller bias (<−10 events) at most of the stations (Figure 8c,d). For post-monsoon and winter season, MERRA-NC overestimated, and MERRA-C underestimated the total frequency of very wet events (Figure 8e–h). Overall, MERRA-NC shows large positive bias and tends to overestimate the frequency of very wet events, while MERRA-C shows more consistent performance to reproduce the very wet events (smaller bias) across the country.

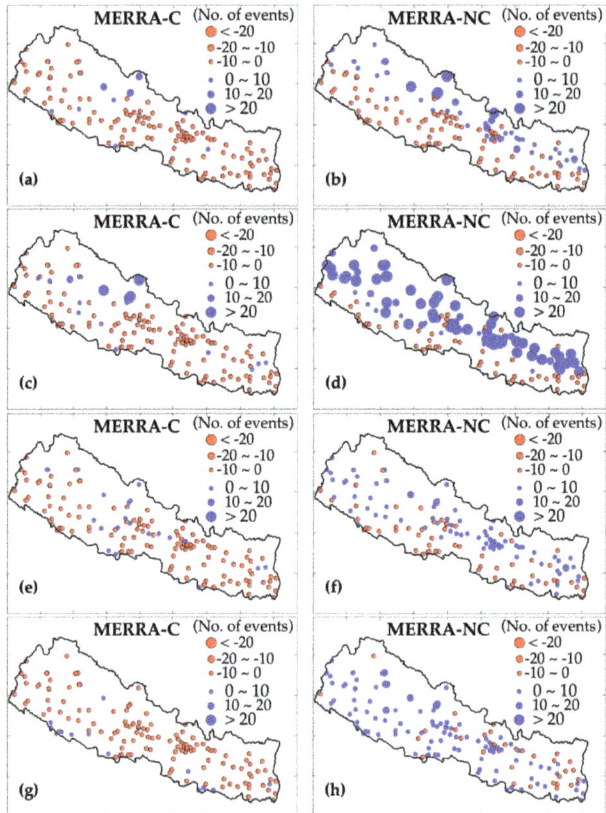

Figure 8. Spatial distribution of bias in very wet events (R95p) at each station for (**a**,**b**) pre-monsoon, (**c**,**d**) monsoon, (**e**,**f**) post-monsoon, (**g**,**h**) winter between observed and MERRA-2 datasets, respectively. Red and blue circles represent the underestimated and overestimated of R95p events across the country, respectively.

Figure 9 shows the bias in total low intensity related extreme events (R5p) for different season at each station over the country. Both datasets show very similar bias distribution of dry events (mostly underestimated by >−10 events) during the pre-, post-monsoon, and winter season. In contrast, MERRA-C largely overestimated (with larger bias >20 events) at most of the station during the monsoon season (Figure 9c,d). In this season, MERRA-NC shows very consistent performance with smaller bias (>−10 events). It is worth noting that MERRA-NC shows large error for R95p events, while MERRA-C shows larger error for R5p during the monsoon season at most of the stations across the country.

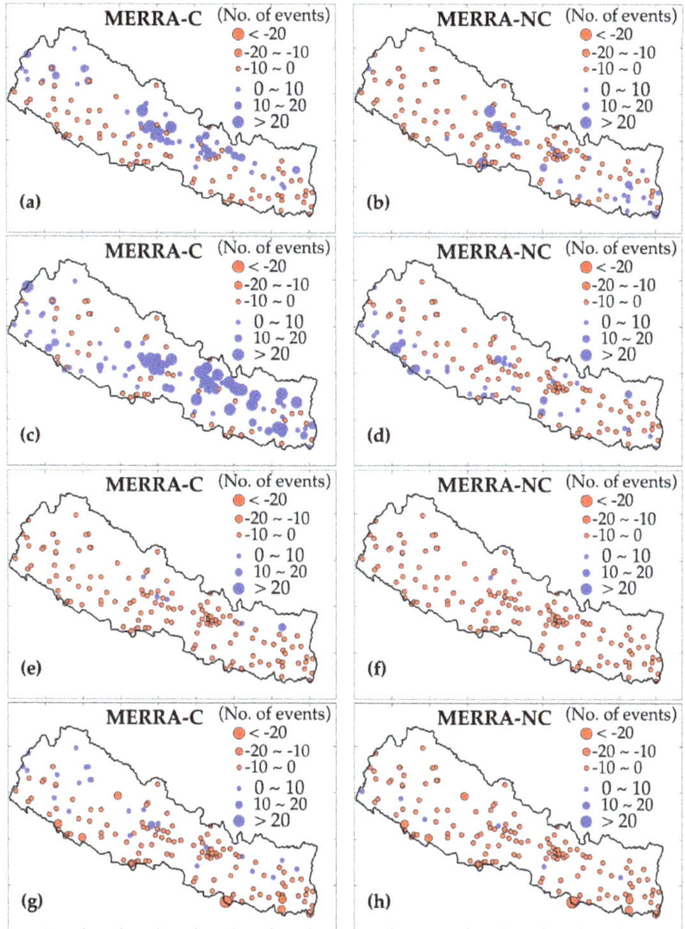

Figure 9. Spatial distribution of bias in very wet events (R5p) at each station for (**a**,**b**) pre-monsoon, (**c**,**d**) monsoon, (**e**,**f**) post-monsoon, (**g**,**h**) winter between observed and MERRA-2 datasets, respectively. Red and blue circles represent the underestimated and overestimated of R5p events across the country, respectively.

3.4. Precipitation Based on the Elevation Gradient

Nepal is a mountainous country, and the precipitation distribution mainly depends on its orography characteristics. Therefore, we examined the variability of mean annual precipitation for individual rain gauge stations and partitioned the station data into a 500 m elevation range over the study area. The elevation dependency was only calculated below 3000 m due to a limited number of stations in high-elevation areas. Figure 10 presents the elevation dependency pattern of mean annual precipitation over 19 years (2000–2018) from observed and both MERRA-2 datasets with the respective number of stations at each interval. The observed data showed the annual precipitation slightly increased from ~1800 mm/year to 2000 mm/year for elevations range 500–1000 m, and decrease to 1700 mm/year for elevations between 1000 and 1500 m with maximum precipitation of ~2200 mm/year at an elevation range of 1500–2000 m. The pattern shows precipitation initially increases up to 2000 m elevation and decreases with increasing elevation. The MERRA-NC precipitation distribution showed a sharp increase in the range between ~1400 mm/year and ~3200 mm/year for elevations below 2500 m, while the MERRA-C data showed the initial increase in precipitation up to 2000 m and subsequent decrease with elevation, with the highest precipitation of ~1200 mm/year between 1500 and 2000 m elevation. Nevertheless, MERRA-C was able to reproduce the observed elevation dependency pattern, and MERRA-NC fails to capture the pattern. It is worth to note that model-only MERRA-NC datasets fail to reproduce the spatial pattern of precipitation (Figure 5), which might be the reason for inconsistent elevation dependency pattern with observation. Some of these DHM gauges might be the reason for the better overall performance of MERRA-C datasets. The result suggests that after gauge calibration MERRA-C is able to reproduce the evident elevation dependency pattern; however, significant underestimation was observed. The orographic effect on these datasets is further discussed in Section 4.

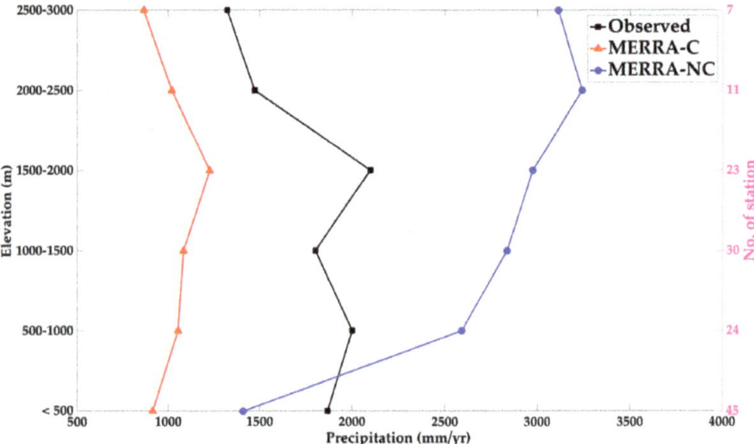

Figure 10. Mean annual precipitation in observed and MERRA-2 datasets at different elevation range with a respective number of the station during the study period.

4. Discussion

This study evaluates and quantifies the spatial and temporal performance of gauge calibrated (MERRA-C) and model-only (MERRA-NC) gridded precipitation datasets from MERRA-2 product using 141-gauge stations of Nepal. We found that MERRA-C datasets underestimated the monthly precipitation cycle and MERRA-NC was largely overestimated, particularly in wet months. In terms of seasonal performance, both the datasets performed poorly in monsoon season (when precipitation amount was higher) than that of the other three seasons. Due to the complex mountainous terrain and interference by wind, DHM gauges may not capture the actual precipitation in the study region. Meanwhile, the lack of solid precipitation measurement in DHM station probably has an impact on the performance of the MERRA-2 product during the winter season. The large uncertainties in the reanalysis product for the wettest season are also reported in previous studies conducted in Assiniboine River Basin (Canada–US border) by Xu et al. [55], Central Africa by Nkiaka et al. [56], and in India by Shah and Mishra [57]. In contrast, Wang et al. [58] found that the reanalysis product showed large bias during the winter season in the Eastern Fringe of Tibetan Plateau. Such uncertainties are primarily related to topographic nature and precipitation distribution of the region. Furthermore, the variational accuracy in interannual performance (Figures 3 and A1) is subject to errors from considerable changes in the number gauge network over the reanalysis period and errors in the input gauge measurements [3].

The spatial distribution of observed precipitation showed large scale precipitation variability with the highest precipitation in the central region (Figure 5). Due to the spatially heterogeneous pattern of precipitation with complex physiography, model datasets have difficulty in resolving the orography effect [5]. The errors over the areas of complex orography may be partially related to the weakness of reanalysis models in simulating the effects of complex terrain [55,58,59]. To minimize such uncertainties, MERRA-C precipitation datasets utilize the several daily CPC gauge-based precipitation datasets from different countries to calibrate the model-only MERRA-NC datasets. The CPC precipitation datasets are derived from ground-based observation data, which is used to adjust MERRA-C datasets. The performance of MERRA-C highly depends on the performance of CPC datasets and adjusted gauge density within each grid box. Hence, the interpolation techniques to generate MERRA gridded datasets from the point and sparse gauge measurements can introduce a considerable level of uncertainty into the gridded dataset, particularly in the high-elevation and remote regions, where a sufficient number of rain gauge stations are usually not available [13,60]. Meanwhile, 45 DHM gauge datasets are also used to generate CPC datasets [61], although information of these assimilated gauge datasets is minimal. The previous study also mentioned that the quality and temporal range of assimilated rain gauge also significantly influence the performance of gauge calibrated precipitation product [62]. DHM rain gauge stations are denser in lowland areas, while too sparse in high-elevation areas of the study region. When these sparse gauges stations were used to calibrate model-only datasets, which may sometimes deteriorate the accuracy of the calibrated product (Tables 1 and 2), it is worth noting that, after gauge correction, MERRA-C moderately captured the spatial distribution of observed precipitation. Similar to this result, most of the previous study revealed the improved performance of gauge calibrate product over the denser gauge network in the larger basin area of West Africa by Nicholson et al. [63] and Malaysia by Mahmud et al. [64]. Meanwhile, Balcutt et al. [65] evaluated the potential bias correction techniques to improve the rainfall representations with spatially aggregated precipitation for driving hydrological simulation in Nepal. These studies also mentioned that a higher number of gauges at each grid during the interpolation would enhance the performance of the calibrated product. Further, the dense gauge network is more representative to provide the actual distribution of precipitation and also more able to capture the orographic effect than the sparse gauge network, thus improving product accuracy [66]. Moreover, there are other various factors in the construction process of MERRA-2, such as boundary layer parameterization, land–atmosphere interactions, and/or convective precipitation parameterization, and this may cause the uncertainty to correctly reflecting the observation precipitation pattern [67].

The evident elevation dependency pattern (i.e., precipitation initially increases with elevations up to 2000 m and decrease with increasing elevation) is similar to previous studies conducted in the same study region [8]. The high mountain blocks the large-scale monsoon flow moving upward and makes the windward (leeward) side of the central region very wet (dry). As mentioned earlier, reanalysis models have difficulty in resolving the orographic effect of precipitation; this might be the reason for the highest precipitation above 2000 m in model-only MERRA-NC datasets (Figures 5 and 10). However, after the gauge calibration, MERRA-C was able to reproduce the observed elevation dependency pattern. The scatter plots with relative statistical metrics were further discussed to demonstrate the overall performances of MERRA-2 product and observed datasets (Figure 11). The statistics showed that MERRA-C relatively performed better with higher correlation (R = 0.71), and lower RMSE (163.85 mm/month) than MERRA-NC; although the estimated precipitation amount by MERRA-NC was more consistent (lower MB) with observed datasets. Besides, NSE also shows the MERRA-C outperformed MERRA-NC during the study period. As similar to Section 3.1, MERRA-C overestimated the observed precipitation, especially in wettest months (>500 mm/month) when precipitation amount is generally higher (Figure 11a). Meanwhile, MERRA-NC overestimated the lower precipitation value (Figure 11b). Overall, MERRA-C were more consistent with reproducing the spatial pattern of observed datasets, while MERRA-NC datasets were more reliable to estimate the observed precipitation amount. The results are in line with the inclusion of the weather stations into the reanalysis gridded datasets that significantly increased the correlation between observed and the gridded products [68]. It is important to mention while interpreting the results, the CPC was derived from point ground-based measurements, and some of these DHM gauges might be the reason for the better overall performance of MERRA-C.

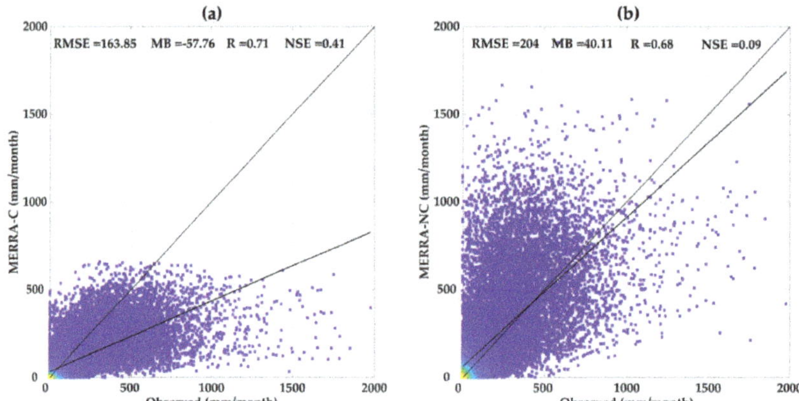

Figure 11. Scatterplot between observed, (**a**) MERRA-C, and (**b**) MERRA-NC generated by monthly datasets at all stations between 2000 and 2018 across the study area. The unit of the figure metrics is in mm/month.

Moreover, MERRA-C and MERRA-NC showed similar tendencies (large error during the monsoon season) with different magnitude of errors (Figures 6–9 and Table 2). Our results provide evidence that interpolation techniques and the sparseness of the gauge measurements can create uncertainties on the performance of gauge calibrated MERRA-2 product. On the other hand, difficulty in resolving the orography effect by the reanalysis model also influenced the overall accuracy of MERRA-2 product [67,68]. Further, the inclusion of a diverse set of station data may increase the accuracy of the reanalysis of data [5]. In conclusion, this study provides shreds of evidence that interpolation techniques and limited or sparseness of the gauge measurements create uncertainties on the performance of reanalysis products, especially in the complex terrain like Nepal. Thus, there is a necessity of adequate gauge stations in the high elevation areas, which not only reduces the uncertainty in the production and validation of gridded datasets but will also be useful for climate and environmental change studies.

5. Conclusions

In this study, we evaluated the performance of MERRA-2 (MERRA-C and MERRA-NC) datasets based on rain-gauge observation over Nepal between 2000 and 2018. The study region features complex terrain and has a significant impact on the distribution of precipitation. Based on the above results, the following conclusions were drawn:

The average seasonal cycle of precipitation during the study period shows the peak precipitation from June to September. MERRA-C and MERRA-NC underestimated and overestimated the observed seasonal cycle of precipitation, respectively. However, MERRA-C and MERRA-NC can reproduce the overall precipitation pattern, but the accuracy is poor in the wettest months for both MERRA-2 datasets.

Gauge calibrated MERRA-C depicted broadly similar spatial distributions of annual precipitation, i.e., precipitation decreased from east to west, with maximum and minimum precipitation in mid-elevation areas of the central region and high-elevation areas of the central and western region, respectively. Both MERRA-2 datasets performed better in winter, post-monsoon, and pre-monsoon than in summer monsoon. For the spatial performance, MERRA-C achieved a higher correlation than MERRA-NC, while both datasets showed a similar magnitude of RMSE. Furthermore, a high correlation in MERRA-C and lower MB in MERRA-NC indicates the better spatial distribution and reliable estimation of precipitation, respectively. MERRA-NC and MERRA-C datasets overestimated the very wet events (R95p) and very dry events (R5p) during the monsoon season, respectively. Whereas, both datasets underestimated the R95p and R5p events for other three seasons at most of the station across the country.

The variability of mean annual precipitation is observed with a 500 m elevation range over the study area. The pattern shows precipitation initially increases up to 2000 m elevation and decreases with increasing elevation. Nevertheless, MERRA-C was able to reproduce the observed elevation dependency pattern, and MERRA-NC fails to capture the observed elevation dependency pattern. The result suggests that MERRA-C datasets have useful implications in Nepal. However, further, improvement is still needed in the MERRA-2 reanalysis product, particularly in the case of mountainous areas such as Nepal.

Evaluation of MERRA-2 precipitation datasets is especially important for understanding the spatio-temporal distribution of precipitation in Nepal. This will eventually benefit from understanding the hydrological processes and water resource management, as it affects the output accuracy of the hydrological model. However, to enhance the precipitation measurement accuracy, the allocation of new measuring stations is still needed to further evaluate the different gridded precipitation products in Nepal.

Supplementary Materials: The following are available online at http://www.mdpi.com/2306-5338/7/3/40/s1, Figure S1: The number of rain gauge with complete data series in each year during the study period (2000–2018).

Author Contributions: Conceptualization, S.S., D.S.; methodology, S.S.; formal analysis, K.H., S.S., N.K.; investigation, S.S., K.H.; funding, B.D., T.X.; data curation, S.S.; writing—original draft preparation, K.H., S.S.; N.K.; writing—review and editing, B.B., M.A., M.S.S., T.X., D.S., B.D.; supervision, D.S., B.D. All authors have read and agreed to the published version of the manuscript.

Funding: The APC was funded by Kathmandu Center for Research and Education, Chinese Academy of Sciences-Tribhuvan University.

Acknowledgments: K. Hamal is supported by Belt and Road Scholarship, and N. Khadka is supported by CAS-TWAS President's fellowship. Authors would like to acknowledge the DHM for observed precipitation datasets and NASA for providing freely available MERRA-2 product. We also express our thankfulness to the Department of Environmental Science, Patan Multiple Campus, Tribhuvan University, Nepal. The authors would like to thank three anonymous reviewers for their insightful comments and suggestion, which helped to improve the manuscript.

Conflicts of Interest: Authors declare that they have no conflict of interest.

Appendix A

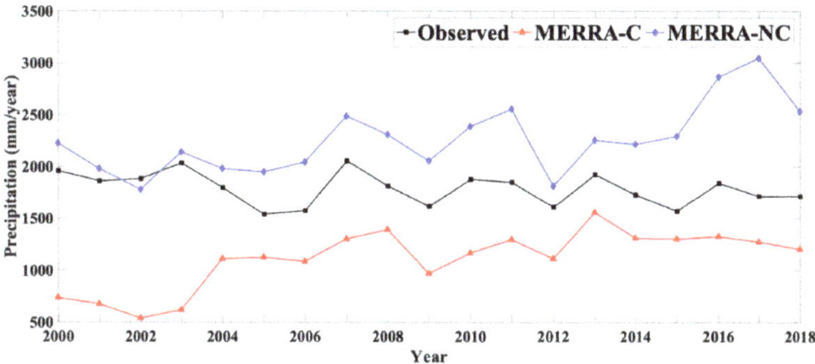

Figure A1. Inter-annual variation of precipitation (mm/year) in observation and both MERRA-2 datasets average over the study area through the study period (2000–2018).

References

1. Kidd, C.; Huffman, G. Global precipitation measurement. *Meteorol. Appl.* **2011**, *18*, 334–353. [CrossRef]
2. Li, D.; Yang, K.; Tang, W.; Li, X.; Zhou, X.; Guo, D. Characterizing precipitation in high altitudes of the western Tibetan plateau with a focus on major glacier areas. *Int. J. Climatol.* **2020**. [CrossRef]
3. Reichle, R.H.; Liu, Q.; Koster, R.D.; Draper, C.S.; Mahanama, S.P.; Partyka, G.S. Land surface precipitation in MERRA-2. *J. Clim.* **2017**, *30*, 1643–1664. [CrossRef]
4. Shrestha, D.; Singh, P.; Nakamura, K. Spatiotemporal variation of rainfall over the central Himalayan region revealed by TRMM Precipitation Radar. *J. Geophys. Res. Atmos.* **2012**, *117*. [CrossRef]
5. Hu, Z.; Hu, Q.; Zhang, C.; Chen, X.; Li, Q. Evaluation of reanalysis, spatially interpolated and satellite remotely sensed precipitation data sets in central Asia. *J. Geophys. Res. Atmos.* **2016**, *121*, 5648–5663. [CrossRef]
6. Reichstein, M.; Bahn, M.; Ciais, P.; Frank, D.; Mahecha, M.D.; Seneviratne, S.I.; Zscheischler, J.; Beer, C.; Buchmann, N.; Frank, D.C. Climate extremes and the carbon cycle. *Nature* **2013**, *500*, 287. [CrossRef]
7. Trenberth, K.E.; Dai, A.; Van Der Schrier, G.; Jones, P.D.; Barichivich, J.; Briffa, K.R.; Sheffield, J. Global warming and changes in drought. *Nat. Clim. Change* **2014**, *4*, 17. [CrossRef]
8. Kansakar, S.R.; Hannah, D.M.; Gerrard, J.; Rees, G. Spatial pattern in the precipitation regime of Nepal. *Int. J. Climatol. J. R. Meteorol. Soc.* **2004**, *24*, 1645–1659. [CrossRef]
9. Talchabhadel, R.; Karki, R.; Thapa, B.R.; Maharjan, M.; Parajuli, B. Spatio-temporal variability of extreme precipitation in Nepal. *Int. J. Climatol.* **2018**, *38*, 4296–4313. [CrossRef]
10. Pokharel, B.; Wang, S.Y.S.; Meyer, J.; Marahatta, S.; Nepal, B.; Chikamoto, Y.; Gillies, R. The east–west division of changing precipitation in Nepal. *Int. J. Climatol.* **2019**. [CrossRef]
11. Shrestha, S.; Yao, T.; Kattel, D.B.; Devkota, L.P. Precipitation characteristics of two complex mountain river basins on the southern slopes of the central Himalayas. *Theor. Appl. Climatol.* **2019**, *138*, 1159–1178. [CrossRef]
12. Hou, A.Y.; Skofronick-Jackson, G.; Kummerow, C.D.; Shepherd, J.M. Global precipitation measurement. In *Precipitation: Advances in Measurement, Estimation and Prediction*; Springer: Cham, Switzerland, 2008; pp. 131–169.
13. Sun, Q.; Miao, C.; Duan, Q.; Ashouri, H.; Sorooshian, S.; Hsu, K.L. A review of global precipitation data sets: Data sources, estimation, and intercomparisons. *Rev. Geophys.* **2018**, *56*, 79–107. [CrossRef]
14. Sharma, S.; Khadka, N.; Hamal, K.; Baniya, B.; Luintel, N.; Joshi, B.B. Spatial and Temporal Analysis of Precipitation and Its Extremities in Seven Provinces of Nepal (2001–2016). *Appl. Ecol. Environ. Sci.* **2020**, *8*, 64–73.
15. Islam, M.; Das, S.; Uyeda, H. Calibration of TRMM derived rainfall over Nepal during 1998–2007. *Open Atmos. Sci. J.* **2010**, *4*, 12–23. [CrossRef]
16. Tapiador, F.J.; Turk, F.J.; Petersen, W.; Hou, A.Y.; Garcia-Ortega, E.; Machado, L.A.T.; Angelis, C.F.; Salio, P.; Kidd, C.; Huffman, G.J.; et al. Global precipitation measurement: Methods, datasets and applications. *Atmos. Res.* **2012**, *104*, 70–97. [CrossRef]
17. Ayugi, B.; Tan, G.; Ullah, W.; Boiyo, R.; Ongoma, V. Inter-comparison of remotely sensed precipitation datasets over Kenya during 1998–2016. *Atmos. Res.* **2019**, *225*, 96–109. [CrossRef]
18. Hou, A.Y.; Kakar, R.K.; Neeck, S.; Azarbarzin, A.A.; Kummerow, C.D.; Kojima, M.; Oki, R.; Nakamura, K.; Iguchi, T. The global precipitation measurement mission. *Bull. Am. Meteorol. Soc.* **2014**, *95*, 701–722. [CrossRef]
19. Huffman, G.J.; Bolvin, D.T.; Braithwaite, D.; Hsu, K.; Joyce, R.; Xie, P.; Yoo, S.-H. NASA global precipitation measurement (GPM) integrated multi-satellite retrievals for GPM (IMERG). *Algorithm Theor. Basis Doc. (ATBD) Vers.* **2015**, *4*, 26.
20. Huffman, G.J.; Bolvin, D.T.; Nelkin, E.J.; Wolff, D.B.; Adler, R.F.; Gu, G.; Hong, Y.; Bowman, K.P.; Stocker, E.F. The TRMM multisatellite precipitation analysis (TMPA): Quasi-global, multiyear, combined-sensor precipitation estimates at fine scales. *J. Hydrometeorol.* **2007**, *8*, 38–55. [CrossRef]
21. Kubota, T.; Shige, S.; Hashizume, H.; Aonashi, K.; Takahashi, N.; Seto, S.; Hirose, M.; Takayabu, Y.N.; Ushio, T.; Nakagawa, K. Global precipitation map using satellite-borne microwave radiometers by the GSMaP project: Production and validation. *IEEE Trans. Geosci. Remote Sens.* **2007**, *45*, 2259–2275. [CrossRef]

22. Joyce, R.J.; Janowiak, J.E.; Arkin, P.A.; Xie, P. CMORPH: A method that produces global precipitation estimates from passive microwave and infrared data at high spatial and temporal resolution. *J. Hydrometeorol.* **2004**, *5*, 487–503. [CrossRef]

23. Parker, W.S. Reanalyses and observations: What's the difference? *Bull. Am. Meteorol. Soc.* **2016**, *97*, 1565–1572. [CrossRef]

24. Bosilovich, M.G.; Chen, J.Y.; Robertson, F.R.; Adler, R.F. Evaluation of global precipitation in reanalyses. *J. Appl. Meteorol. Climatol.* **2008**, *47*, 2279–2299. [CrossRef]

25. De Leeuw, J.; Methven, J.; Blackburn, M. Evaluation of ERA-Interim reanalysis precipitation products using England and Wales observations. *Quart. J. R. Meteorol. Soc.* **2015**, *141*, 798–806. [CrossRef]

26. Behrangi, A.; Christensen, M.; Richardson, M.; Lebsock, M.; Stephens, G.; Huffman, G.J.; Bolvin, D.; Adler, R.F.; Gardner, A.; Lambrigtsen, B.; et al. Status of High latitude precipitation estimates from observations and reanalyses. *J. Geophys. Res. Atmos.* **2016**, *121*, 4468–4486. [CrossRef]

27. Kishore, P.; Jyothi, S.; Basha, G.; Rao, S.V.B.; Rajeevan, M.; Velicogna, I.; Sutterley, T.C. Precipitation climatology over India: Validation with observations and reanalysis datasets and spatial trends. *Clim. Dynam.* **2016**, *46*, 541–556. [CrossRef]

28. Cuo, L.; Zhang, Y. Spatial patterns of wet season precipitation vertical gradients on the Tibetan Plateau and the surroundings. *Sci. Rep.* **2017**, *7*, 5057. [CrossRef]

29. Essou, G.R.C.; Brissette, F.; Lucas-Picher, P. The Use of Reanalyses and Gridded Observations as Weather Input Data for a Hydrological Model: Comparison of Performances of Simulated River Flows Based on the Density of Weather Stations. *J. Hydrometeorol.* **2017**, *18*, 497–513. [CrossRef]

30. Ma, L.; Zhang, T.; Frauenfeld, O.W.; Ye, B.; Yang, D.; Qin, D. Evaluation of precipitation from the ERA-40, NCEP-1, and NCEP-2 Reanalyses and CMAP-1, CMAP-2, and GPCP-2 with ground-based measurements in China. *J. Geophys. Res. Atmos.* **2009**, *114*. [CrossRef]

31. Tong, K.; Su, F.; Yang, D.; Zhang, L.; Hao, Z. Tibetan Plateau precipitation as depicted by gauge observations, reanalyses and satellite retrievals. *Int. J. Climatol.* **2014**, *34*, 265–285. [CrossRef]

32. Ashouri, H.; Sorooshian, S.; Hsu, K.-L.; Bosilovich, M.G.; Lee, J.; Wehner, M.F.; Collow, A. Evaluation of NASA's MERRA Precipitation Product in Reproducing the Observed Trend and Distribution of Extreme Precipitation Events in the United States. *J. Hydrometeorol.* **2016**, *17*, 693–711. [CrossRef]

33. Chen, G.; Iwasaki, T.; Qin, H.; Sha, W. Evaluation of the warm-season diurnal variability over East Asia in recent reanalyses JRA-55, ERA-Interim, NCEP CFSR, and NASA MERRA. *J. Clim.* **2014**, *27*, 5517–5537. [CrossRef]

34. Barros, A.P.; Lang, T.J. Monitoring the monsoon in the Himalayas: Observations in central Nepal, June 2001. *Mon. Weather Rev.* **2003**, *131*, 1408–1427. [CrossRef]

35. Ichiyanagi, K.; Yamanaka, M.D.; Muraji, Y.; Vaidya, B.K. Precipitation in Nepal between 1987 and 1996. *Int. J. Climatol. J. R. Meteorol. Soc.* **2007**, *27*, 1753–1762. [CrossRef]

36. Karki, R.; Schickhoff, U.; Scholten, T.; Böhner, J. Rising precipitation extremes across Nepal. *Climate* **2017**, *5*, 4. [CrossRef]

37. Dimri, A. Relationship between ENSO phases with Northwest India winter precipitation. *Int. J. Climatol.* **2013**, *33*, 1917–1923. [CrossRef]

38. Hamal, K.; Sharma, S.; Baniya, B.; Khadka, N.; Zhou, X. Inter-annual variability of Winter Precipitation over Nepal coupled with ocean-atmospheric patterns during 1987-2015. *Front. Earth Sci.* **2020**, *8*, 161. [CrossRef]

39. Baniya, B.; Tang, Q.; Xu, X.; Haile, G.G.; Chhipi-Shrestha, G. Spatial and temporal variation of drought based on satellite derived vegetation condition index in Nepal from 1982–2015. *Sensors* **2019**, *19*, 430. [CrossRef]

40. Talchabhadel, R.; Karki, R.; Parajuli, B. Intercomparison of precipitation measured between automatic and manual precipitation gauge in Nepal. *Measurement* **2017**, *106*, 264–273. [CrossRef]

41. Llansó, P. *Guidelines on Climate Observation: Networks and Systems*; World Meteorological Organization: Geneva, Switzerland, 2003.

42. Rienecker, M.M.; Suarez, M.J.; Gelaro, R.; Todling, R.; Bacmeister, J.; Liu, E.; Bosilovich, M.G.; Schubert, S.D.; Takacs, L.; Kim, G.-K. MERRA: NASA's modern-era retrospective analysis for research and applications. *J. Clim.* **2011**, *24*, 3624–3648. [CrossRef]

43. Gelaro, R.; McCarty, W.; Suarez, M.J.; Todling, R.; Molod, A.; Takacs, L.; Randles, C.A.; Darmenov, A.; Bosilovich, M.G.; Reichle, R.; et al. The Modern-Era Retrospective Analysis for Research and Applications, Version 2 (MERRA-2). *J. Clim.* **2017**, *30*, 5419–5454. [CrossRef] [PubMed]

44. Rienecker, M.; Suarez, M.; Todling, R.; Bacmeister, J.; Takacs, L.; Liu, H.-C.; Gu, W.; Sienkiewicz, M.; Koster, R.; Gelaro, R. *The GEOS-5 Data Assimilation System—Documentation of Versions 5.0. 1, 5.1. 0, and 5.2. 0*; Technical Report Series on Global Modeling and Data Assimilation; NASA: Washington, DC, USA, 2008.

45. Molod, A.; Takacs, L.; Suarez, M.; Bacmeister, J. Development of the GEOS-5 atmospheric general circulation model: Evolution from MERRA to MERRA2. *Geosci. Model Dev.* **2015**, *8*, 1339–1356. [CrossRef]

46. Monsieurs, E.; Kirschbaum, D.B.; Tan, J.; Maki Mateso, J.-C.; Jacobs, L.; Plisnier, P.-D.; Thiery, W.; Umutoni, A.; Musoni, D.; Bibentyo, T.M. Evaluating TMPA rainfall over the sparsely gauged East African Rift. *J. Hydrometeorol.* **2018**, *19*, 1507–1528. [CrossRef]

47. Liechti, T.C.; Matos, J.P.; Boillat, J.L.; Schleiss, A.J. Comparison and evaluation of satellite derived precipitation products for hydrological modeling of the Zambezi River Basin. *Hydrol. Earth Syst. Sci.* **2012**, *16*, 489–500. [CrossRef]

48. Thiemig, V.; Rojas, R.; Zambrano-Bigiarini, M.; Levizzani, V.; De Roo, A. Validation of Satellite-Based Precipitation Products over Sparsely Gauged African River Basins. *J. Hydrometeorol.* **2012**, *13*, 1760–1783. [CrossRef]

49. Bai, P.; Liu, X.M. Evaluation of Five Satellite-Based Precipitation Products in Two Gauge-Scarce Basins on the Tibetan Plateau. *Remote Sens.* **2018**, *10*, 1316. [CrossRef]

50. Wang, W.; Lu, H.; Zhao, T.J.; Jiang, L.M.; Shi, J.C. Evaluation and Comparison of Daily Rainfall From Latest GPM and TRMM Products Over the Mekong River Basin. *IEEE J. Stars* **2017**, *10*, 2540–2549. [CrossRef]

51. Feidas, H. Validation of satellite rainfall products over Greece. *Theor. Appl. Climatol.* **2010**, *99*, 193–216. [CrossRef]

52. Li, Z.; Yang, D.W.; Hong, Y. Multi-scale evaluation of high-resolution multi-sensor blended global precipitation products over the Yangtze River. *J. Hydrol.* **2013**, *500*, 157–169. [CrossRef]

53. Casanueva Vicente, A.; Rodríguez Puebla, C.; Frías Domínguez, M.D.; González Reviriego, N. Variability of extreme precipitation over Europe and its relationships with teleconnection patterns. *Hydrol. Earth Syst.* **2014**. [CrossRef]

54. Duncan, J.M.; Biggs, E.M.; Dash, J.; Atkinson, P.M. Spatio-temporal trends in precipitation and their implications for water resources management in climate-sensitive Nepal. *Appl. Geogr.* **2013**, *43*, 138–146. [CrossRef]

55. Xu, X.; Frey, S.K.; Boluwade, A.; Erler, A.R.; Khader, O.; Lapen, D.R.; Sudicky, E. Evaluation of variability among different precipitation products in the Northern Great Plains. *J. Hydrol. Reg. Stud.* **2019**, *24*, 100608. [CrossRef]

56. Nkiaka, E.; Nawaz, N.; Lovett, J. Evaluating global reanalysis precipitation datasets with rain gauge measurements in the Sudano-Sahel region: Case study of the Logone catchment, Lake Chad Basin. *Meteorol. Appl.* **2017**, *24*, 9–18. [CrossRef]

57. Shah, R.; Mishra, V. Evaluation of the reanalysis products for the monsoon season droughts in India. *J. Hydrometeorol.* **2014**, *15*, 1575–1591. [CrossRef]

58. Wang, G.; Zhang, X.; Zhang, S. Performance of Three Reanalysis Precipitation Datasets over the Qinling-Daba Mountains, Eastern Fringe of Tibetan Plateau, China. *Adv. Meteorol.* **2019**, *2019*. [CrossRef]

59. Tarek, M.; Brissette, F.P.; Arsenault, R. Evaluation of the ERA5 reanalysis as a potential reference dataset for hydrological modeling over North-America. *Hydrol. Earth Syst. Sci. Discuss.* **2019**, 1–35. [CrossRef]

60. Jiang, S.; Ren, L.; Hong, Y.; Yong, B.; Yang, X.; Yuan, F.; Ma, M. Comprehensive evaluation of multi-satellite precipitation products with a dense rain gauge network and optimally merging their simulated hydrological flows using the Bayesian model averaging method. *J. Hydrol.* **2012**, *452*, 213–225. [CrossRef]

61. Becker, A.; Schneider, U.; Meyer-Christoffer, A.; Ziese, M.; Finger, P.; Stender, P.; Heller, A.; Breidenbach, J. *GPCC Report for Years 2009 and 2010*; DWD: Offenbach, Germany, 2011.

62. Sharma, S.; Chen, Y.; Zhou, X.; Yang, K.; Li, X.; Niu, X.; Hu, X.; Khadka, N. Evaluation of GPM-Era Satellite Precipitation Products on the Southern Slopes of the Central Himalayas Against Rain Gauge Data. *Remote Sens.* **2020**, *12*, 1836. [CrossRef]

63. Nicholson, S.E.; Some, B.; McCollum, J.; Nelkin, E.; Klotter, D.; Berte, Y.; Diallo, B.; Gaye, I.; Kpabeba, G.; Ndiaye, O. Validation of TRMM and other rainfall estimates with a high-density gauge dataset for West Africa. Part I: Validation of GPCC rainfall product and pre-TRMM satellite and blended products. *J. Appl. Meteorol.* **2003**, *42*, 1337–1354. [CrossRef]

64. Mahmud, M.R.; Numata, S.; Matsuyama, H.; Hosaka, T.; Hashim, M. Assessment of effective seasonal downscaling of TRMM precipitation data in Peninsular Malaysia. *Remote Sens.* **2015**, *7*, 4092–4111. [CrossRef]
65. Müller, M.F.; Thompson, S.E. Bias adjustment of satellite rainfall data through stochastic modeling: Methods development and application to Nepal. *Adv. Water Resour.* **2013**, *60*, 121–134. [CrossRef]
66. Blacutt, L.A.; Herdies, D.L.; de Gonçalves, L.G.G.; Vila, D.A.; Andrade, M.J.A.R. Precipitation comparison for the CFSR, MERRA, TRMM3B42 and Combined Scheme datasets in Bolivia. *Atmos. Res.* **2015**, *163*, 117–131. [CrossRef]
67. Robertson, F.R.; Bosilovich, M.G.; Chen, J.; Miller, T.L. The effect of satellite observing system changes on MERRA water and energy fluxes. *J. Clim.* **2011**, *24*, 5197–5217. [CrossRef]
68. Zandler, H.; Haag, I.; Samimi, C. Evaluation needs and temporal performance differences of gridded precipitation products in peripheral mountain regions. *Sci. Rep.* **2019**, *9*, 15118. [CrossRef] [PubMed]

 *hydrology*

Article

Sensitivity Analysis of the Rainfall–Runoff Modeling Parameters in Data-Scarce Urban Catchment

Héctor A. Ballinas-González [1,*], Víctor H. Alcocer-Yamanaka [2], Javier J. Canto-Rios [3] and Roel Simuta-Champo [1]

[1] Mexican Institute of Water Technology, Paseo Cuauhnáhuac 8532, Progreso, Jiutepec, Morelos 62550, Mexico; roel_simuta@tlaloc.imta.mx

[2] National Water Commission, Insurgentes Sur 2416, Copilco El Bajo, Coyoacan, CDMX 04340, Mexico; yamanaka@conagua.gob.mx

[3] Autonomous University of Yucatán, Industrias no contaminantes S/N, Mérida, Yucatán 150, Mexico; javierj.canto@correo.uady.mx

* Correspondence: hballinas@tlaloc.imta.mx

Received: 9 September 2020; Accepted: 1 October 2020; Published: 5 October 2020

Abstract: Rainfall–runoff phenomena are among the main processes within the hydrological cycle. In urban zones, the increases in imperviousness cause increased runoff, originating floods. It is fundamental to know the sensitivity of parameters in the modeling of an urban basin, which makes the calibration process more efficient by allowing one to focus only on the parameters for which the modeling results are sensitive. This research presents a formal sensitivity analysis of hydrological and hydraulic parameters—absolute–relative, relative–absolute, relative–relative sensitivity and R^2—applied to an urban basin. The urban basin of Tuxtla Gutiérrez, Chiapas, in Mexico is an area prone to flooding caused by extreme precipitation events. The basin has little information in which the records (with the same time resolution) of precipitation and hydrometry match. The basin model representing an area of 355.07 km^2 was characterized in the Stormwater Management Model (SWMM). The sensitivity analysis was performed for eight hydrological parameters and one hydraulic for two precipitation events and their impact on the depths of the Sabinal River. Based on the analysis, the parameters derived from the analysis that stand out as sensitive are the Manning coefficient of impervious surface and the minimum infiltration speed with $R^2 > 0.60$. The results obtained demonstrate the importance of knowing the sensitivity of the parameters and their selection to perform an adequate calibration.

Keywords: sensitivity analysis; rainfall–runoff model; parameters model

1. Introduction

Rainfall–runoff phenomena are one of the main processes within the hydrological cycle. In urban zones, the increases in imperviousness cause increased runoff, originating floods. Therefore, to protect the population and movable and immovable property, hydraulic structures that make up urban drainage (storm hydrants, collectors, emitters, and retention works, among others) are analyzed and designed. According to Jha et al. [1], the number of reported flood events affected people and associated economic damage has been significantly increasing over the past two decades. One tool in the analysis and design of these structures is the use of the urban drainage models, which have been developed for the last 30 years to contribute to the management and planning of stormwater [2,3]. The SWMM (Storm Water Management Model) was developed by the Environmental Protection Agency [4] to simulate the rainfall–runoff process in urban watersheds and is widely used for urban planning, analysis and design related to drainage systems [5–7]. Numerous studies have investigated the use of this model to describe different phenomena related to runoff in urban basins; for example, Randall et al. [8]

implemented SWMM to assess the behavior of runoff under Low Impact Development (LID) scenarios. Chang et al. [9] evaluated the DENFIS (Dynamic Evolving Neural Fuzzy Inference System) model performance compared with the physically based model SWMM. Agarwal and Kumar [10] implemented a runoff model to determine flood impact using the Green-Ampt Infiltration model in SWMM.

Technological advances in hydrological modeling have incorporated the use of data with greater detail in terms of spatial and temporal resolution. On the other hand, remote measurement techniques allow information to be obtained quickly in large areas through sensors operating in different spectral bands, which opens the door to the use of large amounts of data applied to hydrological models, which thereby become robust [11]. The ability to incorporate spatially distributed digital information is then hampered by a lack of data on the same time scale (precipitation measurement and runoff) [12,13]. However, modeling hydrological processes can be challenging, particularly in highly heterogeneous urbanized areas (land-use variation, slope, coverage) that produce multiple interactions between urban drainage structures and system (for wastewater and stormwater) at different temporal and spatial scales, which increases data requirements and complexity [14]. These complexities, in addition to the data shortage at the required level, make it difficult to define a universal methodology for reproducing urban flows at the catchment scale.

Hydrological models are approximations of natural systems, which create a substantial discrepancy between the results of the model and reality [15]. The results of the models need to be adjusted by means of parameter calibration, which helps to match the predictions with the corresponding observations [16–18]. The increase in the number of parameters that are adjusted in a model leads to a greater workload in the calibration process [19,20]. Therefore, to increase the speed of the process, it is important to perform a sensitivity analysis to learn the set of parameters to which the models are sensitive, to understand their behavior against their values' variation and to use this information to limit the number of parameters in the calibration [21,22]. It is therefore recommended to perform a sensitivity analysis before starting with hydrological modeling [23,24]. This analysis has been applied in different levels of watersheds. Shin and Choi [25] found that the size of the catchment makes a difference in the parameter sensitivity between rainfall events.

Some researchers, such as Mannina and Viviani [26], considered sensitivity analysis, identifying the model's most sensitive parameters. They applied the analysis to 17 parameters that influence the results of the discharges and concentrations of urban drainage, managing to reduce the number of parameters to 12, which were subsequently used for the calibration of the model. Kleidorfer et al. [27] analyzed the impact of uncertainty on modeling input data considering the parameters with greater sensitivity, using the Metropolis–Hastings algorithm for the assessment of sensitivity in the calibration parameters. Bárdossy [28] mentions that hydrological parameters cannot be identified as a single set of values, and changes to a parameter can be absorbed by the remaining parameters of the model. Other researchers, such as Thorndahl et al. [29], performed a sensitivity analysis of a set of parameters, by comparing the conditions of the conceptual model and the general model, finding that the parameter of greater sensitivity is the hydrological reduction factor. Bajracharya et al. [30] performed a global sensitivity analysis (using the Variogram Analysis of Response Surfaces technique) of the parameters that govern the behavior of runoffs of the Nelson Churchill River basin, represented in the Hydrological Predictions for the Environment model (HYPE). Other studies that perform sensitivity analysis applied to the SWMM model reveal the behavior of the parameters according to the study area and its characterization [31–34].

Due to data being scarce in most of the world (especially in developing countries), research should focus on the reliability of hydrological models. This can be achieved by comparing different sensitivity analysis approaches in data-scarce regions.

This research presents a methodology that uses four common expressions to assess the sensitivity of hydrological model parameters in an urban basin in Mexico. The expressions used are absolute–relative sensitivity, relative–absolute sensitivity, relative–relative sensitivity, and correlation coefficient R^2. This use of expressions aims to show their performance together to determine the sensitivity of parameters in a well-known open-source model (SWMM). This application will show the benefits of implementing such an analysis in applications with limited data. This work builds on our previous study [35], in which a storm with a single peak was evaluated as input data and a sensitivity index was calculated. In this work, a more robust sensitivity analysis is performed, as parameter influence is evaluated with more than one equation. In addition, two different storms were evaluated, allowing us to observe the difference in behavior of the riverbed depths due to the parameters under these differing conditions. The study comprises the analysis of nine parameters that are used for modeling and are difficult to estimate because of their complexity and variability in surface coverage in an urban context.

The document is organized as follows: Section 2 illustrates the materials and methods used in sensitivity analysis; Section 3 presents the results of the analysis; Section 4 presents the discussion on the results. Finally, Section 5 presents the main conclusions arising from this work.

2. Materials and Methods

2.1. Study Area

The Sabinal River basin has 355.07 km^2 of surface, is located between the coordinates 16°42′ and 16°54′ north and between the latitude 93°20′ and 9°02′ west, with an elevation in the range of 384 to 1064 m above sea level and an average elevation of 724 m above sea level. This basin is characterized by 42.31% permeable soil and 57.69% waterproof soil, the latter representing urban areas. In general, the basin has an average slope of 6.89% and a concentration time of 328.80 min. The city of Tuxtla Gutiérrez, Chiapas in México is located within the Sabinal River basin and is crossed from west to east by the main riverbed, 21 km long. It has flooding problems caused by extreme precipitation events, which can occur mainly from May to October. Therefore, sensitivity analysis was carried out using the extreme precipitation events of 07/10/2011 (Event 1) and 07/27/2011 (Event 2), recorded every 10 min by seven automatic stations. Figure 1 shows the automatic stations (yellow circles), as well as the urban basin of the Sabinal River, which was divided into 96 sub-basins for modeling. Tables 1 and 2 show the properties of the precipitation events mentioned above, in view of Figure 2a,b.

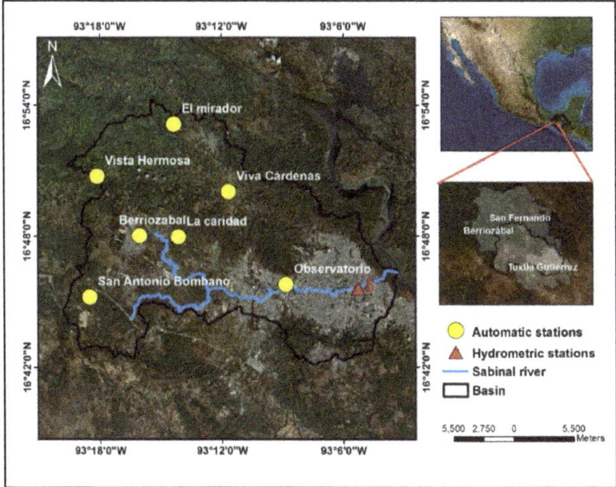

Figure 1. Urban basin of Tuxtla Gutiérrez, Chiapas, México.

Table 1. Precipitation characteristics, Event 1.

Station	Duration (min)	Δt (min)	Maximum Intensity, (mm/h)	Accumulated Precipitation (mm)
Berriozábal	490.00	10.00	100.50	45.50
Caridad	490.00	10.00	117.00	53.75
Mirador	490.00	10.00	81.00	27.00
Observatorio	490.00	10.00	115.50	23.75
San Antonio Bombanó	490.00	10.00	78.00	32.25
Vista Hermosa	490.00	10.00	36.00	33.25
Viva Cárdenas	490.00	10.00	79.50	35.75

Table 2. Precipitation characteristics, Event 2.

Station	Duration (min)	Δt, (min)	Maximum Intensity, (mm/h)	Accumulated Precipitation (mm)
Berriozábal	1820.00	10.00	57.00	22.50
Caridad	1820.00	10.00	70.50	43.75
Mirador	1820.00	10.00	84.00	44.25
Observatorio	1820.00	10.00	81.00	58.75
San Antonio Bombanó	1820.00	10.00	19.50	16.50
Vista Hermosa	1820.00	10.00	51.00	32.00
Viva Cárdenas	1820.00	10.00	40.50	22.25

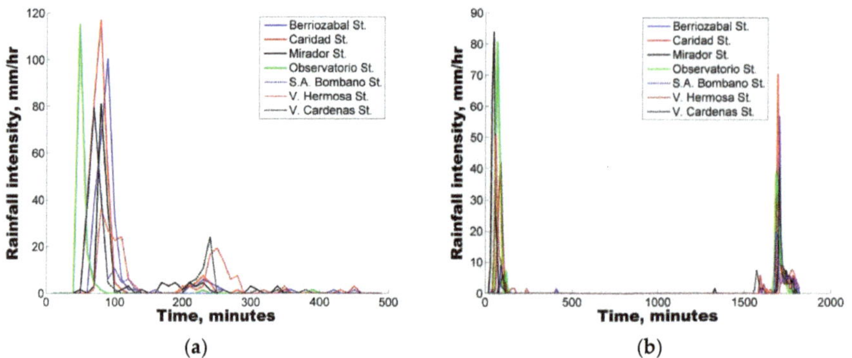

Figure 2. (a) Precipitation Event 1 and (b) Precipitation Event 2.

Like precipitation data, data from the Parque del Oriente hydrometric station are available at the exit of the basin, which has depth information for the month of July with registration every 10 min in the river section. Figure 1 shows the location of the hydrometric station with the number 5 and with the triangle symbol in red. Figure 3a,b show the depths recorded by that station caused by precipitation events 1 and 2.

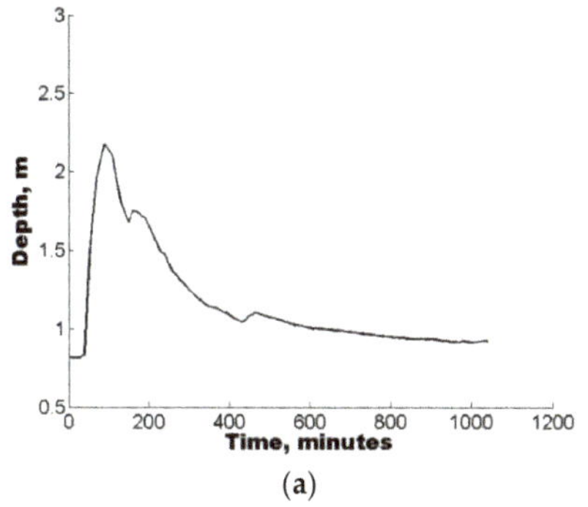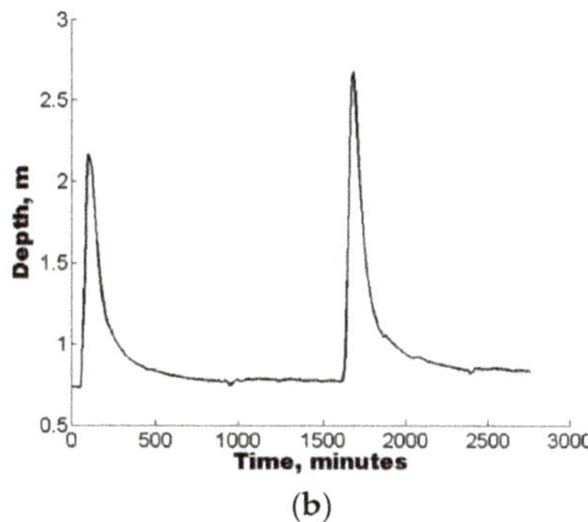

Figure 3. Sabinal river hydrometry in Parque del Oriente station; (**a**) Event 1 and (**b**) Event 2.

2.2. Hydrological Model

The integrated Stormwater Management Model (SWMM) was generated by the Environmental Protection Agency in the United States [5]. SWMM is a dynamic hydrology–hydraulic water quality simulation model that can be used for single-event or long-term (continuous) simulation of runoff quantity and quality from primarily urban areas [4,5,8]. The runoff component operates on a collection of sub-catchment areas that receive precipitation and generate runoff. Each one of the sub-basins is considered as a nonlinear reservoir, where the contribution of flow to the deposits comes from precipitation, snow or releases from upstream stores. SWMM also simulates route flows from the system, infiltration, evaporation and surface runoff. The surface runoff of a given area is determined when the water depth within a catchment exceeds the maximum storage value, and the outflow is determined by Manning (1), which integrates the continuity Equation (2), considering friction through the incorporation of the Manning coefficient (n). The major loss considered in the rainfall and runoff modeling is infiltration loss. In this study, infiltration loss is calculated with the Horton Equation (3) [36]. The infiltration losses are considered only from the pervious areas of a sub-catchment [5].

$$Q_M = \frac{1}{n} L \left(y - y'\right)^{5/3} s^{1/2} \tag{1}$$

Here, Q_M is the flow by Manning, n is the Manning coefficient, L is the width sub-basin, y is the water depth, y' is the lowering of height storage and s is the slope.

$$A \frac{\partial y}{\partial t} = Ai - Q \tag{2}$$

Here, Q is the flow, A is the area of the basin, i is the intensity of the rain, y is the depth of storage in depressions and t is the time.

$$f_p = f_0 - (f_0 - f_\infty) e^{-\propto_d (t - t_w)} \tag{3}$$

Here, f_p is the soil infiltration capacity, f_∞ is the minimum or end value of f_p (*in* $t = \infty$), f_0 is the maximum or initial value of f_p (*in* $t = 0$), t_w is the start time of the storm and $\propto_d$ is the decay coefficient.

SWMM requires the input of parameters related to catchment characteristics, sewer network and soil type. The values range of the parameters was derived from the following (Table 3, [37,38]): Manning's roughness for overland surfaces and conduits, soil infiltration parameters and surface depression storage. Manning's roughness is the measure of resistance to the runoff flow. The value of the roughness coefficient depends on the type of soil, surface cover and vegetation in pervious areas, and in impervious areas it depends on the type of the material used in the construction of streets and building roofs. Other parameters depend on the soil type and the slope; for example, the impervious area depression storage, which is defined as water stored in depressions on impervious areas (depleted only by evaporation), and pervious area depression storage is defined as water stored in depressions on pervious areas (subject to infiltration and evaporation). Likewise, the measure of urbanization is given as the percentage of imperviousness for each sub-catchment or area depression storage, where the urban catchments are composed of pervious and impervious areas (which increases when the urban area develops). For Horton's infiltration equation, the values of minimum or maximum infiltration rate and decay coefficient depend on the soil, vegetation and initial moisture content; these parameters should be estimated using results from the field or from specialized literature. Finally, Manning's roughness coefficient for conduits is one of the parameters used to calculate flow in a pipe or open channel and depends on the material type.

Table 3. SWWM model parameters.

Parameter	Abbreviation	Maximum Value	Minimum Value
Manning's duct coefficient	ManN	0.010	0.030
Manning's N for impervious area	Nimperv	0.001	0.200
Manning's N for pervious area	Nperv	0.010	0.200
Depth of depression storage on impervious area, mm	Simperv	0.000	10.000
Depth of depression storage on pervious area, mm	Sperv	0.000	20.000
Percentage of impervious area with no depression storage, %	PctZero	0.000	100.000
Maximum infiltration rate, mm/hr	MaxRate_fa	1.000	200.000
Minimum infiltration rate, mm/hr	MinRate_fe	1.000	25.000
Decay coefficient, 1/hr	Decay_k	1.000	30.000

2.3. Sensitivity Analysis

Parameter sensitivity analysis is a necessary background for any deeper analysis and helps to improve the understanding of the model's behavior. Its goal is to explore the change in model output resulting from a change in model parameters or model inputs and to separate influential from non-influential parameters. Sensitivity analysis investigates the sensitivity of a parameter with respect to the simulation results at a certain parameter value. The following expressions calculate different sensitivity indexes for each of the possible model parameters [39,40]:

$$s_{i,j}(\Theta_M) = \Theta_{M,j}\frac{\partial f(\Theta_{M,j})}{\partial \Theta_{M,j}} \tag{4}$$

$$s_{i,j}(\Theta_M) = \frac{1}{f(\Theta_{M,j})}\frac{\partial f(\Theta_{M,j})}{\partial \Theta_{M,j}} \tag{5}$$

$$s_{i,j}(\Theta_M) = \frac{\Theta_{M,j}}{f(\Theta_{M,j})}\frac{\partial f(\Theta_{M,j})}{\partial \Theta_{M,j}} \tag{6}$$

where $f(\Theta_{M,j})$ represents the n output variables of the model, and $\Theta_{M,j}$ represents the *j*th independent parameters of the model.

Expression (4) represents the absolute–relative sensitivity, which describes the absolute change in the results for a relative change in parameters. Expression (5) is the relative–absolute sensitivity, which describes the relative change in the results for an absolute change of the parameter. Finally, expression (6) is the relative–relative sensitivity, which describes the relative change in results for a relative change in parameters [39,40].

The gradient term $\partial f(\Theta_{M,j})/\partial\Theta_{M,j}$ is solved numerically by using runs of the model with slightly different values of Θ_M. Then the gradient term can be approximated by expression (7):

$$\frac{\partial f(\Theta_{M,j})}{\partial\Theta_{M,j}} = \frac{f(\Theta_{M,j} + \Delta\Theta_{M,j}) - f(\Theta_{M,j} - \Delta\Theta_{M,j})}{2\Delta\Theta_{M,j}} \tag{7}$$

where $\Delta\Theta_{M,j}$ is a small increment in the parameter value.

In addition to the three equations above, the coefficient of determination R^2 calculation is used. This is defined as the squared value of the coefficient of correlation [41].

$$R^2 = \left(\frac{\sum_{i=1}^{n}(O_i - \overline{O})(P_i - \overline{P})}{\sqrt{\sum_{i=1}^{n}(O_i - \overline{O})^2}\,\sqrt{\sum_{i=1}^{n}(P_i - \overline{P})^2}} \right)^2 \tag{8}$$

Here, O are the observed values and P are the modeling values.

The range of R^2 lies between 0 and 1 and describes how much of the observed dispersion is explained by the prediction. A value of 0 means no correlation, whereas a value of 1 means that the dispersion of the prediction is equal to that of the observation [41].

2.4. Methodology

The general methodology set out in this work for sensitivity analysis is shown in Figure 4. The methodology indicates that the sensitivity analysis process begins with the characterization of the urban basin in which the runoffs (depths in the riverbed) promoted by the precipitation event are evaluated. Topographic, hydrographic and land-use characteristics were configured.

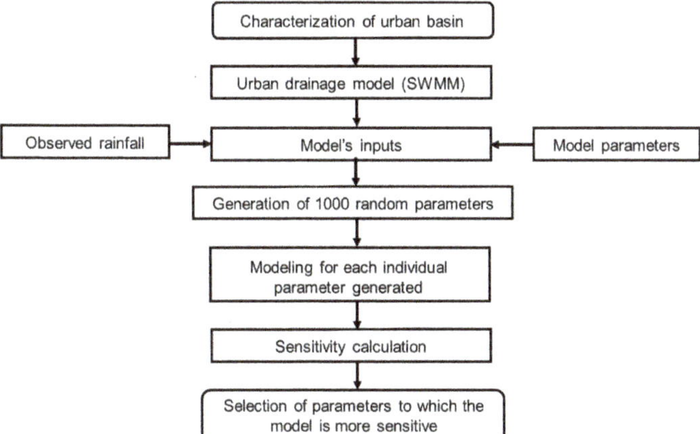

Figure 4. Sensitivity analysis methodology.

The hydrological model SWMM is then selected to perform runoff modeling. The input data required for modeling are then entered—in this case, precipitation events and parameters.

The next step is the individual sensitivity analysis of the input, hydrological and hydraulic parameters. The analysis is performed by evaluating the runoff outputs of the model, generated by 1000 individual values of one of the modeling parameters (randomly generated with the Monte Carlo Simple method), and keeping the values of the other parameters fixed. The process is repeated for each of the parameters. In the next step, the sensitivity associated with the modeling results generated by the variation of individual parameters is calculated with Equations (4)–(6).

Obtaining sensitivity of the parameters includes doing the following:

- Obtaining box plot from the results of accumulated depths and calculation of mean, standard deviation and coefficient of variation in order to evaluate the dispersion of the data.
- Obtaining box plot of parameters and calculation of mean, standard deviation and variance in order to evaluate data dispersion.

Once the sensitivity is calculated, R^2 is obtained in order to evaluate the performance of the results due to the parameters (Equation (8)).

Finally, validation of sensitivity analysis results is carried out using event 2 and repeating the steps of this methodology.

3. Results

3.1. Sensitivity Analysis Event 1

This section shows the results of the sensitivity analysis carried out in the urban basin of the Sabinal River using the Equations (4)–(6) and (8) described above. The model's simulations were carried out using input data (hydrological and hydraulic parameters and precipitation Event 1) in SWMM 5.0. In addition, hydrometric station 5 is represented in the model, from which the results of the cumulative depth series corresponding to the variation of the parameters (1000 embodiments per parameter) were taken.

Figure 5 shows the variation of the accumulated depths with respect to the parameters. The accumulated output depth values for MinRate_fe are in a range of 167.03 to 218.56 m, and Nimperv from 169.94 to 243.66 m, approximately. The accumulated depth values, generated by the MinRate_fe parameter, within quartile two (179.03) are the closest to the accumulated depth (183.33 m) of the hydrometric series observed in the Sabinal. The Nimperv parameters that generate depth values closer to those observed (Figure 5) are observed between the lower limit and quartile one (169.94 to 210.25). On the other hand, the parameters Decay_k, MaxRate_fa, PctZero, Sperv, Simperv and Nperv generate output results with less variation between them. As for the variation of the ManN parameter, it does not generate a change in the modeling results.

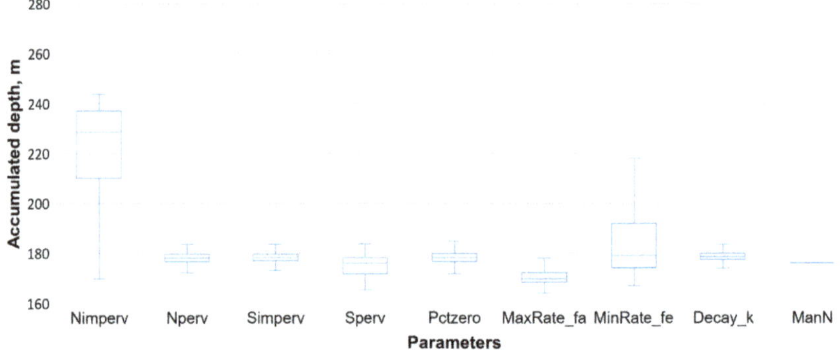

Figure 5. Variation of the accumulated depth on the Sabinal River with respect to the parameters.

A key feature of Figure 5 is that MinRate_fe has a positive bias of the accumulated depths, since the case's whisker is longer towards the high values (there is greater dispersion of quartile three at the upper limit of the data), while Nimperv has a negative bias. This is because the box whiskers are long towards the low values (greater dispersion of the lower limit to the quartile one data).

In Table 4, it can be noted that the results obtained from accumulated depths have greater variation with respect to the MinRate_fe and Nimperv parameters since the standard deviation and coefficient of variation are 20.09 m and 10.73%, respectively, for MinRate_fe, and for Nimperv, the standard deviation is 19.85 m and coefficient of variation is 8.97%.

Table 4. Mean µ, standard deviation σ, and coefficient of variation Cv of the depth values with respect to the parameters, Event 1.

Parameter	µ (m)	σ (m)	Cv (%)
Nimperv	**221.30**	**19.85**	8.97
Nperv	178.05	2.15	1.21
Simperv	178.31	1.96	1.10
Sperv	175.10	3.89	2.22
PctZero	177.75	3.23	1.82
MaxRate_fa	170.61	3.34	1.96
MinRate_fe	**187.25**	**20.09**	**10.73**
Decay_k	178.57	1.85	1.04
ManN	176.08	0.00	0.00

Once the analysis of the accumulated depths had been done, sensitivity analysis was carried out for each of the expressions presented. Figure 6 shows that the rate of change of the random parameters used in this analysis generated different values of absolute–relative sensitivity indexes, except ManN, the value of which was zero because it had no variation in accumulated depth. As a result, in quartile three at the upper limit, the parameters MinRate_fe and Nimperv have greater dispersion of the high values (positive bias) and of absolute-relative sensitivity values: 0.0810 to 0.1890 and 0.0761 to 0.1767, respectively.

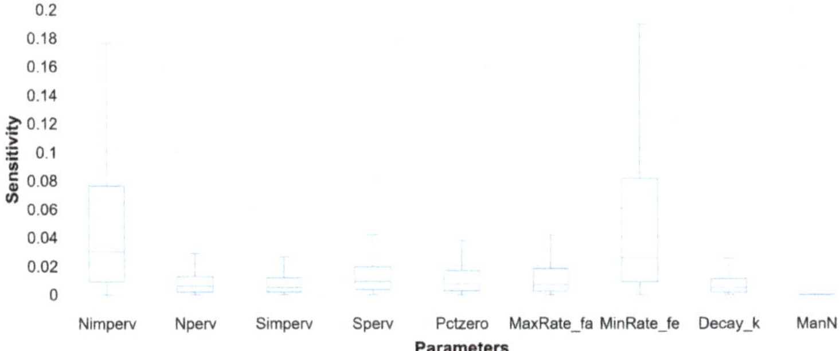

Figure 6. Comparison of parameters with respect to the absolute–relative sensitivity index, Event 1.

On the other hand, Figure 7 shows the results of the relative–absolute sensitivity. The Nimperv and Nperv parameters have high dispersion values: Nimperv from 1.1232 to 2.1777 (quartile three at upper limit) and Nperv from 0.2231 to 0.4943 (quartile three to upper limit), both parameters with positive bias. All other parameters have low sensitivity values with less dispersion.

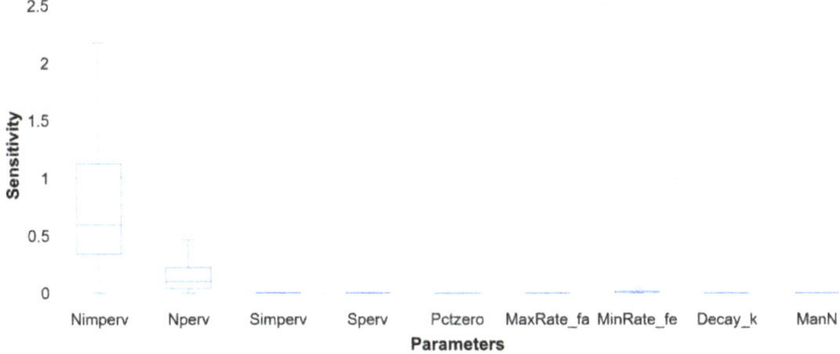

Figure 7. Comparison of parameters with respect to the relative–absolute sensitivity index, Event 1.

Finally, Figure 8 shows the sensitivity values of the parameters obtained from the relative–relative relationship: the Nimperv and MinRate_fe parameters are those with high sensitivity values and with the highest dispersion ranging from quartile two to the upper limit. The sensitivity values for Nimperv range from 0.0482 to 0.1635 and for MinRate_fe from 0.0554 to 0.2180. As in calculations with previous sensitivity calculation expressions, the other parameters have less bias in the obtained values.

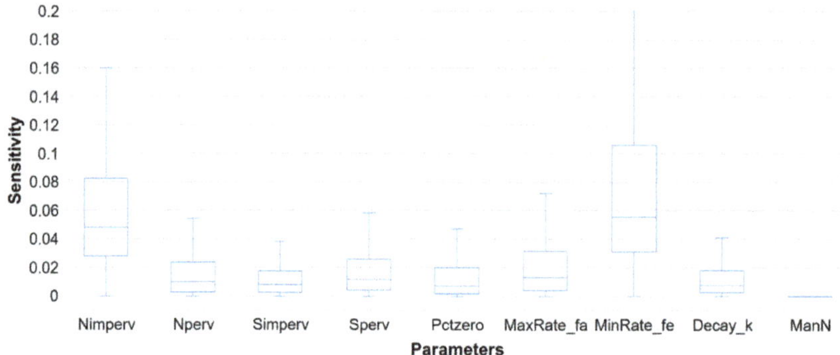

Figure 8. Comparison of parameters with respect to the relative–relative sensitivity index, Event 1.

Table 5 shows the statistical characteristics of each of the above figures, which correspond to the different methods to obtain the sensitivity of the model results according to the parameters. For example, the absolute–relative sensitivity values of the Nimperv parameter are 0.1047 and 0.0110 of standard deviation and variance, respectively; 0.5723 for standard deviation and 0.3276 for variance in the case of relative–absolute sensitivity; and for relative–relative sensitivity, 0.04870 for standard deviation and 0.00237 for variance. In the case of MinRate_fe for the relative–absolute sensitivity, the values are 0.2223 and 0.00494 for standard deviation and variance, respectively, and for the relative–relative sensitivity they are 0.08626 for standard deviation and 0.00744 for variance. In the case of relative–absolute sensitivity, instead of MinRate_fe, it was the Nperv parameter with values of 0.3961 for standard deviation and 0.1569 for variance.

Table 5. Mean μ, standard deviation σ and variance, σ^2 of sensitivity values, Event 1.

Parameter	Absolute–Relative			Relative–Absolute			Relative–Relative		
	μ	σ	σ^2	μ	σ	σ^2	μ	σ	σ^2
Nimperv	0.0670	**0.1047**	**0.0110**	0.7875	**0.5723**	**0.3276**	0.05945	**0.04870**	**0.00237**
Nperv	0.0139	0.0491	0.0024	0.2332	0.3961	0.1569	0.02644	0.06645	0.00442
Simperv	0.0116	0.0241	0.0006	0.0061	0.0262	0.0007	0.02112	0.05493	0.00302
Sperv	0.0184	0.0424	0.0018	0.0041	0.0192	0.0004	0.02677	0.05586	0.00312
Pctzero	0.0187	0.0419	0.0018	0.0009	0.0066	0.0000	0.02659	0.07668	0.00588
Maxrate_fa	0.0207	0.0695	0.0048	0.0007	0.0042	0.0000	0.03378	0.06498	0.00422
Minrate_fe	0.0970	**0.2223**	**0.0494**	0.0202	**0.0660**	**0.0044**	0.08293	**0.08626**	**0.00744**
Decay_k	0.0109	0.0240	0.0006	0.0026	0.0106	0.0001	0.02447	0.07148	0.00511
ManN	0.0000	0.0000	0.0000	0.0000	0.0000	0.0000	0.00000	0.00000	0.00000

Based on the above analysis, Figure 9a (medium sensitivity) and Figure 9b (maximum sensitivity, according to the upper limit of the whisker box) show the most sensitive parameters in the hydrological simulation model.

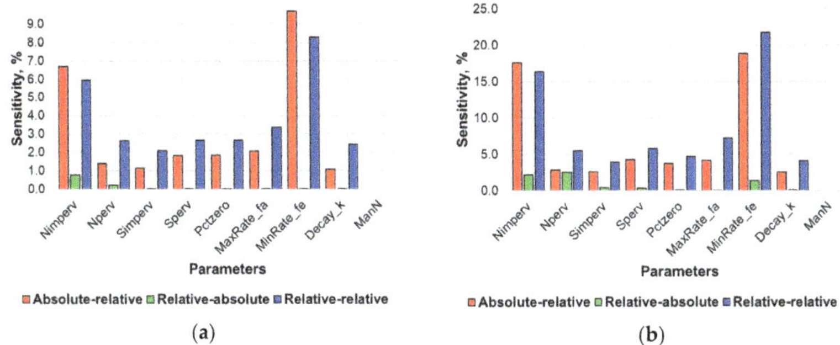

Figure 9. (**a**) Average sensitivity of modeling parameters and (**b**) maximum sensitivity of modeling parameters, Event 1.

R^2 was calculated according to the methodology. In Figure 10, R^2 efficiency results are shown: it can be seen that the R^2 values for Nimperv are in the range of 0.31 to 0.82, having greater amplitude in the whiskers box. The next parameter with greater amplitude in the whiskers box is MinRate_fe, with a range of R^2 from 0.62 to 0.79. The other parameters have less bias in the obtained values.

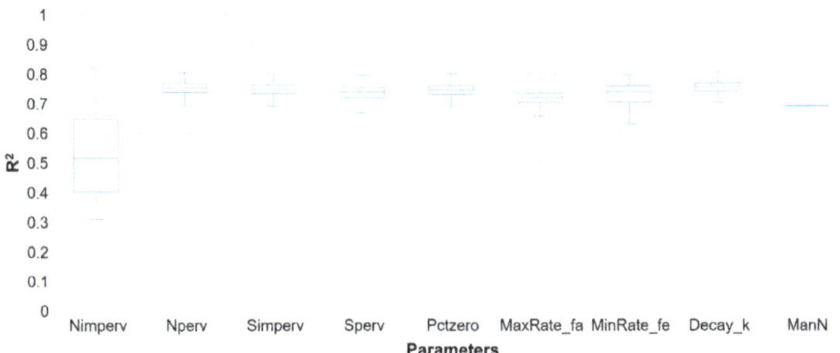

Figure 10. Comparison of parameters with respect to R^2, Event 1.

Based on the sensitivity and R^2 calculation, Figure 11a,b show the behavior of the depths generated by Event 1. This shows the change of the depth hydrogram according to the maximum, minimum and mean value and the parameter value with greater R^2, for parameters with higher Nimperv sensitivity and MinRate_fe.

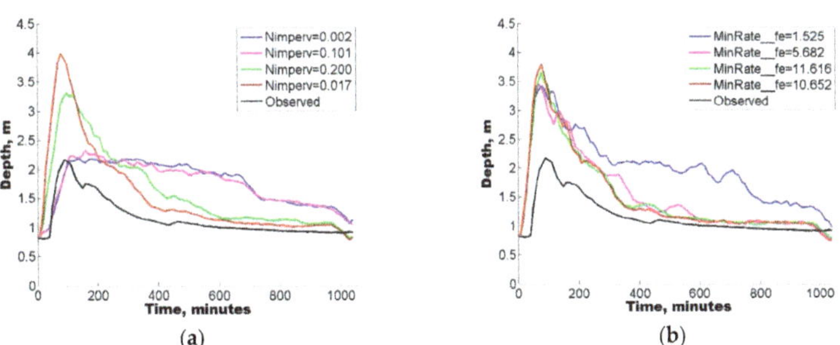

Figure 11. Depth levels of Sabinal river using (**a**) Nimperv parameter and (**b**) MinRate_fe parameter. The blue line is a minimum value, the magenta line is an average value, the green line is a maximum value, the red line is a parameter with a maximum value of R^2 and the black line is depth levels in hydrometric station.

In contrast to Figure 12a,b, the change in the depth hydrogram is observed according to the maximum, minimum, mean value and the parameter value with the highest R^2 for two of the parameters with lower Simperv sensitivity and Decay_k.

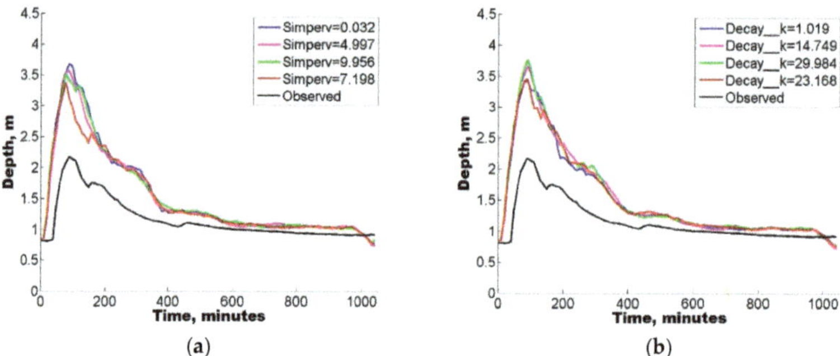

Figure 12. Depth levels of Sabinal river using (**a**) Simperv parameter and (**b**) Decay_k parameter. The blue line is a minimum value, the magenta line is an average value, the green line is a maximum value, the red line is a parameter with maximum value of R^2 and the black line is depth levels in hydrometric station.

3.2. Validation of Results, Event 2

This section shows the results of the sensitivity analysis carried out in the urban basin of the Sabinal River according to the methodology proposed (expressions (4), (5), (6) and (8)). The modeling was carried out with the input data (hydrological and hydraulic parameters and precipitation Event 2). In this case, the accumulated depth of the observed hydrometry was 201.25 m.

Figure 13 shows the variation of the accumulated depths with respect to the parameters, and it can be observed that the parameter that generates the greatest variation in the accumulated output depth values is Nimperv, which is in a range of 371.58 to 496.41 m. The accumulated depth values, generated by the Nimperv parameter, within the lower bound and quartile two (371.58 to 437.20), are the most dispersed and close to the observed accumulated depth. The MinRate_fe parameter generated depth values close to observed and less dispersion than the previous parameter (Figure 13) are observed between the lower bound and the upper bound (367.84 to 384.52). On the other hand, the parameters MaxRate_fa, PctZero, Decay_k, Sperv, Simperv and Nperv generate output results with less variation between them (compact boxes). As for the variation of the ManN parameter, these do not result in a change in modeling results.

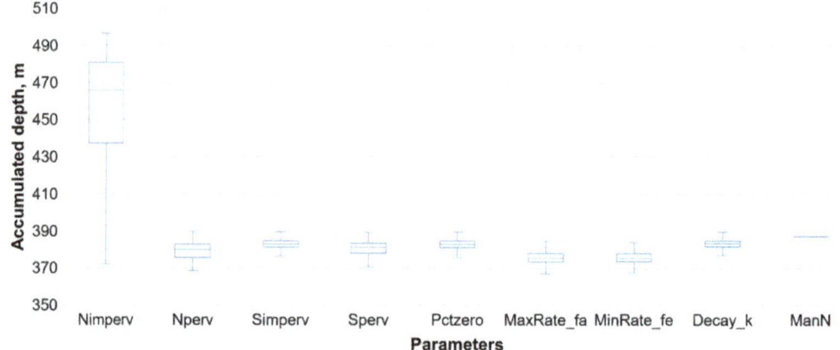

Figure 13. Variation of the accumulated depth on the Sabinal River with respect to the parameters.

One key feature in Figure 13 is that for Nimperv, it has a negative bias of the accumulated depths, since the whisker of the box is longer towards the low values (there is greater dispersion of the lower limit to quartile one of the data). With respect to the above, in Table 6, it can be noted that the results obtained from accumulated depths have greater variation with respect to the Nimperv parameter, since the standard deviation and coefficient of variation are 32.47 m and 7.12%. For MinRate_fe, the standard deviation is 4.89 m and the coefficient of variation is 1.30%.

Table 6. Mean μ, standard deviation σ, and coefficient of variation Cv of the depth values with respect to the parameters, Event 2.

Parameter	μ (m)	σ (m)	Cv (%)
Nimperv	**455.74**	**32.47**	**7.12**
Nperv	379.46	4.38	1.15
Simperv	383.17	2.60	0.68
Sperv	380.89	3.85	1.01
PctZero	383.11	2.66	0.69
MaxRate_fa	376.21	3.44	0.91
MinRate_fe	376.43	4.89	1.30
Decay_k	383.45	2.38	0.62
ManN	387.46	0.00	0.00

On the other hand, the sensitivity results indicate that parameter Nimperv has greater dispersion (positive bias) in quartile three and the upper limit (Figure 14), where the values of absolute–relative sensitivity range from 0.0654 to 0.1514, respectively. In contrast, the parameters MinRate_fe, Decay_k, MaxRate_fa, PctZero, Sperv, Simperv and Nperv have sensitivity values with less dispersion.

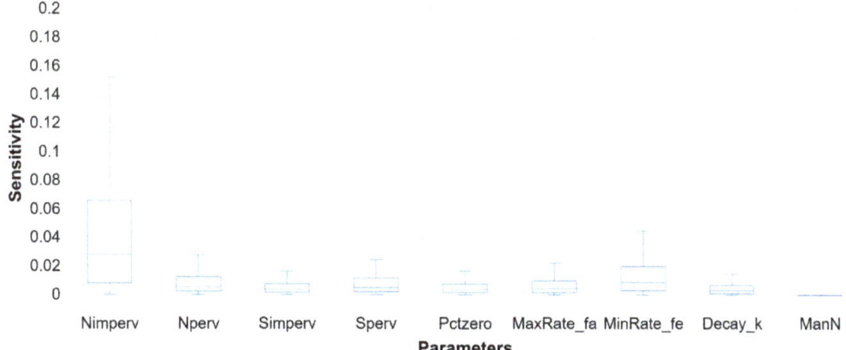

Figure 14. Comparison of parameters with respect to the absolute–relative sensitivity index, Event 2.

Figure 15 shows the results of the relative–absolute sensitivity: the Nimperv and Nperv parameters are the ones with the highest sensitivity values with respect to the model outputs; in this case, Nimperv from 0.7863 to 1.4513 (quartile three at upper limit) and Nperv from 0.2231 to 0.4943 (quartile three to upper limit), both parameters with positive bias. All other parameters have low sensitivity values with less dispersion.

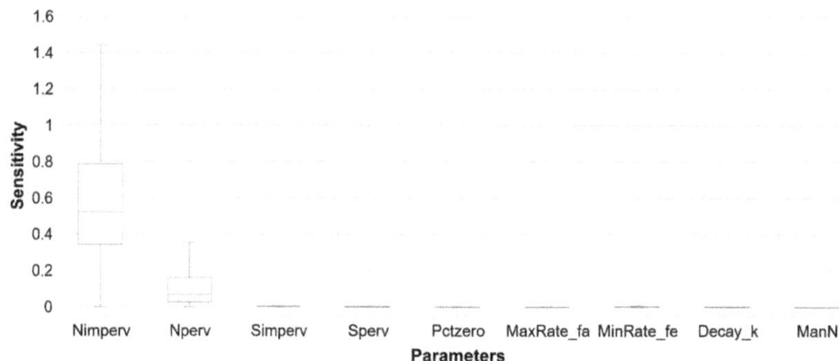

Figure 15. Comparison of parameters with respect to the relative–absolute sensitivity index, Event 2.

Finally, Figure 16 shows the sensitivity values of the parameters obtained from the relative–relative relationship: the Nimperv parameter is the one with high sensitivity values with the highest dispersion ranging from quartile three to the upper limit. In this case, the sensitivity values for Nimperv range from 0.0679 to 0.1326. As in calculations with previous sensitivity calculation expressions, the other parameters have less dispersion in the obtained values.

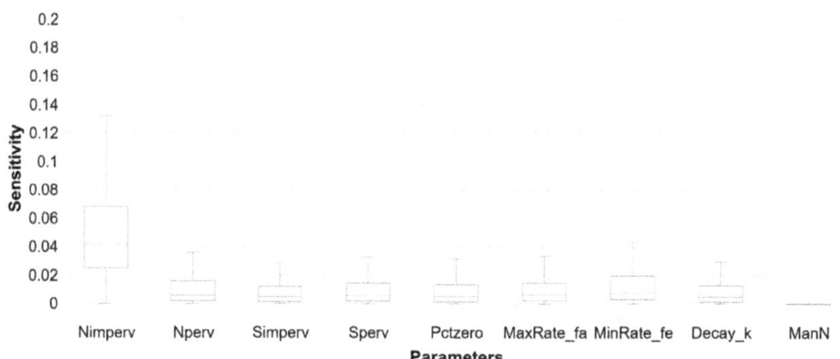

Figure 16. Comparison of parameters with respect to the relative–relative sensitivity index, Event 2.

Table 7 shows the statistical characteristics of each of the above figures, which correspond to the different methods of obtaining the sensitivity of the model results according to the parameters. For example, the Nimperv parameter for relative–absolute sensitivity has values of 0.1031 and 0.0106 of standard deviation and variance, respectively. On the other hand, these values are 0.6241 for the standard deviation and 0.3895 of variance with the relative–absolute sensitivity. Finally, for relative–relative sensitivity, these values are 0.05722 for standard deviation and 0.00327 for variance. In this case, the Nperv and Simperv parameters are not considered, because having a large amount of outlier data causes the standard deviation and sensitivity variance values to increase.

Based on the above analysis, Figure 17a,b show the medium sensitivity and maximum sensitivity (according to the upper limit of the whisker box) of parameters that generate the sensitivity in the hydrological simulation model.

Table 7. Mean, μ, standard deviation, σ, and variance, σ^2, of sensitivity values, Event 2.

Parameter	Absolute–Relative			Relative–Absolute			Relative–Relative		
	μ	σ	σ^2	μ	σ	σ^2	μ	σ	σ^2
Nimperv	0.0596	**0.1031**	**0.0106**	0.6875	**0.6241**	**0.3895**	0.05345	**0.05722**	**0.00327**
Nperv	0.0107	0.0221	0.0005	0.2182	0.6057	0.3668	0.02482	0.10692	0.01143
Simperv	0.0090	0.0445	0.0020	0.0059	0.0323	0.0010	0.03140	0.20057	0.04023
Sperv	0.0107	0.0294	0.0009	0.0023	0.0090	0.0001	0.02197	0.10039	0.01008
Pctzero	0.0090	0.0415	0.0017	0.0005	0.0028	0.0000	0.02181	0.07978	0.00637
Maxrate_fa	0.0108	0.0321	0.0010	0.0002	0.0008	0.0000	0.02257	0.10334	0.01068
Minrate_fe	0.0490	0.3743	0.1401	0.0057	0.0305	0.0009	0.03501	0.23318	0.05437
Decay_k	0.0086	0.0340	0.0012	0.0017	0.0108	0.0001	0.02140	0.10890	0.01190
ManN	0.0000	0.0000	0.0000	0.0000	0.0000	0.0000	0.00000	0.00000	0.00000

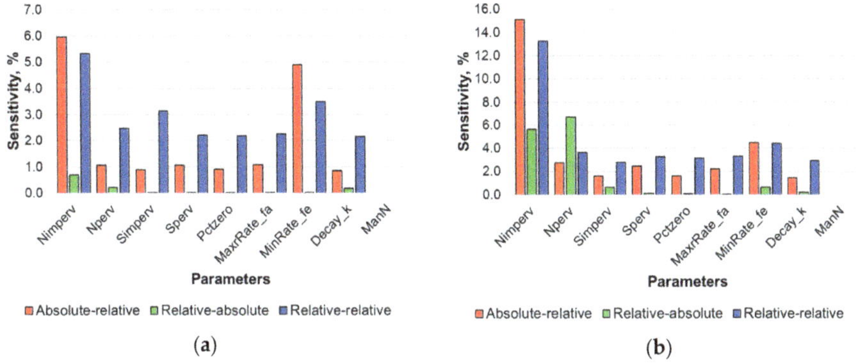

Figure 17. (**a**) Average sensitivity of modeling parameters and (**b**) maximum sensitivity of modeling parameters, Event 2.

R^2 was calculated according to the methodology. In Figure 18, R^2 efficiency results are shown, where the R^2 values for Nimperv are in the range of 0.35 to 0.74, having greater amplitude in the whiskers box. The other parameters have less bias in the obtained values.

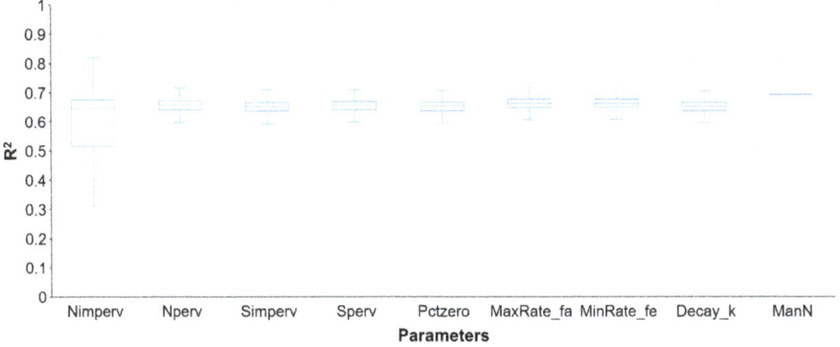

Figure 18. Comparison of parameters with respect to R^2, Event 2.

Based on the sensitivity and R^2 calculation, Figure 19a,b show the behavior of the depths generated by Event 1. This shows the change of the depth hydrogram according to the maximum, minimum, and mean values and the parameter value with greater R^2. These four parameters have higher Nimperv sensitivity and one of the lowest sensitivity parameters, MinRate_fe.

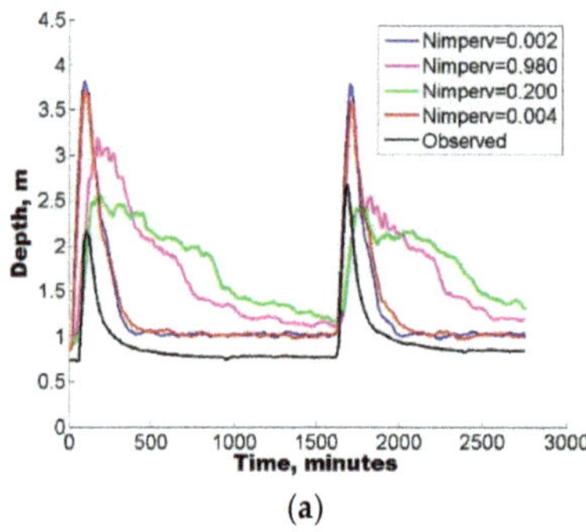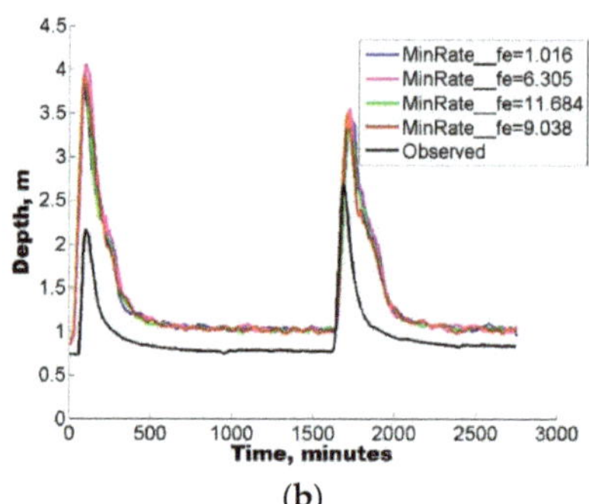

Figure 19. Depth levels of Sabinal river using (**a**) Nimperv parameter and (**b**) MinRate_fe parameter. The blue line is a minimum value, the magenta line is an average value, the green line is a maximum value, the red line is a parameter with maximum value of R2 and the black line are depth levels in hydrometric station.

4. Discussion

Some authors, such as Baek et al. and Sharifan et al. [42,43], suggest Nimperv and ManN as influential parameters in modeling with SWMM. Other researchers like Randall et al. [8] identified Horton's infiltration speeds as relatively sensitive parameters. Temprano et al. [44] also found the parameter with the highest sensitivity is the percentage of impervious surface. Guan et al. [45–47] report that the sensitive parameters in modeling in SWMM are those related to waterproofness, specifically in areas with depressions, Manning n of surfaces, and ducts.

This work implements the use of scarce data. Two different precipitation events and nine different input parameters were used. Each event was evaluated to show the robustness of the sensitivity analysis methodology in the hydrological modeling of urban basins, due to each individually analyzed parameter. The results show differences in the sensitivity of the model to the calculated parameters, where there is the accumulated depth as the comparison target.

For Event 1, it was found that the parameters with the highest sensitivity are Nimper and MinRate_fe. For Nimperv, cumulative depth values decrease while Nimperv values decrease and increase their values in the same way the parameter values grow. In the case of MinRate_fe, as the value of the parameter is lowered, the accumulated depth increases, and as the value of the parameter increases, the accumulated depth decreases. The above has a direct impact on the peak flow and runoff volume of the output hydrogram. For example, a high Nimperv value produces a dimmed output hydrogram with a higher runoff volume. A small Nimperv value produces a hydrogram with higher peak flow and lower runoff volume. In the case of MinRate_fe, a low parameter value produces an output hydrogram with high peak flow and higher runoff volume, and a high parameter value results in a hydrogram with a lower runoff volume and high peak flow.

In the case of validation with Event 2, the highest sensitivity parameter is Nimperv. The behavior of the parameter values and the values of the output hydrograms behave in the same way as in the previous case: its impact is on the volume of runoff and peak flow. MinRate_fe is less sensitive because the precipitation event is made up of two storms, and as a result, there is greater saturation in the ground at the end of the first peak of the storm. The remaining parameters for both Event 1 and Event 2 have a lower sensitivity relative to the result spectrum of the accumulated depths and have less influence on runoff volume and peak flow. In addition to the configuration of the basin and the location of the analysis point, the ManN parameter does not generate sensitivity in the model outputs. This may be due to the amount of base flow, the runoff generated by the nearby sub-basins, and the velocity.

According to the results, these parameters can be applied to carry out the calibration process. In turn, these parameters can be used individually or together. For example, in this case, R^2 efficiencies greater than 0.60 were found, indicating that there is more than one possible Nimperv parameter value that calibrates the model with a reliable fit, that is, forming sets of two or more parameters and thus having a better fit to the observed hydrogram [48]. The results show that with the methodology used, it is possible to have reliable calibration with scarce data. According to this study, having more than one hydrometric station would be calibrated according to the area of influence of the station, applying the methodology used in this research.

For expressions for calculating the sensitivity of modeling results with respect to the reason for changing the parameters, it was found that they correctly identify the parameters that cause the most variation in the results. SWMM correctly represents the rainfall–runoff phenomenon, and the sensitivity of the input parameters depends on the characterization of the basin under study, precipitation and antecedent moisture, so the order of sensitive parameters is different for each area under study.

5. Conclusions

This study reflects the importance of sensitivity analysis of hydrological and hydraulic parameters interacting in a hydrological model of an urban basin, mainly because these parameters allow one to perform an adequate calibration of a model in which the runoff time series best fit the observed data obtained from hydrometry.

Therefore, an effort should be made to perform this type of analysis and give certainty to the results in modeling. Thus, it is also important to perform sensitivity analysis with different methodologies that fit the needs of the modeler, since each case study is different.

For the case study, it can be concluded that satisfactory results were obtained, achieving the objective of characterizing the sensitivity of the modeling parameters under a framework of hydrometric data shortages and obtaining the result that the most important parameters of this basin are Nimperv and MinRate. In addition, the analysis was not only performed using the Equations (4)–(6) but complemented by the calculation of the standard deviation, variance and R^2 of the result. Implemented R^2 shows that there are several parameters that provide a good representation of the system. The results show that the sensitivity of the parameters depends on the basin under study and the effects of secondary interactions between the model parameters. It is also shown that the most sensitive parameters in a simulation vary according to the storm and the accumulated precipitation. As Knighton [21] suggested, SWMM is well parameterized for the calculation of the rainfall–runoff ratio, so care must be taken when identifying sensitive parameters and the order in which they are applied.

In cases where data scarcity is high, the implementation of methods that enable the quantification of sensitivity in model predictions permit more reliable results.

Finally, subjectivity in rainfall–runoff modeling should be considered, since it depends mainly on the expertise decision-making of the modeler; in future studies, the uncertainty analysis of such models should be considered as well.

Author Contributions: H.A.B.-G. performed most of the analysis and numerical simulations shown in the study and also contributed to the manuscript preparation; V.H.A.-Y. contributed to the design of the study and discussion of the results; J.J.C.-R. revised the methodology, participated in the discussion of results and contributed to the final manuscript. R.S.-C. revised the methodology, participated in the discussion of results and contributed to the final manuscript. All authors have read and agreed to the published version of the manuscript.

Funding: This research received no external funding.

Acknowledgments: The authors express their gratitude to the Department of Urban Hydraulics of the Mexican Institute of Water Technology (IMTA) for their support in this investigation.

Conflicts of Interest: The authors declare no conflict of interest.

References

1. Jha, A.K.; Bloch, R.; Lamond, J. *Cities and Flooding: A Guide to Integrated Urban Flood Risk Management for the 21st Century*; World Bank Publications: Washington, DC, USA, 2012.
2. Zhao, D.; Chen, J.; Wang, H.; Tong, Q. Application of a sampling based on the combined objectives of parameter identification and uncertainty analysis of an urban rainfall-runoff model. *J. Irrig. Drain. Eng.* **2013**, *139*, 66–74. [CrossRef]
3. Dotto, C.; Kleidorfer, M.; Deletic, A.; Rauch, W.; McCarthy, D.; Fletcher, T. Performance and sensitivity analysis of stormwater models using a Bayesian approach and long-term high resolution data. *Environ. Modell. Softw.* **2011**, *26*, 1225–1239. [CrossRef]
4. Huber, W.; Dickinson, R. *Stormwater Management Model (SWMM) Version 4.0: User's Manual*; United States Environmental Protection Agency: Athens, GA, USA, 1988.
5. James, W.; Rossman, L.E.C.; James, W.R. *User's Guide to SWMM5*, 13th ed.; CHI: Guelph, ON, Canada, 2010; pp. x1–x905.
6. Li, J.; Li, Y.; Li, Y. SWMM-based evaluation of the effect of rain gardens on urbanized areas. *Environ. Earth Sci.* **2016**, *75*, 17. [CrossRef]
7. Gumbo, B.; Munyamba, N.; Sithole, G.; Savenije, H.H.G. Coupling of digital elevation model and rainfall-runoff model in storm drainage network design. *Phys. Chem. Earth Parts A/B/C* **2002**, *27*, 755–764. [CrossRef]
8. Randall, M.; Sun, F.; Zhang, Y.; Jensen, M.B. Evaluating Sponge City volume capture ratio at the catchment scale using SWMM. *J. Environ. Manag.* **2019**, *246*, 745–757. [CrossRef]
9. Chang, T.K.; Talei, A.; Chua, L.H.C.; Alaghmand, S. The Impact of Training Data Sequence on the Performance of Neuro-Fuzzy Rainfall-Runoff Models with Online Learning. *Water* **2019**, *11*, 52. [CrossRef]
10. Agarwal, S.; Kumar, S. Applicability of SWMM for semi Urban Catchment Flood modeling using extreme Rainfall Events. *Int. J. Recent. Technol. Eng.* **2019**, *8*, 245–251.
11. Rodriguez-Rincon, J.P.; Pedrozo-Acuña, A.; Breña-Naranjo, J.A. Propagation of hydro-meteorological uncertainty in a model cascade framework to inundation prediction. *Hydrol. Earth Syst. Sci.* **2015**, *19*, 2981–2998. [CrossRef]
12. Petrucci, G.; Bonhomme, C. The dilemma of spatial representation for urban hydrology semi-distributed modelling: Trade-offs among complexity, calibration and geographical data. *J. Hydrol.* **2014**, *517*, 997–1007. [CrossRef]
13. Sun, N.; Hall, M.; Hong, B.; Zhang, L.J. Impact of SWMM catchment discretization: Case study in Syracuse, New York. *J. Hydrol. Eng.* **2014**, *19*, 223–234. [CrossRef]
14. Hamel, P.; Fletcher, T.D. Modelling the impact of stormwater source control infiltration techniques on catchment baseflow. *Hydrol. Process.* **2014**, *28*, 5817–5831. [CrossRef]
15. Sorooshian, S.; Chu, W. Review of Parameterization and Parameter Estimation for Hydrologic Models. In *Land Surface Observation, Modeling and Data Assimilation*; Liang, S., Li, X., Xie, X., Eds.; World Scientific: Singapore, 2013; Volume 5, pp. 127–138.
16. Gupta, H.V.; Sorooshian, S.; Yapo, P.O. Status of Automatic Calibration for Hydrologic Models: Comparison with Multilevel Expert Calibration. *J. Hydrol. Eng.* **1999**, *4*, 135–143. [CrossRef]
17. McMillan, H.; Clark, M. Rainfall-runoff model calibration using informal likelihood measures within a Markov chain Monte Carlo sampling scheme. *Water Resour. Res.* **2009**, *45*, 1–12. [CrossRef]
18. Gou, J.; Miao, C.; Duan, Q.; Tang, Q.; Di, Z.; Liao, W.; Wu, J.; Zhou, R. Sensitivity analysis-based automatic parameter calibration of the VIC model for streamflow simulations over China. *Water Resour. Res.* **2020**, *56*, 1–19. [CrossRef]
19. Muleta, M.K.; Nicklow, J.W. Sensitivity and uncertainty analysis coupled with automatic calibration for a distributed watershed model. *J. Hydrol.* **2005**, *306*, 127–145. [CrossRef]
20. Peel, M.C.; Blöschl, G. Hydrological modelling in a changing world. *Prog. Phys. Geogr.* **2011**, *35*, 249–261. [CrossRef]
21. Knighton, J.; Lennon, E.; Bastidas, L.; White, E. Stormwater detention system parameter sensitivity and uncertainty analysis using SWMM. *J. Hydrol. Eng.* **2016**, *21*, 1–15. [CrossRef]

22. Wan, B.; James, W. SWMM Calibration using Genetic Algorithms. Global Solutions for Urban Drainage. In Proceedings of the Ninth International Conference on Urban Drainage (9ICUD), Portland, OR, USA, 8–13 September 2002.

23. Fraga, I.; Cea, L.; Puertas, J.; Suárez, J.; Jiménez, V.; Jácome, A. Global sensitivity and GLUE-based uncertainty analysis of a 2D-1D dual urban drainage model. *J. Hydrol. Eng.* **2016**, *21*, 1–11. [CrossRef]

24. Razavi, S.; Gupta, H.V. What do we mean by sensitivity analysis? The need for comprehensive characterization of "global" sensitivity in Earth and Environmental systems models. *Water Resour. Res.* **2015**, *51*, 3070–3092. [CrossRef]

25. Shin, M.-J.; Choi, Y.S. Sensitivity Analysis to Investigate the Reliability of the Grid-Based Rainfall-Runoff Model. *Water* **2018**, *10*, 1839. [CrossRef]

26. Mannina, G.; Viviani, G. An urban drainage stormwater quality model: Model development and uncertainty quantification. *J. Hydrol.* **2010**, *381*, 248–265. [CrossRef]

27. Kleidorfer, M.; Deletic, A.; Fletcher, T.D.; Rauch, W. Impact of input data uncertainties on urban stormwater model parameters. *Water Sci. Technol.* **2009**, *60*, 1545–1554. [CrossRef] [PubMed]

28. Bárdossy, A. Calibration of hydrological model parameters for ungauged catchments. *Hydrol. Earth Syst. Sci.* **2007**, *11*, 703–710. [CrossRef]

29. Thorndahl, S.; Beven, K.J.; Jensen, J.B.; Schaarup-Jensen, K. Event based uncertainty assessment in urban drainage modelling, applying the GlUE methodology. *J. Hydrol.* **2008**, *357*, 421–437. [CrossRef]

30. Bajracharya, A.; Awoye, H.; Stadnyk, T.; Asadzadeh, M. Time Variant Sensitivity Analysis of Hydrological Model Parameters in a Cold Region Using Flow Signatures. *Water* **2020**, *12*, 961. [CrossRef]

31. Zaghloul, N.A. Sensitivity analysis of the SWMM Runoff-Transport parameters and the effects of catchment discretization. *Adv. Water Resour.* **1983**, *6*, 214–226. [CrossRef]

32. Jamali, B.; Bach, P.M.; Deletic, A. Rainwater harvesting for urban flood management—An integrated modelling framework. *Water Res.* **2020**, *171*, 1–11. [CrossRef]

33. Luan, B.; Yin, R.; Xu, P.; Wang, X.; Yang, X.; Zhang, L.; Tang, X. Evaluating Green Stormwater Infrastructure strategies efficiencies in a rapidly urbanizing catchment using SWMM-based TOPSIS. *J. Clean. Prod.* **2019**, *223*, 680–691. [CrossRef]

34. Babaei, S.; Ghazavi, R.; Erfanian, M. Urban flood simulation and prioritization of critical urban sub-catchments using SWMM model and PROMETHEE II approach. *Phys. Chem. Earth* **2018**, *105*, 3–11. [CrossRef]

35. Ballinas-González, H.A.; Alcocer-Yamanaka, V.H.; Pedrozo-Acuña, A. Uncertainty Analysis in Data-Scarce Urban Catchments. *Water* **2016**, *8*, 524. [CrossRef]

36. Horton, R.E. An Approach Toward a Physical Interpretation of Infiltration Capacity. *Soil Sci. Soc. Am. J.* **1940**, *5*, 399–417. [CrossRef]

37. Rawls, W.J.; Ahuja, L.R.; Brakensiek, D.L.; Shirmohammadi, A. Infiltration and soil water movement. In *Handbook of Hydrology*; Maidment, D.R., Ed.; McGraw-Hill, Inc.: New York, NY, USA, 1993; pp. 1–51. Available online: https://www.mheducation.co.uk/handbook-of-hydrology-9780070397323-emea (accessed on 20 April 2020).

38. Rossman, L.A. *Storm Water Management Model: User's Manual Version 5.0*; US EPA, National Risk Management Research Laboratory: Cincinnati, OH, USA, 2010.

39. Kleidorfer, M. Uncertain Calibration of Urban Drainage Models: A Scientific Approach to Solve Practical Problems. Ph.D Thesis, Universität Innsbruck, Innsbruck, Austria, 2009.

40. Reichert, P. Environmental System Analysis. In *Department of System Analysis, Integraated Assessment and Modelling*; eawag. Swiss Federal Institute of Aquatic Science and Technology: Dübendorf, Switzerland, 2009; pp. 40–65.

41. Krause, P.; Boyle, D.P.; Bäse, F. Comparison of different efficiency criteria for hydrological model assessment. *Adv. Geosci.* **2005**, *5*, 89–97. [CrossRef]

42. Baek, S.-S.; Choib, D.-H.; Jung, J.-W.; Lee, H.-J.; Lee, H.; Yoon, K.-S.; Cho, K.-H. Optimizing low impact development (LID) for stormwater runoff treatment in urban area, Korea: Experimental and modeling approach. *Water Res.* **2015**, *86*, 122–131. [CrossRef] [PubMed]

43. Sharifan, R.A.; Roshan, A.; Aflatoni, M.; Jahedi, A.; Zolghadr, M. Uncertainty and sensitivity analysis of SWMM model in computation of manhole water depth and subcatchment peak flood. *Proced. Soc. Behav. Sci.* **2010**, *2*, 7739–7740. [CrossRef]

44. Temprano, J.; Arango, O.; Cagiao, J.; Suárez, J.; Tejero, I. Stormwater quality calibration by SWMM: A case study in Northern Spain. *Water SA* **2006**, *32*, 55–63. [CrossRef]

45. Guan, M.; Sillanpää, N.; Koivusalo, H. Modelling and assessment of hydrological changes in a developing urban catchment. *Hydrol. Process.* **2015**, *29*, 2880–2894. [CrossRef]

46. Krebs, G.; Kokkonen, T.; Valtanen, M.; Setälä, H.; Koivusalo, H. Spatial resolution considerations for urban hydrological modelling. *J. Hydrol.* **2014**, *512*, 482–497. [CrossRef]

47. Barco, J.; Wong, K.M.; Stenstrom, M.K. Automatic Calibration of the U.S. EPA SWMM Model for a Large Urban Catchment. *J. Hydraul. Eng.* **2008**, *134*, 466–474. [CrossRef]

48. Beven, K.; Binley, A. The future of distributed Models: Model Calibration and Uncertainty Prediction. *Hydrol. Process.* **1992**, *6*, 279–298. [CrossRef]

Article

Rainfall Intensity-Duration-Frequency Relationship. Case Study: Depth-Duration Ratio in a Semi-Arid Zone in Mexico

Ena Gámez-Balmaceda [1,2], Alvaro López-Ramos [3], Luisa Martínez-Acosta [2,3], Juan Pablo Medrano-Barboza [3], John Freddy Remolina López [4], Georges Seingier [1], Luis Walter Daesslé [1] and Alvaro Alberto López-Lambraño [1,2,5,6,*]

[1] Instituto de Investigaciones Oceanólogicas, Universidad Autónoma de Baja California, Baja California 22860, Mexico; ena.gamez@uabc.edu.mx (E.G.-B.); georgess@uabc.edu.mx (G.S.); walterr@uabc.edu.mx (L.W.D.)

[2] Faculty of Engineering, Architecture and Design, Universidad Autónoma de Baja California, Baja California 22860, Mexico; lmartinez18@uabc.edu.mx

[3] GICA Group, Faculty of Civil Engineering, Universidad Pontificia Bolivariana campus Montería, Montería 230002, Córdoba, Colombia; alvaro.lopezr@upb.edu.co (A.L.-R.); juan.medrano@upb.edu.co (J.P.M.-B.)

[4] ITEM Group, Faculty of Electronic Engineering, Universidad Pontificia Bolivariana campus Montería, Montería 230002, Córdoba, Colombia; john.remolina@upb.edu.co

[5] Hidrus S.A. de C.V., Ensenada 22760, Mexico

[6] Grupo Hidrus S.A.S., Montería 230002, Colombia

[*] Correspondence: altoti@gmail.com or alopezl@hidrusmx.com; Tel.: +521-442-194-6654 or +521-646-134-5766

Received: 9 September 2020; Accepted: 12 October 2020; Published: 15 October 2020

Abstract: Intensity–Duration–Frequency (IDF) curves describe the relationship between rainfall intensity, rainfall duration, and return period. They are commonly used in the design, planning and operation of hydrologic, hydraulic, and water resource systems. Considering the intense rainfall presence with flooding occurrences, limited data used to develop IDF curves, and importance to improve the IDF design for the Ensenada City in Baja California, this research study aims to investigate the use and combinations of pluviograph and daily records, to assess rain behavior around the city, and select a suitable method that provides the best results of IDF relationship, consequently updating the IDF relationship for the city for return periods of 10, 25, 50, and 100 years. The IDF relationship is determined through frequency analysis of rainfall observations. Also, annual maximum rainfall intensity for several duration and return periods has been analyzed according to the statistical distribution of Gumbel Extreme Value (GEV). Thus, Chen's method was evaluated based on the depth-duration ratio (R) from the zone, and the development of the IDF relationship for the rain gauges stations was focused on estimating the most suitable (R) ratio; chosen from testing several methods and analyzing the rain in the region from California and Baja California. The determined values of the rain for one hour and return period of 2 years (P_1^2) obtained were compared to the values of some cities in California and Baja California, with a range between 10 and 16.61 mm, and the values of the (R) ratio are in a range between 0.35 and 0.44; this range is close to the (R) ratio of 0.44 for one station in Tijuana, a city 100 km far from Ensenada. The values found here correspond to the rainfall characteristics of the zone; therefore, the method used in this study can be replicated to other semi-arid zones with the same rain characteristics. Finally, it is suggested that these results of the IDF relationship should be incorporated on the Norm of the State of Baja California as the recurrence update requires it upon recommendation. This study is the starting point to other studies that imply the calculation of a peak flow and evaluation of hydraulic structures as an input to help improve flood resilience in the city of Ensenada.

Keywords: hydrologic statistics; flood design; extreme rainfall intensity time series

---

1. Introduction

Intensity–Duration–Frequency (IDF) relationship, or IDF curves, is a representation of intense rainfall events that allows for calculation of a peak flow needed to design hydraulic structures (e.g., storm sewers, culverts, drainage systems), to assess and predict flood hazard, and design flood protection structures [1–3]. Most of these structures were designed in many developing countries a long time ago without an updated IDF—remarkable, since rain is a variable that changes with space and time. Consequently, any update from the IDF relationship in urban catchments will necessarily imply revision and modification of the local standard structural designs [4]. The updated IDF design would also help to predict flood risk occurrences and map flood hazards of the expected peak flow. This is especially relevant in arid and semiarid zones where rain characteristics indicate that yearly variation in storms is very large, and the intensity of rare storms is always very high for a brief period, and so therefore the flooding is of sudden occurrence and rapid rise [5]. Northwest Mexico is a semi-arid region with low annual average rainfall; however, with the presence of rainfall intensities associated with climate variability, that has caused flooding [6–8].

The Intensity-Duration-Frequency relationship design is based on the measurement of peak rainfall events, regarding duration and developed for a certain recurrence interval or return period [9], where accuracy depends on the rainfall characteristics, such as magnitude, frequency, and duration [10,11]. The analyzed data is the precipitation time series, modeled for future projections at a regional scale. These projections indicate that the precipitation return period tends to increase if the climate conditions do not change [12,13]. However, due to Climate Change, there are uncertainties of intense rainfall occurrence affecting the return period of the IDF design, which in turn tends to decrease in some global regions [14]. This issue makes it necessary for trend analysis of the extreme rainfall events to design or update the IDF relationship.

The best estimation of the precipitation intensity (unit/time) is directly obtained from the automatic (pluviograph) weather station that measures the sub-hourly rain, and for which the records are automatically transmitted every five or ten minutes [15,16]. There is a low density of this kind of weather station in the Mexican territory, though. The separation between stations should be between 5 and 30 km [17,18] according to the Manual on the Global Observing System proposed by the World Meteorological Organization. However, distance between the stations in Mexico is 70 km on average; thus, there is a lack of information between stations along space and time. In the country, and only since 1999, the available sub-hourly rain records have been registered through automatic weather stations (EMAs), by the National Meteorological Service (SMN).

The Ensenada's city has two IDF designs: one published as an Official Norm by the State of Baja California, with daily rainfall records from 1948 to 2008, just for one station [19]; and the other, by the Federal Communication and Transportation Ministry (SCT) published in 2000, and whose analyzed time series are unknown [20]. These are the only IDF studies found in the literature for the study area, and both serve as official Mexican documents [21,22]. Due to a lack of automatic stations, none of these studies used the pluviograph records from the last two decades (2000–2020), to assess extreme rainfall events necessary to develop the IDF curves. This problem of a lack of data and weather stations is evident in several cities in Mexico; however, to minimize the inconvenience of the periodicity of measurements and the distance between weather stations, most studies in Mexico have used different empirical methods to estimate the IDF relationship based on daily rainfall from standard rain gauges (pluviometric), suggesting that the Chen method [23] is the most appropriate for the IDF estimation [24,25].

Considering flood occurrence caused by intense rains, the insufficient studies on the IDF relationship, lack of pluviographic information, and other important aspects for planning hydraulic structures in Ensenada city and its surroundings, it is necessary to carry out a detailed analysis of the Intensity–Duration–Frequency relationship for the area. Therefore, the purpose of this study is to analyze, estimate, and propose the Depth–Duration–Ratio (R), appropriate for the characteristics of the rains that occur in the study area, to obtain representative IDF curves. For this, it is necessary to estimate the rainfall of one hour, and the return period (P_1^2) of two years. (P_1^2) is traditionally derived from the proposals made by Hershfield [26], Reich [27], and Bell [28] in areas where only pluviometric information is available. However, based on pluviographic information obtained in automatic stations, it was possible to adjust the expression of Bell [28] that represents the characteristics of the rainfall occurring in the study area. Based on this, the Chen method [23] is used to obtain the IDF curves and consequently their updating for the return periods of 10, 25, 50, and 100 years. Additionally, the spatial distribution of the rainfall-duration ratio (R) was obtained, which facilitates obtaining the IDF relationship with the Chen method [23] in places where there are no pluviographic and standard rain gauge stations.

Finally, the adequate estimation of IDF curves will allow and guarantee the planning and optimal design of hydraulic structures. Future flood hazard studies and consequently disaster risk management will also benefit directly from this work.

2. Materials and Methods

The urban zone of Ensenada and its surroundings has been selected as the study area, located on the Pacific coast of Baja California (31°30′–31°60′N; 116°50′–116°10′W. See Figure 1). This zone, in the northwest of Mexico, has a semiarid climate, with convective type rains characterized for being intense and of short term. The average annual rainfall is 273 mm, where the rainy season usually occurs between November and April. Topographic landscape is variable, with steep slopes, alluvial valleys, and an alluvial coastal plain [29]. This area is divided into four urban and seven rural subbasins that drain toward the Pacific coast. There are 11 weather stations inside the basins located at different elevations, that contain the main data to achieve the purpose of the study (Figure 1).

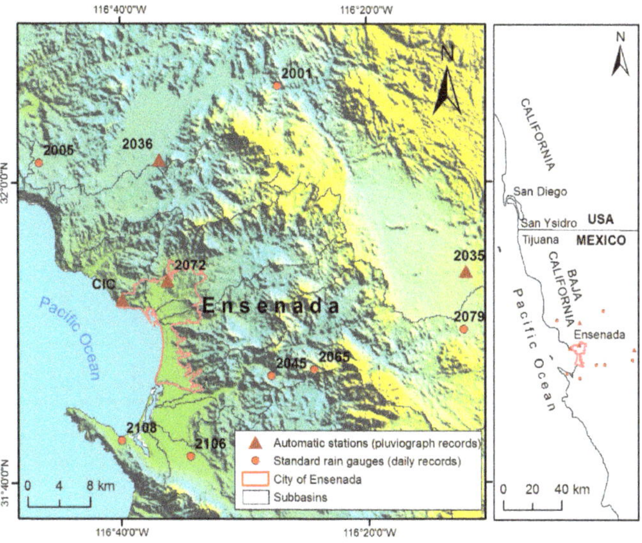

Figure 1. Study area location showing sub-basins and distribution of weather stations.

The required data to develop the IDF relationship in the study zone were the maximum rainfall events, annually recorded at different durations; obtained from four automatic stations (pluviograph records). On the other hand, historical rainfall data of ten standard rain gauges (daily rainfall records) were collected and included in the calculations (Table 1).

Table 1. Rain gauge network from the study zone.

Station Name	Station ID	Elevation above Sea Level (M)	Type of Records	Lat	Long	Record Length	Source of Data
Emilio Lopez Zamora	2072	43	10 min 24 h	31.89	−116.60	1999–2019 1940–2019	CONAGUA CONAGUA/CICESE
CICESE	CIC	60	5 min	31.86	−16.66	2007–2019	CICESE
Guadalupe	2036	361	5 min 24 h	32.02	−116.61	2009–2019 1954–2019	CICESE CONAGUA
Ojos Negros	2035	680	5 min 24 h	31.91	−116.23	2009–2019 1948–2019	CICESE CONAGUA
Agua Caliente	2001	400	24 h	32.10	−116.45	1969–2011	CONAGUA
Boquilla de Santa Rosa	2005	250	24 h	32.02	−116.77	1969–2011	CONAGUA
San Carlos	2045	164	24 h	31.78	−116.46	1969–2011	CONAGUA
Santo Tomas	2065	180	24 h	31.79	−116.40	1969–2011	CONAGUA
El Alamar	2079	710	24 h	31.83	−116.20	1969–2011	CONAGUA
Maneadero	2106	50	24 h	31.69	−116.57	1969–2011	CONAGUA
Punta Banda	2108	15	24 h	31.71	−116.66	1969–2010	CONAGUA

5-min: automatic stations, 24 h (daily): rain gauge stations. Lat: latitude, Long: longitude.

Based on the characteristics of the study area and available precipitation data, the methodology was defined with the steps shown in Figure 2.

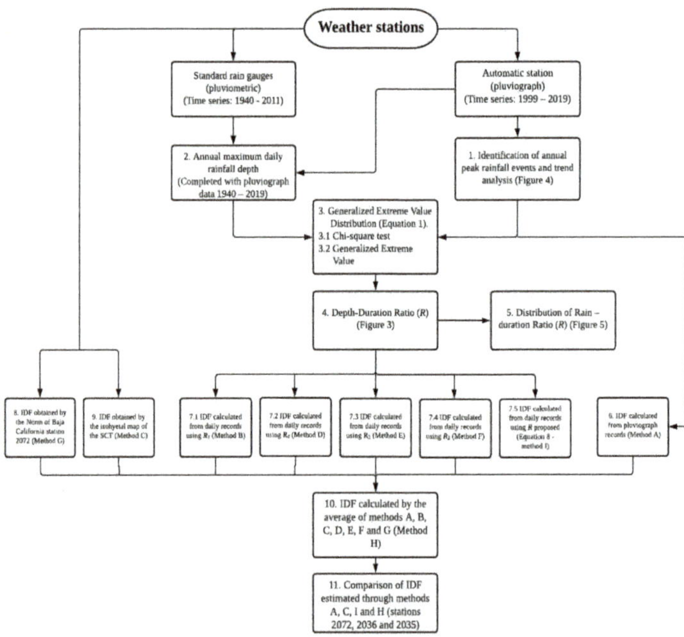

Figure 2. General diagram of the methodology.

Table 1 introduces four automatic stations. The Emilio Lopez Zamora station is the automatic station with more recorded data. It has 20 years of records, with measures from every ten minutes. This station is managed by a governmental agency called Comisión Nacional del Agua (CONAGUA). The other three automatic stations scarcely have 10 years of records, with measures for every five minutes. They are CICESE (CIC), Guadalupe (2036), and Ojos Negros (2035); all three were administered by Centro de Investigación Científica y de Educación Superior de Ensenada (CICESE). From these data, the maximum rainfall events of each year were identified and classified by their magnitude and duration in order to estimate the IDF curves.

Identification of each event was based on the continuous pluviograph records, from the rain start, until it stopped for more than one hour; e.g., if the rain ended at any time, but it suddenly started before an hour passed, this record was considered to be part of the previous event. Nonetheless, it was also considered a separate event during the day if, after one event had occurred, there were two hours without rain, and it started to rain again. Once all events were identified, they were classified by duration and magnitude to select the most intense of every year. Intensity is defined by the rain magnitude in mm, registered for a determined time; e.g., there are events with the same magnitude and different durations in hours, where the most intense is always the one with the shortest duration. Measures of the events recorded by the four automatic stations had different durations. In some automatic stations, the longest event recorded (not the most intense), was one of 180 min. Other intense events had a duration of only 10 minutes. However, most of the intense events had a duration of 120 min; therefore, this duration was selected to identify the intense events of each year. López–Lambraño developed this method to identify and classify the intense events [30], and it has been satisfactorily applied by Maldonado et al. [31].

Regarding rainfall information, there are 10 stations around the city of Ensenada (Table 1), which have records of maximum rainfall in 24 h, from 1940 to 2011. In the same location of these stations there are three automatic stations (2072, 2036 and 2035), which have data from 1999 to 2019. Thus, the time series of the pluviometric data of these three stations were completed from the pluviograph records (automatic stations); e.g., for station 2072 a time series of 79 years was obtained. This was achieved through the daily accumulation of rainfall recorded every 10 min in that station until 2019. This same procedure was performed for stations 2036 and 2035.

A trend analysis was carried out on the pluviographic and pluviometric data, to determine the trend lines and verify if the registered rainfall tends to increase or decrease in the area. The stationarity in all the stations was also verified through the Augmented Dickey–Fuller test, which is a unit root test that allows accepting or rejecting the stationarity hypothesis in a time series [32]. Subsequently, the Chi square goodness and fit test was performed for the precipitation data analyzed with the probability distribution functions of Gumbel type I, Log-normal, Frechet and double Frechet to verify which of them presented the best fit.

The development of the IDF relationship is a procedure that starts with data availability, type of records, years of records, quality, and coverage, to choose an efficient method and claim the best results. Two methods, for different conditions, were applied in this case to estimate the IDF relationship for the return periods of 10, 25, 50, and 100 years. These periods, proposed by the National Agency for Prevention of Disasters, are suitable for the design of minor urban hydraulic structures and flood hazard assessments [21]. The first method to estimate the IDF relationship is applied to pluviograph data (from automatic stations). These data provide the real intensity that occurred for each rain event through the years and can be extrapolated applying a probability distribution. The Generalized Extreme Value Distribution (GEV) is usually applied to estimate IDF curves, like the extrapolation of the available data to estimate the peak value of the sample. The Gumbel's method, based on the theory of extreme values, is the first method applied here, where the probability of an event of a determined magnitude not being equaled or exceeded, can safely be adopted and have been widely used [30,33]. This is expressed by the Equation (1).

$$F(X) = e^{[-e^{(-\frac{X-\mu}{\alpha})}]} \quad (-\infty < X \leq \infty) \tag{1}$$

where $0 < \alpha < \infty$ is the scale parameter, $-\infty < \mu < \infty$ is the location parameter or central value, e is the base of the Naperian logarithms, α, X, and μ; corresponding to the parameters of the statistic moments of the distribution. The distribution derivative provides the probability density function where the values of X, for different return periods (T), are estimated by means of:

$$X_T = \mu + \alpha Y_T \tag{2}$$

$$Y_T = -\ln\left[\ln\left(\frac{T}{T-1}\right)\right] \tag{3}$$

Confidence intervals are important to estimate the return period and data accuracy. Generally, the data to calculate the IDF curves have a standard error of 10% for short return periods and 20% for periods of 50 and 100 years [28]. By Gumbel's method, applied to the automatic data, the IDF curves can be estimated with a confidence value for a return period of 20 years, considering the 10 and 20 years of records from the three automatic stations. The IDF curves for a return period of 20 years are very useful for minor urban structures design. However, to develop the IDF curves for large return periods, such as 50 or 100 years, there would be uncertainty when applying the Gumbel method. Therefore, the requirement to apply Gumbel's method with a good confidence value, would be to have a long series of precipitation data.

Pluviograph records were useful to have real data of the events to estimate IDF for short return periods but were not enough for extrapolating rainfall intensities for the larger return periods. Therefore, the use of daily rainfall depth, from the standard rain gauges, was necessary to make the calculation of the rainfall intensity for all the return periods proposed. This allows us to counteract the calculations with the automatic data and disseminate better results.

Consequently, a second method to develop the IDF curves recommended for urban hydrological design in the Mexican Republic [24] is applied to daily rainfall records (from the 10 standard rain gauges). This method was developed by Chen as an alternative of the absent pluviograph records [23], by the following equation:

$$P_t^{Tr} = \frac{a P_1^{10} \log\left(10^{2-X} Tr^{X-1}\right) t}{60(t+b)^c} \tag{4}$$

where, P_t^{Tr} is the intensity of precipitation in mm/h, $P_1^{10} = R(P_{24}^{10})$, P_1^{10} is the rain in mm, generated in one hour for a return period of 10 years, (X) is the ratio of the rain-return period $X = \frac{P_t^{100}}{P_t^{10}}$, P_t^{100} and P_t^{10} is the rain of 24 h and return period of 100 and 10 years respectively. (Tr) is the return period in years, (t) is the duration in minutes, (a), (b), and (c) are parameters of regional characterization of the rain defined by the (R) ratio.

The depth-duration ratio (R) is the most important parameter to estimate the IDF relationship, from daily records, by using the Chen equation for a specific geographic location, because this ratio is related to rain characteristics of the zone [34]. As shown before, Equation (4) is based on (R) ratio to determine the rain of one hour and a return period of ten years P_1^{10}, and therefore, the rain for a given return period. The (R) ratio for any average condition of rainfall over any geographical areas, also, has been proposed by Chen [23], through the following equation:

$$R = \frac{P_1^2}{P_{24}^2} \tag{5}$$

where, (P_1^2) is the rain in mm, generated in one hour for a return period of 2 years. (P_{24}^2) is the rain in mm, generated in 24 h for a return period of 2 years. This value is easily estimated by applying Gumbel distribution for each standard rain gauge. (P_1^2) can be estimated from pluviograph records applying the Gumbel distribution for each station. However, the issue to apply formula 5 is finding the value of the rain of one hour and a return period of two years (P_1^2), when there are only daily records, not pluviograph records. Therefore, to validate Chen's method in this region, it was considered to make a good estimation of (R) ratio through the value of (P_1^2).

Several alternatives to apply Equation (5) for any geographic area, based only on daily records, have been given by Hershfield and Wilson through a diagram that relates the mean of maximum annual observations of precipitation days with the mean annual number of thunderstorm days; Reich, through a proposed world isopluvial map of 2-year/1-h maximum precipitation; and Bell, based on Hershfield method through the following equation:

$$P_1^2 = 0.17MN^{0.33} \tag{6}$$

Equation (6) could be applied if $0 < M \le 2.0$, $1 < N \le 80$, in which (P_1^2) = 2-yr, 1-h rainfall in inches: M = mean of maximum annual observational—precipitation day in inches: and N = mean annual number of thunderstorm days.

The purpose was to find the R-value to 10 rain gauge stations applying Equation (5), where (P_{24}^2) was easily estimated using daily records, by applying Gumbel distribution (Equations (2) and (3)). In the case of (P_1^2) there are three stations with pluviograph records and seven stations with daily records, the issue was to find (P_1^2) for the seven stations with daily records. Consequently, it was decided to evaluate the methods available for estimating the value of (P_1^2) and choosing the best way of finding the accurate (P_1^2) to calculate the (R) ratio.

The process started by knowing the approximated value of the (P_1^2) along the regional area to compare them with the estimated values through the mentioned methods, started the process. Approximated values of (P_1^2) in the regional area were examined through a literature review of regional studies. This review, with the same rainfall characteristics, includes California in the US (CA), near to the international border, and Baja California in Mexico, as shown in Figure 1.

The first method evaluated to calculate (P_1^2) was the Gumbel distribution using pluviograph records from stations: CIC, 2072, 2036, and 2035. Then, a weighted average of (P_1^2) was estimated from the four automatic stations, through the Thiessen polygon. The averaged value was assigned to the seven rain gauges stations, to replace it in Equation (5) and have the first (R) value in the zone, as (R_1).

The next step was to calculate (P_1^2) for the 10 rain gauges stations through the methods previously described methods; Hershfield, Bell, and Reich [26–28]. Each value was replaced in Equation (5) to have different values of (R) like (R_2) and (R_3). Three different (R) values were estimated. Nevertheless, from the different values of (P_1^2), it was considered to search for a relationship to provide values of (P_1^2) for the rain gauges stations close to the values of (P_1^2) estimated from pluviograph records. The Hershfield and Bell methods provide the same results. This way, a new method to find (P_1^2) was derived from Bell´s formula (Equation (6)). After several tests of parameters variation, the new relationship resulted from modifying Equation (6) in the following equation:

$$P_1^2 = 0.12MN^{0.33} \tag{7}$$

The (P_1^2) values estimated from Equation (7) were replaced in Equation (5) to have new values of (R), such as (R_4). Therefore, the evaluated methods provided four different results of $(R_{1\,\dots\,4})$. The challenge was to choose the best method of finding (R); thus, to address this challenge, an average of all calculated (R) was made, and this average was proposed as the best option to determine the (R)

ratio for the city of Ensenada (Equation (8)). A summary of the process to find (R) ratio for each station in the zone is represented in Figure 3.

$$\overline{R} = \frac{\sum\limits_{i=1}^{n} R}{n} \tag{8}$$

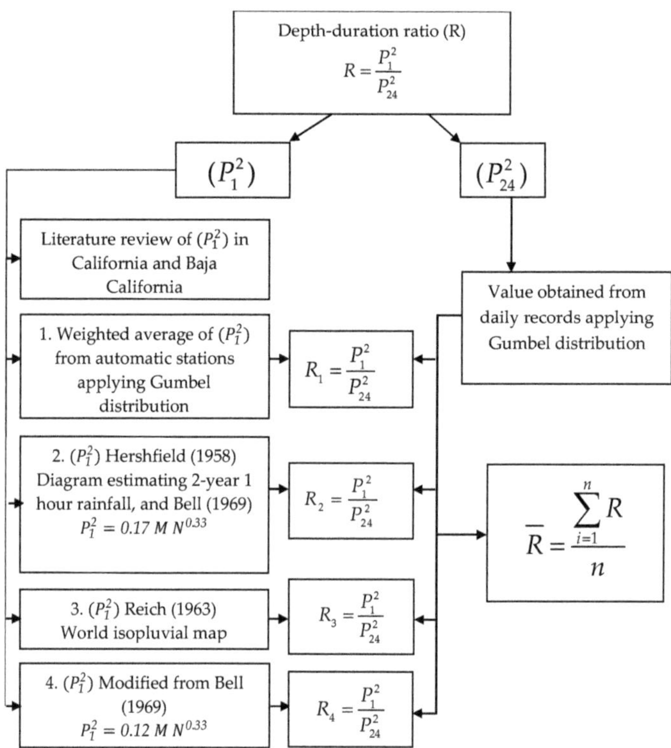

Figure 3. Procedure to determine the (R) ratio proposed. $\overline{R}$ is the average ratio and n is the number of $(R_{1\ldots4})$ calculated for each (P_1^2).

Once the (R) ratio was determined for each rain gauge station, the spatial distribution of the (R) ratio was drawn on a map by interpolation, using the Inverse Distance Weighting (IDW) method. The error analysis of this spatial interpolation was carried out between actual and interpolated values using three statistical parameters: Mean Absolute Error (MAE), Root Mean Square Error (RMSE) and coefficient of determination (r^2). Consequently, the distribution of (R) ratio for the study area can be successfully used to calculate the IDF curves in the zone.

The procedure to develop the IDF relationship for the proposed return periods includes the combination of pluviograph and daily rainfall records, through the methods described above, testing different values of estimated (R), and comparing their results, to choose the most suitable, according to rain characteristics.

The first step was the calculations of the IDF relationship in three automatic stations (2072, 2036, and 2035), using the maximum rainfall events for each year, and applying the Gumbel method (Equation (1)).

The second step was the estimation of the IDF relationship for the 10 rain gauges stations, with daily records for more than 40 years, applying the Chen method (Equation (4)). There are three pluviographic stations (2072, 2036, and 2035) on the same location from the previous stations. The calculations using Chen equation included the use of the (R) ratios found with the different methods already mentioned, for all stations, followed by the comparison of the different results of IDF developments with the IDF values by the (SCT), and the comparison with the IDF relationship developed in 2011 by the by Norm of the State of Baja California, that only has results for station 2072.

Therefore, for each rain gauge station, with daily records for more than 40 years, there are several IDF calculations to enhance the comparison between each other, and have more options to choose the best results, developed through the following methods:

A.   IDF calculated from pluviograph records using Gumbel distribution (Equation (1)); this included only three stations (2072, 2036, and 2035).

B.   IDF calculated from daily records of ten stations (standard rain gauge), using the Chen equation, where (R) was calculated through the weighted average of the (P_1^2) from automatic stations, applying Gumbel distribution, implementing (R_1).

C.   IDF obtained by the isohyetal map of the (SCT) only for stations 2072, 2036, 2035, 2001, 2005, and 2045.

D.   IDF calculated from daily records of ten stations (standard rain gauge), using the Chen equation, where (R) was calculated through the (P_1^2) by the modification of Bell equation [28], implementing (R_4).

E.   IDF calculated from daily records of ten stations (standard rain gauge), using the Chen equation, where (R) was calculated through the (P_1^2) by the world isopluvial map of Reich [27], implementing (R_3).

F.   IDF calculated from daily records of ten stations (standard rain gauge), using the Chen equation, where (R) was calculated through the (P_1^2) by Hershfield and Bell [26,28], implementing (R_2), both methods provide the same results of (P_1^2).

G.   IDF obtained by the Norm of the State of Baja California only for station 2072.

H.   IDF calculated by the average of IDF developed through options A, B, C, D, E, F, and G.

I.   IDF calculated from daily records of ten stations (standard rain gauge), using Chen equation, where (R) is the ratio proposed calculated by Equation (8).

Once the IDF relationships were estimated from the previously established methods, two comparisons were made. The first comparison was made from the use of methods A, B, C, D, E, F, and G. The second comparison was made from the use of methods A, C, H, and I. From the first comparison method A and C were chosen again, considering that method A is very important because it was developed with pluviograph information (2072, 2036, and 2035 stations), representing the continuous measure of rainfall intensity. However, the rainfall time series data at stations do not exceed 20 years of record (Table 1). Thus, the method C was chosen because it corresponds to the official IDF relationship from the country. The chosen method and the corresponding results of the IDF curve for the city of Ensenada and its surroundings are shown in the results and discussion section.

3. Results and Discussion

Peak rainfall events of each year were identified from the records of the four automatic stations, following the previously described method. Table 2 provides the events extracted from the Emilio Lopez Zamora station (2072). Since this station has the most extensive time series of rainfall data (20 years), it will be taken to illustrate the maximum precipitation events.

Table 2. Peak rainfall events for ELZ station (2072).

| | Peak Rainfall Events (Mm) at Different Durations (ELZ -2072) | | | | | |
Years	10 min	20 min	30 min	60 min	120 min	180 min
1999	3.30	5.08	5.58	5.58	7.35	9.12
2000	6.35	8.13	8.12	10.40	14.23	20.60
2001	5.59	6.85	8.12	9.14	10.90	13.40
2002	3.05	4.32	4.57	7.11	10.16	11.70
2003	2.54	3.55	5.09	6.60	11.18	12.40
2004	4.83	7.11	9.14	10.90	18.79	24.60
2005	3.56	5.34	6.61	11.70	13.72	13.70
2006	4.32	7.37	9.14	12.40	18.29	18.30
2007	4.32	8.13	12.19	18.03	24.37	35.30
2008	3.56	6.35	6.60	11.40	18.27	21.60
2009	4.83	6.86	9.14	11.40	17.00	18.80
2010	6.86	8.89	9.65	12.20	19.56	23.40
2011	3.56	5.33	7.11	11.40	16.00	20.10
2012	4.83	5.34	6.86	9.15	16.52	16.50
2013	3.05	6.10	6.61	7.61	12.95	19.60
2014	6.60	10.92	12.19	12.70	13.19	13.40
2015	6.80	9.91	12.45	13.20	21.34	22.90
2016	4.32	7.62	8.38	9.90	14.21	17.80
2017	5.84	8.89	9.14	13.00	20.07	22.40
2018	3.00	4.57	6.00	7.40	9.89	13.20
2019	4.40	5.60	7.40	14.60	17.40	27.60

It is clearly seen that the highest precipitation depth of each event occurs during the first 10 and 20 min. These observations confirm the short periods of rainfall events, as a local rain characteristic. Moreover, to analyze the increase or decrease of extreme events throughout the years, a rainfall trend has been drawn. The line trend of time series was analyzed from data of the station that have pluviograph and daily rainfall record. Figure 4 shows the line trend for data of station (2072). Figure 4a shows the trend of the high daily rainfall depth of every year, and Figure 4b presents the trend of the extreme rainfall events of the years at different durations, not necessarily the most intense (pluviograph records). The series presents a positive trend. In both cases, the trend indicates that maximum daily rainfall depth and peak rainfall events have been increasing with the years and will continue occurring in the future. In the trend analysis shown in Figure 4a, the slope establishes that the average increase in the maximum daily rainfall is 18%, i.e., if the current conditions that govern the occurrence of rainfall in the study area were to be maintained over time, then precipitation would increase at a rate of 1.8 mm per decade. For the case of Figure 4b there would be an increase of approximately 1.2 mm. The above magnitudes are considered significant if we consider that precipitation falls over a given coverage area, which translates into the generation of more direct runoff volume over the city of Ensenada.

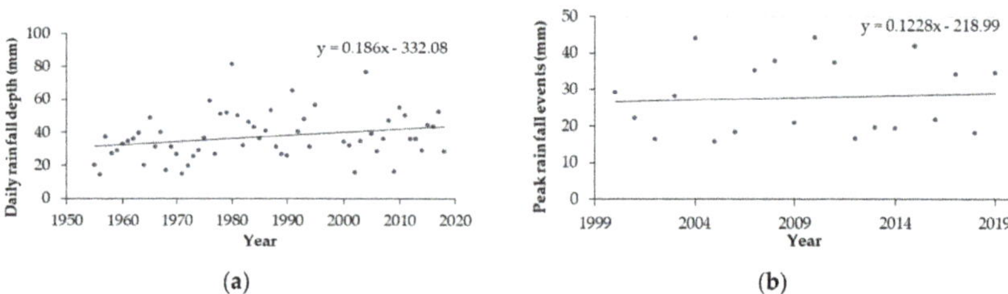

Figure 4. Trend analysis for station 2072 using automatic and daily records. (**a**) Annual maximum daily rainfall depth. (**b**) Annual Peak rainfall events.

These results can also be complemented by a detailed analysis of the rainfall time series for the State of Baja California [35].

The traditional methods used in hydrology to estimate rainfall and extreme flows for different return periods are based on the stationarity hypothesis in the probability distribution function of the series. According to Poveda et al. [36], this hypothesis could not be valid given the effects of climate change, climate variability, changes in land use, and the records of hydrological variables. Therefore, the Augmented Dickey–Fuller test was performed to evaluate the rainfall time series in the city of Ensenada fulfilling the hypothesis of stationarity or non-stationarity (Table 3).

Table 3. Augmented Dickey–Fuller hypothesis test for precipitation data from stations used in the study area.

Data		Alternative Hypothesis (Significance 0.05)	Result	p-Value
2072 (Rainfall Data)	10 min	Stationary	Stationary	0.0382
	20 min	Stationary	Non stationary	0.2505
	30 min	Stationary	Non stationary	0.5175
	60 min	Stationary	Non stationary	0.5984
	120 min	Stationary	Non stationary	0.7600
	180 min	Stationary	Non stationary	0.6485
2001		Stationary	Non stationary	0.2235
2005		Stationary	Non stationary	0.2138
2035		Stationary	Non stationary	0.5035
2036		Stationary	Non stationary	0.1968
2045		Stationary	Non stationary	0.0870
2065		Stationary	Non stationary	0.4364
2072		Stationary	Non stationary	0.3331
2079		Stationary	Stationary	0.0380
2106		Stationary	Non stationary	0.0603
2108		Stationary	Non stationary	0.0535

Table 3 shows the results of the Augmented Dickey–Fuller test. The analysis was carried out for a 5% significance level, finding that none of the time series corresponding to weather stations comply with the stationarity assumption because the p-value is greater than 0.05 [32]. Given the previous analysis, the Chi-square goodness and fit test was used for the precipitation data analyzed with the Gumbel type I, Log-normal, Frechet and double Frechet probability distribution functions to verify which of them had the best fit. In the case of the rain gauge stations, eight were better adjusted to the Gumbel type I distribution function, and two stations to the double Frechet. The previous analysis was carried out with a 0.05 confidence level. In the case of the automatic station 2072, both the Gumbel type I distribution function and the double Frechet distribution function presented equivalent results. Given the above, it was decided to use the Gumbel distribution function for the analysis of the relationship Intensity-Duration-Frequency of rainfall in the study area. In addition, it has been found that the chosen distribution fits well for precipitation in semi-arid areas [11].

The (R) ratio was estimated based on the (P_1^2) following the process shown in Figure 3. The result of the regional review of (P_1^2) is shown in Table 4. The regional review reveals that values of (P_1^2) from Southern California to Northern Baja California are in the range between 10 and 16.61 mm, indicating that values of (P_1^2) for the study area should be estimated in this range. For this reason, the evaluation of the different methods to calculate (P_1^2) and estimate (R) followed by the comparison of their results was carefully analyzed to determine the accurate (R) ratio. The results of (P_1^2) are in the range of the regional review, when $(P_1^2)_1$ is calculated by the average of pluviograph and daily records applying Gumbel, $(P_1^2)_2$ is calculated by Hershfield [26], $(P_1^2)_3$ is calculated by Reich [27], and $(P_1^2)_4$ is calculated by the adjusting and modification of Bell [28]. The results of $(R_{1...4})$ ratios estimated for each station through the different methods of (P_1^2) were averaged to define and support the proposed (R) for each station. These values are shown in Table 5.

Table 4. Literature review of (P_1^2) for California and Baja California.

Location of Weather Stations	Rain (P_1^2) (mm)	Reference
Southern California	12.7	Frevert et al., 1963 [37]
Southern California	12.7	Reich, 1963 [27]
Ensenada B.C	14	Reich, 1963 [27]
San Diego	14.22	Dedrick et al., 1976 [27]
Coronado San Diego, CA	13.13	Hodges et al, 1961 [26]
	Ranges (11.48–16.61)	NOAA, 2020 [38]
Chula Vista San Diego, CA	11.63	NOAA, 2020 [38]
Imperial Beach, CA	10.84	NOAA, 2020 [38]
San Ysidro, CA	Ranges (9.39–13.63)	NOAA, 2020 [38]
Northern of Baja California	10	CENAPRED, 2016 [22]

Table 5. Depth-duration ratio (R) estimated through the average ratios $R_{1\ldots4}$ estimated from (P_1^2) and (P_{24}^2). Where $(P_1^2)_1$ was averaged of pluviograph and daily records applying Gumbel, $(P_1^2)_2$ calculated by Hershfield [26] and Bell [28], $(P_1^2)_3$ calculated by Reich [27], and $(P_1^2)_4$ calculated by the adjusting and modification of Bell [28] (Equation (7)).

Station	$(P_1^2)_1$ (mm)	$(P_1^2)_2$ (mm)	$(P_1^2)_3$ (mm)	$(P_1^2)_4$ (mm)	(P_{24}^2) (mm)	R_1	R_2	R_3	R_4	R
2072	10.56	17.35	14.20	12.29	32.90	0.32	0.53	0.43	0.37	0.41
2036	11.00	21.13	13.80	14.92	38.40	0.29	0.55	0.36	0.39	0.40
2035	10.96	17.13	14.50	12.09	31.20	0.35	0.55	0.47	0.39	0.44
2001	10.20	19.40	13.90	11.62	34.70	0.29	0.56	0.40	0.33	0.40
2005	10.20	23.33	13.70	12.86	41.00	0.25	0.57	0.33	0.31	0.37
2045	10.20	22.38	14.50	13.02	39.90	0.26	0.56	0.36	0.33	0.38
2065	10.20	21.87	14.50	11.05	40.70	0.25	0.54	0.36	0.27	0.35
2079	10.20	20.33	14.70	12.96	35.10	0.29	0.58	0.42	0.37	0.41
2106	10.20	18.34	14.50	10.80	34.00	0.30	0.54	0.43	0.32	0.40
2108	10.20	21.22	14.5	10.16	39.70	0.26	0.53	0.36	0.26	0.35

The (R) ratio calculated for each station varied from 0.35 to 0.44; this range is close to the (R) ratio of 0.44 for one station in Tijuana reported by Campos–Aranda [24], which indicates that both cities share the same rain characteristics. Thus, once the (R) ratio was estimated, the spatial distribution of this ratio was projected in a map for the study area (Figure 5). The error analysis was carried out between actual and interpolated values using three statistical parameters: Mean Absolute Error (MAE = 2.341×10^{-14}), Root Mean Square Error (RMSE = 1.53011×10^{-7}), and coefficient of determination (R^2=1). There is a significant positive correlation between actual and interpolated values estimated for the depth-duration ratio. Values for MAE and RMSE indicate that the IDW method created a good interpolated surface for the whole area of interest based on the observed data. It can be established that in Ensenada City the average value of R is about 0.39 and this value is proposed as (R) ratio for the area.

It is highly important to highlight that in the absence of pluviograph data, the (R) ratio calculated becomes the key to develop a satisfactory IDF curves for the standard rain gauge stations, by applying Chen equation.

IDF curves are commonly developed using historical annual maximum precipitation, this involves utilization of long-term historical rainfall observations. When sub-daily rainfall records are not available, the characteristics of extreme rainfall intensities, and subsequently, their distribution functions corresponding to the short durations might not be captured [39–43]. This problem is clearly addressed by applying Equation (7) to rain duration ratio (R_4) estimated (Table 5), likewise distribution of the depth-duration ratio (Figure 5). In this way, empirical formulas (e.g., Chen) can be used in areas where there are no pluviographs (high temporal resolution), rain gauges, and/or information available.

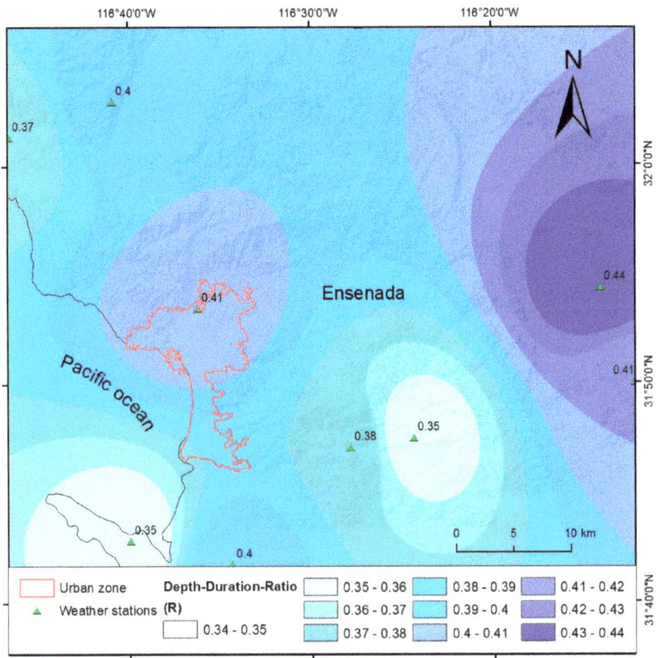

Figure 5. Distribution of the Depth-Duration Ratio (R) for the city of Ensenada.

The IDF relationship were calculated with pluviograph records, by applying Gumbel distribution and, calculated with daily records by applying Chen methods through different options of (R) ratio. The different results allowed us to make comparisons to choose the most suitable method for calculations of the IDF relationship.

The comparisons of results were focused on the three stations that have pluviograph and daily records (2072, 2036, and 2035). However, station 2072 has been chosen to define the methods of comparisons, because it has the widest time series of pluviograph and daily records.

Table 6 presents the results of the first comparison that tested methods A, B, C, D, E, F, and G previously described, and its average likelihood method (H), to assess the IDF relationship. In this table the values of the IDF relationship obtained by the different methods vary considerably, and for this reason it was decided to carry out an average of these methods. However, it is important to highlight that method G presents the highest magnitudes and correspond to the IDF reported in the Baja California State Standards. It should be noted that these magnitudes are far from the average. With these results, it was identified that these IDFs are over-estimated, which implies a greater over determination of dimensions in the designs of hydraulic structures, and so therefore higher construction costs.

From this comparison, another comparison was selected to assess and estimate the IDF relationship for the rest of the rain gauge stations. The comparison defines the selection of the best method of IDF estimation, that includes the IDF estimated using pluviograph data (method A), the IDF reported by the SCT (method C), the IDF estimated with the R ratio proposed in this zone (method I), and the IDF from method H. This comparison is shown in Figures 6–8 for the stations 2072, 2036, and 2035.

Table 6. First comparison of IDF relationship, estimated through different methods for station 2072 for different return periods.

Return Period	Method	IDF (mm/hr) Station 2072				
		10 (min)	20 (min)	30 (min)	60 (min)	120 (min)
10 years	(A)	38.3	28.1	22.3	14.6	10.5
	(B)	45.2	32.2	25.8	17.2	11.2
	(C)	48.0	38.0	26.0	16.0	13.0
	(D)	52.8	37.8	30.1	19.7	12.5
	(E)	59.6	42.7	33.9	21.8	13.5
	(F)	70.6	50.8	40.1	25.1	14.9
	(G)	99.2	59.0	43.5	25.9	15.4
	Average	59.1	41.2	31.7	20.0	13.0
25 years	(A)	44.3	32.4	25.6	16.7	12.0
	(B)	53.3	38.0	30.5	20.3	13.2
	(C)	57.0	40.0	31.0	20.0	15.0
	(D)	62.4	44.6	35.5	23.2	14.7
	(E)	70.3	50.4	40.1	25.7	15.9
	(F)	83.4	60.0	47.3	29.7	17.6
	(G)	119.0	70.0	52.0	31.0	18.0
	Average	70.0	47.9	37.4	23.8	15.2
50 years	(A)	48.7	35.5	28.0	18.2	13.1
	(B)	59.5	42.5	34.0	22.6	14.7
	(C)	64.0	45.0	35.0	22.0	18.0
	(D)	69.6	49.7	39.6	25.9	16.4
	(E)	78.5	56.3	44.7	28.7	17.7
	(F)	93.0	66.9	52.8	33.1	19.7
	(G)	134.7	80.1	59.1	35.1	20.9
	Average	78.3	53.7	41.9	25.8	17.0
100 years	(A)	53.1	38.7	30.4	19.7	14.2
	(B)	65.7	46.9	37.5	25.0	16.3
	(C)	71.0	50.0	38.0	24.0	20.0
	(D)	76.8	54.9	43.8	28.6	18.1
	(E)	86.6	62.1	49.3	31.7	19.6
	(F)	102.7	73.9	58.3	36.6	21.7
	(G)	150.0	89.2	65.8	39.1	23.3
	Average	86.6	59.4	46.2	29.2	19.0

(A) Estimated from pluviograph records using Gumbel (Equation (1)). (B) Estimated from the Chen equation using R_1 from Table 5. (C) Obtained from the isohyetal map of the SCT. (D) Estimated from the Chen equation using R_4 from Table 5 (Equation (7)). (E) Estimated from the Chen equation using R_3 from Table 5 (Reich, 1963). (F) Estimated from the Chen equation using R_2 from Table 5 (Equation (6)). (G) Obtained from the of Norm by the State of Baja California.

The comparisons of IDF relationship for the 2072 station showed similar values calculated with methods (I) and (H). The method (I) is calculated by using Chen with the proposed (R) ratio, and the method (H) is calculated from the average of IDF developed by methods A, B, C, D, E, F, and G described before. On the other hand, the intensities obtained with method C are lower than those obtained with method I. This may be because the IDFs obtained with method C have not been updated since 2000, therefore, they do not involve the analyzes performed in Figure 4a,b.

Figures 6–8 show that the results of the method (I) are closer to methods (C) and (H) than method (A). Therefore, considering the position of the results of (I) on the range of values of all methods, it could be suggested as the most suitable to calculate the IDF curves. The values to estimate Figures 6–8 can be seen in Tables S1–S3 in Supplementary Materials. Method (I) was chosen to calculate IDF for the 10 rain gauge stations like the best option of the methods tested. Table 7 shows the IDF relationship for the 10 stations calculated by the Chen method through the proposed (R) ratio.

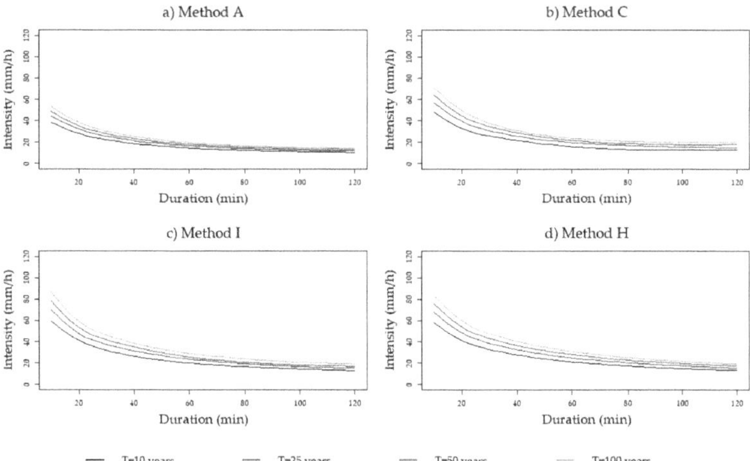

Figure 6. Established comparison of the IDF relationship estimated through different methods for station 2072 for different return periods. (**a**) shows the IDF curves from method (A), estimated from pluviograph records using Gumbel (Equation (1)). (**b**) shows the IDF curves from method (C) estimated from isohyetal map of the SCT. (**c**) shows the IDF curves from method (H) estimated from the average of IDF developed by methods A, B, C, D, E, F, and G, and (**d**) shows the IDF curves from method (I) estimated from Chen equation using the (R) ratio proposed (Equation (8)).

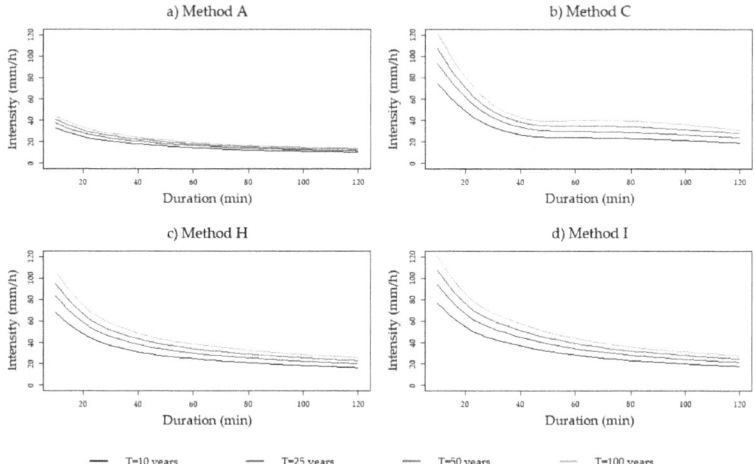

Figure 7. Established comparison of the IDF relationship estimated through different methods for station 2036 for different return periods. (**a**) shows the IDF curves from method (A), estimated from pluviograph records using Gumbel (Equation (1)). (**b**) shows the IDF curves from method (C) estimated from isohyetal map of the SCT. (**c**) shows the IDF curves from method (H) estimated from the average of IDF developed by methods A, B, C, D, E, and F, and (**d**) shows the IDF curves from method (I) estimated from Chen equation using the (R) ratio proposed (Equation (8)).

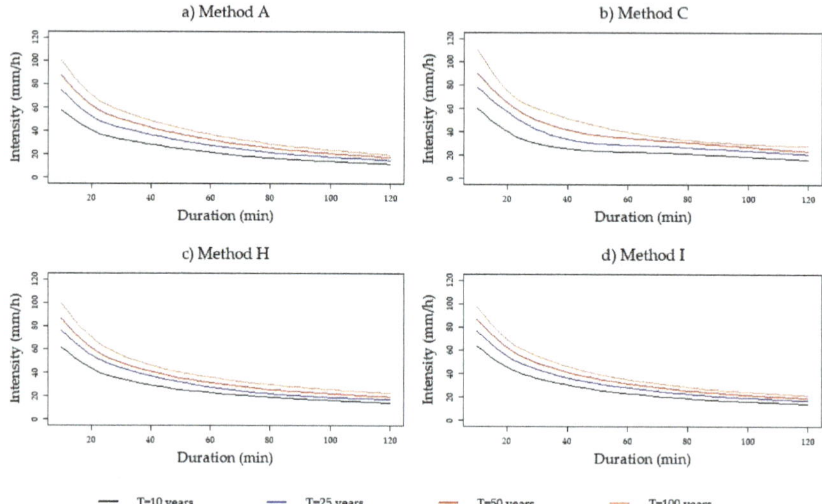

Figure 8. Established comparison of the IDF relationship estimated through different methods for station 2035 for different return periods. (**a**) shows the IDF curves from method (A), estimated from pluviograph records using Gumbel (Equation (1)). (**b**) shows the IDF curves from method (C) estimated from isohyetal map of the SCT. (**c**) shows the IDF curves from method (H) estimated from the average of IDF developed by methods A, B, C, D, E, and F, and (**d**) shows the IDF curves from method (I) estimated from Chen equation using the (*R*) ratio proposed (Equation (8)).

For the stations 2036 and 2035, the results of the IDF relationship were similar by methods (C) and (I). However, when graphing the IDF curves with the C method, it is noted that they do not show the typical fit of said curves. This may be because there are errors in the process of estimating them. A comparison has also been made for the stations 2001, 2005, and 2045, that provided equivalent results by applying methods (C) and (I). From the analysis of the comparison of results, method I is recommended as the most suitable for the IDF design in the evaluated stations.

For the IDF relationship calculation from daily records to be used in Mexico, the Chen's method has been mostly recommended by other authors [24,44]. However, considering that rain changes in space and time, the use of this method has been validated through the accurate calculation of (*R*) ratio, where it depended on the (P_1^2). The results of (P_1^2) obtained by pluviograph data, Reich and the adjusting from Bell equation, are similar; however, the method widely used is Hershfield that provides the highest values. Therefore, it was considered to calculate (*R*) with each value of (P_1^2) of each method to make an average of all estimated (*R*). This average was proposed as the (*R*) ratio of the zone that can be safe to calculate the IDF curves in other area stations.

The differences between the various methods in this study can be attributed to the length of the time series used. Finally, it has been verified that the quality and length of the pluviographic and rainfall precipitation records are important aspects in the study of the intensity-duration-frequency relationship in an area.

Table 7. IDF relationship selected for the rain gauge stations estimated through the Chen Method with the (R) ratio proposed in this study.

Return Period	Stations	Intensity Duration Frequency (mm/h) for Different Durations				
		10 (min)	20 (min)	30 (min)	60 (min)	120 (min)
10 years	2072	57.71	41.35	32.86	21.22	13.21
	2036	76.70	54.92	43.68	28.30	17.69
	2035	63.58	45.63	36.19	23.17	14.22
	2001	68.68	49.18	39.11	25.34	15.84
	2005	76.61	54.76	43.66	28.55	18.12
	2045	66.59	47.62	37.93	24.73	15.61
	2065	68.58	48.97	39.11	25.76	16.51
	2079	71.32	51.10	40.61	26.23	16.32
	2106	61.60	44.11	35.08	22.73	14.21
	2108	74.84	53.44	42.68	28.11	18.02
25 years	2072	68.15	48.83	38.80	25.06	15.59
	2036	93.96	67.28	53.51	34.66	21.67
	2035	76.82	55.14	43.73	28.00	17.18
	2001	83.95	60.12	47.81	30.97	19.36
	2005	94.00	67.19	53.56	35.03	22.23
	2045	79.00	56.51	45.01	29.34	18.53
	2065	83.28	59.47	47.50	31.28	20.05
	2079	87.28	62.54	49.70	32.10	19.97
	2106	73.74	52.81	41.99	27.21	17.01
	2108	93.09	66.48	53.09	34.97	22.42
50 years	2072	76.04	54.48	43.30	27.96	17.40
	2036	107.01	76.63	60.94	39.48	24.68
	2035	86.85	62.33	49.43	31.65	19.43
	2001	95.50	68.39	54.38	35.23	22.03
	2005	107.15	76.59	61.06	39.94	25.34
	2045	88.39	63.22	50.36	32.83	20.73
	2065	94.41	67.41	53.84	35.46	22.73
	2079	99.36	71.19	56.57	36.54	22.74
	2106	82.92	59.38	47.22	30.59	19.13
	2108	106.90	76.34	60.97	40.15	25.74
100 years	2072	83.93	60.14	47.79	30.87	19.21
	2036	120.07	85.98	68.38	44.30	27.70
	2035	96.87	69.52	55.13	35.30	21.67
	2001	107.05	76.66	60.96	39.49	24.69
	2005	120.30	85.99	68.55	44.84	28.45
	2045	97.79	69.94	55.71	36.32	22.93
	2065	105.53	75.35	60.18	39.63	25.41
	2079	111.43	79.84	63.45	40.98	25.50
	2106	92.11	65.96	52.45	33.98	21.25
	2108	120.71	86.20	68.84	45.34	29.07

4. Conclusions

The procedure and methods for developing IDF relationship in Ensenada were carefully tested making effective use of the available rainfall data, including pluviograph and daily records that allowed to achieve the research objectives.

Despite the absence of enough pluviograph data in this study, the IDF relationship was carried out successfully using the combination of Chen's methods, through the average (R) calculated from the average value of the one-hour rainfall and the two-year return period. The values of $\left(P_1^2\right)$ obtained were compared to the values of some cities in California and Baja California, with a range between 10 and 16.61 mm, and the values of the (R) ratio in a range between 0.35 and 0.44; this range is close to the (R) ratio of 0.44 for one station in Tijuana, a city 100 km farther North from Ensenada. The values found here correspond to the rainfall characteristics of the zone; therefore, the method used in this study can be replicated to another semi–arid zones with the same rain characteristics. The parameterization of the IDF relation for different durations, allows better understanding and realization of spatio-temporal analysis of the characteristics of rainfall in an area.

Chen's method is applicable to any zone if the (R) ratio is well defined. This value was carefully reviewed in the region and carefully tested with updated pluviograph and daily data. Therefore, the (R) ratio estimated is proposed to develop IDF curves in the absence of pluviograph data.

After analyzing the results of the IDF relationship for the station 2072, it was observed that the IDF relationship published by the Norm of the State of Baja California presents the highest values of all methods. Although they are also safe for designing, they imply the highest cost of construction—greater sizing—that could be minimized considering the new results. The document review indicates that the IDF relationships were developed with data available up to the year 2011. Therefore, it is suggested that these results from the IDF relationship should be included in the Norm of the State of Baja California, as it requires the recurrence update upon its recommendation.

This study guarantees the following aspects: input to rain–runoff models to improve the information available for an adequate design, planning, estimation of dimensions of civil works, and the integral management of water resources in this semi-arid region. Also, the proposed intensity–duration–frequency relationship will facilitate the evaluation of the flood hazard. The city of Ensenada periodically suffers the effects of floods, is out of electricity service, it is not possible to travel, there are drainage problems, health hazards, and work activities are often stopped. Therefore, proper risk management is important to protect not only the lives of citizens, but also the public and private assets, as well as the progress made in the development process. These results constitute a starting point for risk management and flood resilience.

Supplementary Materials: The following are available online at http://www.mdpi.com/2306-5338/7/4/78/s1, Table S1: Established comparison of the IDF relationship estimated through different methods for station 2072 for different return periods, Table S2: Established comparison of IDF relationship estimated through different methods for the station 2036 for different return periods, Table S3: Established comparison of the IDF relationship estimated through different methods for the station 2035 for different return periods.

Author Contributions: Conceptualization, E.G.-B.; A.A.L.-L.; methodology, E.G.-B., A.A.L.-L., J.P.M.-B., A.L.-R. and J.F.R.L; formal analysis, E.G.-B., G.S., L.W.D., A.A.L.-L., L.M.-A., J.P.M.-B., and A.L.-R.; writing—original draft preparation, E.G.-B., A.A.L.-L., L.M.-A., J.P.M.-B., A.L.-R. and J.F.R.L.; writing—review and editing, E.G.-B., A.A.L.-L., L.M.-A., A.L.-R. and J.F.R.L.; visualization, E.G.-B. and L.M.-A.; supervision, A.A.L.-L.; project administration, E.G.-B., A.A.L.-L., L.M.-A., A.L.-R. and J.F.R.L.; funding acquisition, J.F.R.L. All authors have read and agreed to the published version of the manuscript.

Funding: This research was funded by Universidad Autónoma de Baja California and HIDRUS S.A de C.V, Grupo HIDRUS S.A.S and Universidad Pontificia Bolivariana Campus Montería, the APC was funded by Universidad Pontificia Bolivariana Campus Montería.

Acknowledgments: We would like to thank the Centro de Investigación Científica y de Educación Superior de Ensenada (CICESE), and especially Santiago Higareda from the meteorology laboratory, who provided the precipitation data of automatic station.

Conflicts of Interest: The authors declare no conflict of interest.

References

1. De Paola, F.; Giugni, M.; Topa, M.E.; Bucchignani, E. Intensity-Duration-Frequency (IDF) rainfall curves, for data series and climate projection in African cities. *Springer* **2014**, *3*, 1–18. [CrossRef] [PubMed]

2. Raghunath, H.M. *Hydrology Principles-Analysis-Design*; New Age International: New Delhi, India, 2006; ISBN 9788122423327.

3. Sherif, M.; Chowdhury, R.; Shetty, A. Rainfall and Intensity-Duration-Frequency (IDF) Curves in United Arab Emirates. In Proceedings of the World Environmental and Water Resources Congress, Portland, OR, USA, 1–5 June 2014.

4. Gericke, O.J.; Du Plessis, J.A. Evaluation of the standard design flood method in selected basins in South Africa. *J. S. Afr. Inst. Civ. Eng.* **2012**, *54*, 2–14.

5. Lin, X. *Flash Floods in Arid and Semi-Arid Zones*; UNESCO: Paris, France, 1999.

6. Cruz Aguirre, R.U. Además de "Rosa", Ocho Huracanes más han Pasado a Menos de 200 km de Ensenada. Available online: https://todos.cicese.mx/sitio/noticia.php?n=1215#.XdhI15NKjIV (accessed on 31 August 2020).

7. Pacheco, B. Tiene Baja California Historial de Desastres—El Vigía. Available online: https://www.elvigia.net/general/2015/12/10/tiene-baja-california-historial-desastres-220040.html (accessed on 31 August 2020).

8. Rodríguez Esteves, J.M. La conformación de los "desastres"—Construcción social del riesgo y variabilidad climática en Tijuana, B.C. *Front. Norte* **2007**, *19*, 83–112.

9. Arnell, V. *Analysis of Rainfall Data for Use in Design of Storm Sewer Systems*; International Conference on Urban Storm Drainage: Southampton, UK, 1978; pp. 1–15.

10. Campos-Aranda, D.F. Relación y estimación de predicciones de lluvia horaria-diaria en dos zonas geográficas de México. *Tecnol. Cienc. Agua* **2012**, *3*, 141–152.

11. López-Lambraño, A.; Fuentes, C.; González-Sosa, E.; López-Ramos, A. Pérdidas por intercepción de la vegetación y su efecto en la relación intensidad, duración y frecuencia (IDF) de la lluvia en una cuenca semiárida. *Tecnol. Cienc. Agua* **2017**, *8*, 37–56.

12. DeGaetano, A.T.; Castellano, C.M. Future projections of extreme precipitation intensity-duration-frequency curves for climate adaptation planning in New York State. *Clim. Serv.* **2017**, *5*, 23–35. [CrossRef]

13. Groisman, P.Y.; Knight, R.W.; Karl, T.R. Changes in intense precipitation over the Central United States. *J. Hydrometeorol.* **2012**, *13*, 47–66. [CrossRef]

14. WHO; UNEP. *Managing the Risks of Extreme Events and Disasters to Advance Climate Change Adaptation*; Cambridge University Press: New York, NY, USA, 2012; p. 594.

15. Jaklič, A.; Šajn, L.; Derganc, G.; Peer, P. Automatic digitization of pluviograph strip charts. *Meteorol. Appl.* **2016**, *23*, 57–64. [CrossRef]

16. Llasat, M.-C. An Objective Classification of Rainfall Events on the Basis of their Convective Features. Application to Rainfall Intensity in the North-East of Spain. *Int. J. Climatol.* **2001**, *21*, 1385–1400. [CrossRef]

17. Organización Meteorológica Mundial. *Guía del Sistema Mundial de Observación*; Organización meteorológica mundial: Geneva, Switzerland, 2010; ISBN 9789263304889.

18. Plummer, N.; Allsopp, T.; Lopez, J.A. *Guidelines on Climate Observation Networks and Systems*; Llansó, P., Ed.; World Meteorological Organization: Geneva, Switzerland, 2003.

19. Secretaría de Infraestructura y Desarrollo Urbano. Normas Técnicas para Elaboración de Proyectos de Alcantarillado Pluvial en el Estado de Baja California. 2013. Available online: http://www.ordenjuridico.gob.mx/Documentos/Estatal/Baja%20California/wo83549.pdf (accessed on 31 August 2020).

20. Secretaría Comunicaciones y Transporte. *Isoyetas de Intensidad-Duración-Periodo de Retorno para la República Mexicana—Baja California*; Secretaria Comunicaciones y Transporte: Mexicali, Mexico, 2014. Available online: http://www.sct.gob.mx/carreteras/direccion-general-de-servicios-tecnicos/isoyetas/ (accessed on 31 August 2020).

21. SEDATU. *Términos de Referencia para Elaborar Atlas de Riesgos*; Secretaría de Desarrollo Agrario, Territorial y Urbano: Ciudad de México, Mexico, 2016; p. 121.

22. Secretaría de Gobernación y Cenapred. *Guía Básica para la Elaboración de Atlas Estatales y Municpales de Peligros y Riesgos*; Violeta, R., Ed.; Secretaría de Gobernación y Cenapred: Ciudad de México, Mexico, 2014; ISBN 9706289054.

23. Chen, C. Rainfall intensity-duration-frequency formulas. *J. Hydraul. Eng.* **1983**, *109*, 1603–1621. [CrossRef]

24. Campos-Aranda, D. Intensidades máximas de lluvia para diseño hidrológico urbano en la república mexicana Rainfall Maximum Intensities for Urban Hydrological Design in Mexican Republic. *Ing. Investig. Tecnol.* **2010**, *11*, 179–188.

25. Lorente Castelló, J.; Casas Castillo, M.C.; Rodríguez Sola, R.; Redaño Xipell, Á. Intensidades extremas y precipitación máxima probable. *Fenóm. Meteorol. Advers. Esp.* **2013**, *Volumen 5*, 142–155.

26. Hodges, L.H.; Hershfield, D.M.; Washington, D.C.; Reichelderfer, F.W. *Rainfall Frequency Atlas of the United States for Durations from 30 Minutes to 24 Hours and Return Periods from 1 to 100 Years*; Weather Bureau: Washington, DC, USA, 1961; p. 65.

27. Reich, B.M. Short-Duration Rainfall-Intensity Estimates and Other Design Aids for Regions of Sparse Data. *J. Hydrol.* **1963**, *1*, 3–28. [CrossRef]

28. Bell, F.C. Generalized Rainfall-Duration-Frequency Relationships. *J. Hydraul. Div.* **1969**, *95*, 311–327.

29. Smith, S.V.; Bullock, S.H.; Hinojosa-Corona, A.; Franco-Vizcaíno, E.; Escoto-Rodríguez, M.; Kretzschmar, T.G.; Farfán, L.M.; Salazar-Ceseña, J.M. Soil erosion and significance for carbon fluxes in a mountainous Mediterranean-climate watershed. *Ecol. Appl.* **2007**, *17*, 1379–1387. [CrossRef]

30. López-Lambraño, A. *Elaboración de las Curvas de Intensidad-Duración-Frecuencia de la Lluvia en las Estaciones Pluviográficas Aeropuerto los Garzones, Universidad de Córdoba y Turipana, Localizadas en la Cuenca Media del Río Sinú, Montería*; Universidad Pontificia Bolivariana: Montería, CO, USA, 2001.

31. Maldonado, M.; Martínez, G.; Matajira, F.F. Elaboración curvas IDF estaciones: Cinera Villa Olga y Santa Isabel—Municipio de Cúcuta—Colombia. *Rev. Ambient. Agua Aire Suelo* **2007**, *2*, 80–94.

32. Said, S.E.; Dickey, D.A. Testing for Unit Roots in Autoregressive-Moving Average Models of Unknown Order. *Biometrika* **1984**, *71*, 599–607. [CrossRef]

33. Courty, L.G.; Wilby, R.L.; Hillier, J.K.; Slater, L.J. Intensity-duration-frequency curves at the global scale. *Environ. Res. Lett.* **2019**, *14*, 1–11. [CrossRef]

34. Hersfield, D.M.; Wilson, W.T. Generalizing of Rainfall-Intensity-Frequency Data. *AIHS. Gen. Ass. Toronto* **1957**, *1*, 499–506.

35. López-Lambraño, A.A.; Fuentes, C.; López-Ramos, A.A.; Mata-Ramírez, J.; López-Lambraño, M. Spatial and temporal Hurst exponent variability of rainfall series based on the climatological distribution in a semiarid region in Mexico. *Atmosfera* **2018**, *31*, 199–219. [CrossRef]

36. Poveda, G.; Álvarez, D.M. El colapso de la hipótesis de estacionariedad por cambio y variabilidad climática: Implicaciones para el diseño hidrológico en ingeniería. *Rev. Ing.* **2012**, *36*, 65–76.

37. Frevert, R.; Schwab, G.; Edminster, T.; Barnes, K. *Soil and Water Conservation Engineering*; John Wiley & Sons Ltd.: New York, NY, USA, 1963.

38. Robie, R.B. *Rainfall Analysis for Drainage Design*; Department of Water Resources: Sacramento, CA, USA, 1976.

39. NOAA's, N.W.S. NOAA Atlas 14 Point Precipitation Frequency Estimates: CA. Available online: https://hdsc.nws.noaa.gov/hdsc/pfds/pfds_map_cont.html?bkmrk=ca (accessed on 31 August 2020).

40. Nhat, M.L.; Tachikawa, Y.; Takara, K. Establishment of Intensity-Duration-Frequency Curves for Precipitation in the Monsoon Area of Vietnam. *Annu. Disas. Prev. Res. Inst.* **2006**, *49B*, 93–103.

41. Tfwala, C.M.; van Rensburg, L.D.; Schall, R.; Mosia, S.M.; Dlamini, P. Precipitation intensity-duration-frequency curves and their uncertainties for Ghaap plateau. *Clim. Risk Manag.* **2017**, 1–9. [CrossRef]

42. Koutsoyiannis, D.; Kozonis, D.; Manetas, A. A mathematical framework for studying rainfall intensity-duration-frequency relationships. *J. Hydrol.* **1998**, *206*, 118–135. [CrossRef]

43. Sun, Y.; Wendi, D.; Kim, D.E.; Liong, S.Y. Deriving intensity–duration–frequency (IDF) curves using downscaled in situ rainfall assimilated with remote sensing data. *Geosci. Lett.* **2019**, *6*, 1–12. [CrossRef]

44. Domínguez, R.; Carrizosa, E.; Fuentes, G.E.; Arganis, M.L.; Osnaya, J.; Galván-Torres, A.E. Análisis regional para estimar precipitaciones de diseño en la república mexicana. *Tecnol. Cienc. Agua* **2018**, *9*, 5–29. [CrossRef]

Publisher's Note: MDPI stays neutral with regard to jurisdictional claims in published maps and institutional affiliations.

Article

Intra-Storm Pattern Recognition through Fuzzy Clustering

Konstantinos Vantas and Epaminondas Sidiropoulos *

Department of Rural and Surveying Engineering, Aristotle University of Thessaloniki,
54124 Thessaloniki, Greece; kon.vantas@gmail.com
* Correspondence: nontas@topo.auth.gr

Abstract: The identification and recognition of temporal rainfall patterns is important and useful not only for climatological studies, but mainly for supporting rainfall–runoff modeling and water resources management. Clustering techniques applied to rainfall data provide meaningful ways for producing concise and inclusive pattern classifications. In this paper, a timeseries of rainfall data coming from the Greek National Bank of Hydrological and Meteorological Information are delineated to independent rainstorms and subjected to cluster analysis, in order to identify and extract representative patterns. The computational process is a custom-developed, domain-specific algorithm that produces temporal rainfall patterns using common characteristics from the data via fuzzy clustering in which (a) every storm may belong to more than one cluster, allowing for some equivocation in the data, (b) the number of the clusters is not assumed known a priori but is determined solely from the data and, finally, (c) intra-storm and seasonal temporal distribution patterns are produced. Traditional classification methods include prior empirical knowledge, while the proposed method is fully unsupervised, not presupposing any external elements and giving results superior to the former.

Keywords: rainfall; rainstorm events; inter-event time; intra-storm patterns; fuzzy clustering; clustering analysis; clustering tendency; Greece

Citation: Vantas, K.; Sidiropoulos, E. Intra-Storm Pattern Recognition through Fuzzy Clustering. *Hydrology* **2021**, *8*, 57. https://doi.org/10.3390/hydrology8020057

Academic Editor: Davide Luciano De Luca

Received: 28 February 2021
Accepted: 22 March 2021
Published: 25 March 2021

Publisher's Note: MDPI stays neutral with regard to jurisdictional claims in published maps and institutional affiliations.

1. Introduction

Knowledge of the temporal and spatial distribution of rainfall is essential both for climatological studies, especially regarding climate change, and for purposes of flood studies and water resources planning. Effective and illuminating studies of rainfall data are achieved through the detection of patterns or groupings. Stochastic precipitation models utilize Markov chains to simulate the occurrence of wet or dry days and consequently the daily precipitation depth [1–3]. Numerous models, extensively developed, are based on the concept of rectangular pulses point process (RPPP), which can be categorized into two types, the Bartlett–Lewis [4] and the Neyman–Scott [5,6] precipitation models. Both of them use the assumption that storms arrive according to a Poisson process, the most basic example of continued-time Markov chains, of which a concise description can be found in Onof et al. [7]. The application of RPPP methods concerns the fitting of a small set of parameters that define the distributions that follow different rainfall quantities, such as the rainstorm and rain-cell origins, durations and intensities, a multi-objective problem without analytical solution and, sometimes, erratic results [5,7–9].

A different family of models are the profile-based ones [10,11] that utilize the internal structure of rainstorms in their entirety and not as derived parameters of statistical distributions. These rainfall models employ the construction of Huff or mass curves [12], a probabilistic method for sub-daily precipitation records, in which rainstorm data are represented in the form of normalized time, versus normalized cumulative precipitation depth and classified by the quartile where the maximum intensity occurs. The use of mass curves offers several advantages: (a) adequate stability regarding the sample size of rainstorms [13]; (b) coarser hourly data giving nearly identical results to data with finer time-step in the order of minutes [13]; (c) similarity in very long distances [12–14]; (d) the fact that they are not affected by elevation, type of storm, storm duration or storm precipitation depth [15].

Given that precipitation records contain both wet and dry periods, an important issue is the delineation or the extraction of individual rainstorm events from the contiguous recorded precipitation data. A systematic review [16] of different methods that are based on various rainstorm events characteristics (depth, intensity, generated runoff) reveals the necessity of selecting a criterion that defines objectively the rainless intervals, or minimum inter-event time, and not the arbitrary selection of constant time length. The latter is a common practice: Huff used 6 h to separate storms [12], Yu et al. used 2 h [17] and the calculation of rainfall erosivity in the universal soil loss equation (USLE) and its revisions [18] uses, also, 6 h. A different approach was proposed by Restrepo-Posada and Eagleson [19], commonly used today by many authors [13,20], in which "an empirical and inexact", but easily applied, method was developed for the separation of precipitation timeseries, into statistically independent rainstorms.

Unsupervised learning methods, clustering in particular, are now employed for pattern recognition in hydrological data [21]. In general, clustering analysis is a popular exploratory task aimed at partitioning the content of databases into smaller groups on the basis of inherent similarity criteria [22]. Clustering yields meaningful patterns to be further utilized for understanding of processes or for simulations. Clustering methods are applied in many fields, motivated by the large amounts of accumulated data and the developed needs for better data management through grouping and detection of patterns. The unsupervised nature of clustering analysis is a definite advantage, but it raises additional difficulties regarding suitable choice of methods and metrics. The challenge of clustering also lies in adapting the method and its parameters to the nature of the specific application.

A variety of such unsupervised learning methods are reviewed by Sheikholeslami et al. [23], including descriptions of clustering approaches. Among these, the well-known k-means and k-medoids methods divide the data into distinct clusters, such that each element belongs to exactly one cluster. This type of clustering may be characterized as hard, crisp or non-fuzzy. On the contrary, in fuzzy clustering, each point-element is allowed to belong to more than one cluster, albeit with a varying degree of membership. The membership degree of an element is a number in the interval [0,1]. Points closer to the center of a cluster are assigned a higher membership degree than points further away. One of the most widely used fuzzy clustering algorithms is fuzzy c-means clustering. It is used frequently in pattern recognition, and a related review can be found in Nayak et al. [24]. A major problem in clustering is the determination of the optimal number of clusters, which, in real-world applications, is generally not known in advance. As a result, a number of methods have been developed for the determination of the optimal number of clusters. Many of them use the concept of relative cluster validity [25], where results from different clustering methods are compared using a predefined metric. A number of these methods can be found in Milligan et al. [26] and Charrad et al. [27].

In hydrology, machine learning and clustering methods are being applied with increasing frequency, but, apparently, not to their full potential yet. More specifically, cluster analysis has been applied for the identification of hydrologically homogeneous regions, as noted in several literature publications: (a) hierarchical clustering on monthly rainfall erosivity density [21]; (b) fuzzy c-means clustering on annual precipitation [28] and annual maximum intensities [29]; self-organizing maps on monthly precipitation [30], while less work has appeared in terms of temporal pattern investigations applying: (a) the AL algorithm on rainstorms mass curves [31]; (b) self-organizing maps on design hyetographs; (c) k-means, also, on rainstorms mass curves [32]. On the other hand, rainfall mass curves have been utilized, as previously reported, in the stochastic generation of rainfall events [10,11,33,34], also as design storms [14,35–37], for the simulation of floods [38,39] and in changes in storm properties due to climate change [40].

In view of the above considerations, this paper aims to investigate the presence of intra-storm temporal patterns using timeseries of rainfall data coming from the Greek National Bank of Hydrological and Meteorological Information. Prior to clustering analysis, a more precise statistical analysis is applied in order to define a seasonal model to aid in the extraction of statistically independent rainstorm events, which has never been presented for Greece. The computational process that follows is a custom-developed, domain-specific algorithm that produces temporal rainfall patterns using common characteristics from the data via fuzzy clustering in which (a) every storm may belong to more than one cluster, allowing for some equivocation in the data, (b) the number of the clusters is not assumed known a priori, but is determined solely from the data and, finally, (c) intra-storm and seasonal temporal distribution patterns are produced. The optimal temporal rainfall distribution curves presented recently by the authors employed hierarchical clustering on principal components and utilized data from a specific water region of Greece [41]. Additionally, a preliminary research that utilizes self-organizing maps and crisp clustering has been presented by the authors in a recent European Geosciences Union General Assembly [42]. The present paper (a) adopts the more general approach of fuzzy clustering, which, to our knowledge, has not been applied so far to rainstorm timeseries; (b) the data utilized extend country wide; (c) the advantages of mass curves mentioned in the literature for different parts of the word are materialized for the case of Greece, regarding regionalization and the effect of elevation. Moreover, the overall result of clustering is verified and compared with the established Huff's classification, which has also never been presented for the country, via visualization effected through non-linear projection and topographic maps that are created using emergent self-organizing maps, a method that can reveal patterns in high-dimensional datasets.

2. Materials and Methods

The methodology that was applied in the study is presented in Figure 1 in the form of a flowchart. Precipitation data with a time step of 30 min were imported; rainstorms were extracted from the dataset and preprocessed in order to have an appropriate form as input to clustering analysis. A cluster tendency assessment was applied, so as to examine if clustering is meaningful, and the optimal number of clusters were determined by a custom method that utilizes fuzzy c-means. Additionally, the Huff's curves were compiled for the same dataset. The resulted patterns were visualized using a nonlinear projection of the data in two dimensions, and finally, their characteristics were examined in terms of internal and seasonal structure.

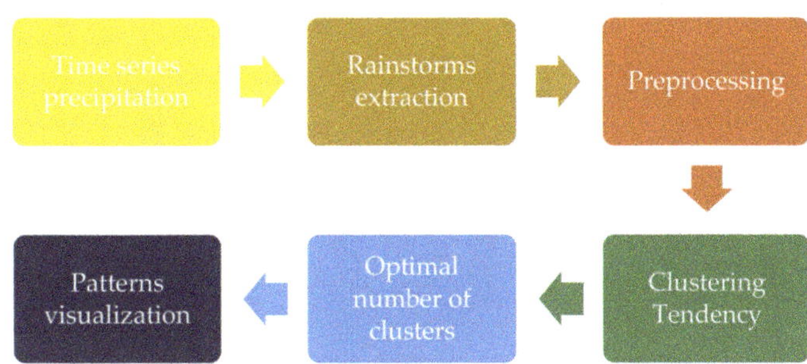

Figure 1. Flowchart of the applied methodology.

2.1. Data Acquisition and Processing

The data used in the analysis were taken from the Greek National Bank of Hydrological and Meteorological Information [43] for 108 meteorological stations across Greece (Figure 2). The timeseries comprised a total of 2926 years of 30 min records for the time period from 1953 to 1997, with a mean length of 26.6 years per station. The timeseries were checked for consistency and errors as follows: (a) in records of repetitive values near zero (i.e., ≤0.01 mm), these values were set to zero and (b) records of aggregated values, where the time step was larger than the reported, were removed. The pluviograph data coverage was 43.2% on average. The missing values in the timeseries are not random points but contiguous data related to entire months or years.

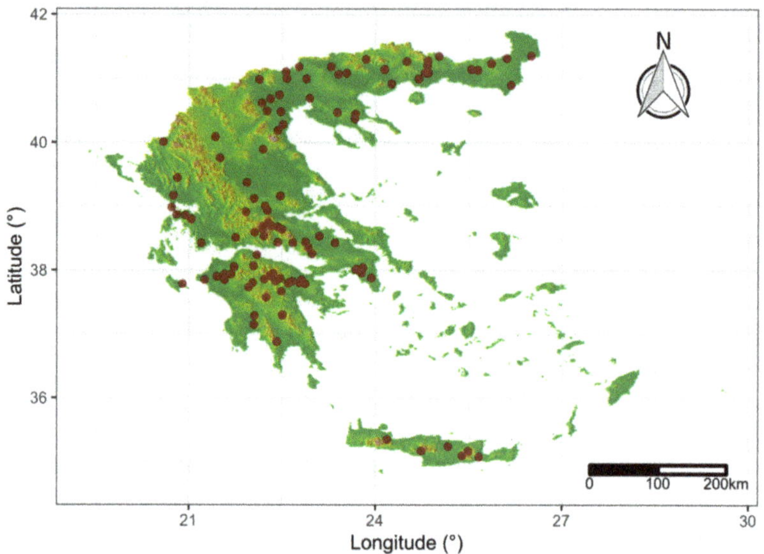

Figure 2. Station locations in Greece used in the analysis obtained from the Greek National Bank of Hydrological and Meteorological Information.

Greece, given its geographic characteristics, the distribution of sea and land and the complex and rich relief, has a mosaic of different micro-climates and regional variations, all in the Mediterranean context. Precipitation varies from its maximum values during winter to a minimum during summer. The highest values are observed westwards and on the mountain range of Pindos and its expansion on Peloponnesus (two to three times higher). This fact reveals the importance of relief on the distribution of rainfall over the country. Furthermore, higher precipitation depth is connected to the movement of Mediterranean depressions that follow a characteristic path from west to east. Northerly, at the valleys of central Macedonia, smaller amounts of precipitation are recorded, due to prevailing dry winds. At the northeastern part, at Thrace, higher values are observed also due to relief. Finally, eastern parts of Greece are drier. During summer months when convective activity prevails and over northern Greece, it produces higher precipitation amounts than in the drier southern parts.

2.2. Extraction of Rainstorms

A Poisson process hypothesis is assumed for the separation of the precipitation timeseries to statistically independent rainstorm events, where:

- The time intervals of rainstorms that come from the same month are distributed exponentially.
- The rainstorms are separated by a minimum critical time duration of no precipitation (**CD**) or inter-event time [19].
- There is a seasonal pattern for **CD** that is assumed to have constant monthly values.

- The probability density function is:

$$f(t_a) = \omega \cdot e^{-\omega \cdot t_a}, \quad t_a \geq 0 \tag{1}$$

where ω is the average storm arrival rate and

$$t_a = t_r + t_b \tag{2}$$

where t_a is the storm interarrival time, t_r is the storm duration, and t_b is the dry time between storms.

The estimation of **CD** that separates rainstorms is based on an iterative procedure of statistical tests (Algorithm 1). Inter-month data per station are used to ensure homogeneity and: (a) a test value of cd is used, coming from a predefined vector of values **CD**, (b) a vector $\mathbf{T}_a$ of t_a values is computed for each cd, (c) ω values for $\mathbf{T}_a$ are estimated by means of the maximum likelihood estimation method, (d) the goodness-of-fit between $\mathbf{T}_a$ and the exponential distribution is estimated via parametric bootstrapping of s samples that utilizes the Kolmogorov–Smirnov test [44], and (e) the cd value with the maximum p-value from the empirical non-parametric distribution is selected. In Algorithm 1, a threshold of 50 t_a values was imposed empirically prior to statistical testing.

Algorithm 1: Temporal model of CD

Input: Stations' precipitation time series P_i where $i = 1,\dots,k$; Critical durations test vector $CD = [120, 180, \dots, 1800]$ (min); Number of samples that are drawn for parametric bootstrapping $s = 50,000$;

1 **for** *station* $i \leftarrow 1$ *to* k **do**
2   **for** *month* $m \leftarrow 1$ *to* 12 **do**
3   **for** *cd in* **CD** **do**
4   Compute the vector of interarrival times t_a using inter-month data and n = length(t_a);
5   **if** $n \geq 50$ **then**
6   Estimate the average storm arrival rate $\hat{\omega}$ from t_a using Maximum Likelihood Estimation;
7   Obtain the Kolmogorov–Smirnov's p-value for the original sample t_a and the estimated distribution;
8   Generate s samples of size n from the estimated distribution;
9   For each sample compute the one-sample Kolmogorov–Smirnov's p-value using the estimated distribution as theoretical;
10   Use the empirical non-parametric distribution of p-values to obtain the p-value for the original sample t_α;
11   Get minimum dry period duration $MDPD_{i,m}$ from $CD[max(p - value)]$;

Result: Monthly values of CD

A first version of the above process was presented in previous publications of the authors [41,45]. The basic hypothesis concerns the application of the probability density function (1). A recent related publication [20] adopted the same exponential distribution to estimate inter-event times in a region of China, but using the "inexact" Restrepo-Posada and Eagleson approach.

2.3. Preprocessing

Prior to clustering analysis, preprocessing of data is necessary, in order to transform them into standardized uniform representations that will facilitate the recognition of the common features. More specifically, each one of the storms is a vector of different length, so that a method must be applied to transform the dataset to one with a fixed number of features that represent time. Thus, the storms were scaled and interpolated to unitless form in which (a) the time expresses the ratio of the rainstorm duration and (b) the rainfall expresses, also, the ratio of total rainstorm depth:

$$p'_i = \frac{p_i}{\sum_{i=1}^{n} p_i} \tag{3}$$

$$t'_i = \frac{t_i}{\sum_{i=1}^{n} t_i} \tag{4}$$

where p_i and t_i are the precipitation depth and time for the timestep i, respectively, and p'_i and t'_i their scaled values.

In the sequel, since the scaled rainstorm vectors have variable length, linear interpolation was applied to compute the unitless rainfall for every 1% of unitless time values. Finally, a matrix of unitless rainstorms **U** was produced with the values of unitless rainfall, in which every row represents a storm and each column the unitless time values. In the clustering analysis, use was made only of the rainstorms with cumulative rainfall greater than 12.7 mm (i.e., with the exception of light rain) that were no single points, due to the timestep of 30 min, and also erosive by definition [18], as in other similar publications [12–14].

2.4. Clustering Tendency

Before the application of a clustering algorithm, it is advisable to have a preliminary look into the dataset in order to detect any existing clustering tendency. This was done in the following ways: (a) a visual assessment of cluster tendency (VAT [46]) and (b) application of the Hopkins index, H [47].

The VAT method creates an image matrix that can be used to visually assess the cluster tendency. In the method, as it was applied, the pairwise dissimilarity values of the rainstorms were computed, and these values were reordered in the form of a matrix and displayed as an intensity image. In this image, the clusters are indicated as dark blocks of pixels along the diagonal [46]. As a dissimilarity measure, the Euclidean distance d was used

$$d(u,v) = \sqrt{\sum_{i=1}^{100} (u_i - v_i)^2} \tag{5}$$

where u and v are two row vectors from the unitless rainstorms **U** matrix. The re-ordering was achieved applying agglomerative hierarchical clustering using the Ward's minimum variance criterion, an algorithm that minimizes the total within-cluster variance [48]. At the beginning of the algorithm, the number of the clusters is equal to the number of data points (all clusters contain a single point). At every step, the algorithm finds the pair of clusters that result after merging to the minimum increase in the total within-cluster variance, which is expressed as the sum of squared differences between the clusters' centers. Finally, all clusters are combined to one cluster that contains all the data using a hierarchical method [21].

The Hopkins index, H, can be used to test the null hypothesis of randomly and uniformly distributed data, generated by a Poisson point process and is calculated with

$$H = \frac{\sum_{j=1}^{m} u_j^d}{\sum_{j=1}^{m} w_j^d + \sum_{j=1}^{m} u_j^d} \tag{6}$$

where X is a collection of n data points that have d dimensions. A random sample from X without replacement with members $x_i (i = 1 \text{ to } m, m \ll n)$ is formed. Y is a set of uniformly random data points, also with d dimensions and members $y_j (j = 1 \text{ to } m)$, u_j in turn is the Euclidean distance from y_j to its nearest neighbor in X, and w_j is also the Euclidean distance from x_i to its nearest neighbor in X. A value of H close to one indicates that the data are highly clustered, 0.5 indicates randomly distributed data, and zero indicates regularly spaced data [21,25].

2.5. Fuzzy Clustering

The unitless rainstorm data were clustered by the fuzzy c-means (FCM) algorithm, which was developed by J. Dunn [49] and improved by J. Bezdek [50]. FCM aims to minimize the objective function

$$\underset{C}{\arg\min} \sum_{i=1}^{n} \sum_{j=1}^{c} w_{ij}^{m} \|x_i - c_j\|^2 \tag{7}$$

where $w_{ij} \epsilon [0, 1]$ is the degree or membership of item x_i from a set of n elements $X = x_1, \ldots, x_n$ that belong to cluster c_j, C is the set of c cluster centers $C = c_1, \ldots, c_c$, and $\| \ldots \|$ denotes any norm that expresses similarity, such as the Euclidean distance.

From an iterative optimization procedure on (7), each element w_{ij} of the partition matrix W is equal to

$$w_{ij} = \frac{1}{\sum_{k=1}^{c} \left(\frac{\|x_i - c_j\|}{\|x_i - c_k\|} \right)^{\frac{2}{m-1}}} \tag{8}$$

where m is the "fuzzifier" that determines the level fuzziness with $m \epsilon \mathbb{R}$ and $m \geq 1$, commonly set to two [51]. The larger the m value is, the fuzzier the membership values of the clustered data points are. The centers c_j of the clusters are the mean of all the elements weighted by their degrees:

$$c_j = \frac{\sum_{i=1}^{n} w_{ij}^{m} x_i}{\sum_{i=1}^{n} w_{ij}^{m}} \tag{9}$$

FCM stops when the maximum number of iterations given a priori is reached, or when the algorithm is unable to reduce the current value of the objective function further to a predefined, usually very small, value.

2.6. Optimal Number of Clusters

The most common and fundamental problem in clustering analysis is the determination of the number of clusters in an unlabeled dataset, as the one used in the analysis. As previously reported and due to the fact that most clustering algorithms, including FCM, require the number c of clusters to be known a priori, the next step, after answering the question of clustering tendency, has to do with the determination of c. The proposed method uses statistical testing (Algorithm 2). Variations of it have been presented by the authors in the context of different clustering algorithms, namely, (a) self-organizing maps [42] and (b) a dendrogram produced by hierarchical clustering on principal components [41].

A different approach has been presented [52], using k-means (crisp) clustering, which involves two different parameters: (a) an initial threshold that indicates similarity, through a Pearson correlation coefficient, between all pairs of the unitless cumulative rainstorms and (b) an additional one, based on the distances matrix of all pairs of the unitless cumulative rainstorms, which creates the initial seeds used in the k-means clustering algorithm. In other studies, the number of clusters was set without any estimation [32] or through a visual, ambiguous method involving the density maps derived from self-organizing maps [53]. Comparing the reported methodologies to the proposed one, Algorithm 2 presents the following advantages: (a) It is easier to implement, and it can be combined with any clustering algorithm that uses as a parameter the number of clusters directly or indirectly, as is for example the *eps* parameter in the DBSCAN algorithm [54]. (b) It utilizes hypothesis testing about the centers of the clusters, and (c) it has only one parameter to choose, the significance level of well-known statistical tests.

In Algorithm 2, firstly, the cumulative values of the unitless rainstorms matrix are computed, as that kind of transformation was found to help the computational procedures of clustering algorithms. At each step of the iteration, FCM is applied using a trial number of clusters, greater than two and smaller than a predefined maximum value n_{max}. Afterwards, the cluster centers that resulted are represented as cumulative rainfall distributions.

These centers, and for all possible pairs, are tested to find out whether they are drawn from the same distribution using the two-sample Kolmogorov–Smirnov test [55]. The test quantifies the distance between the two samples' empirical distribution functions and has been used in similar comparisons [56,57]. Due to the multiple statistical testing that arise, the resulted p-values are adjusted using the Benjamini and Hochberg method [58], which controls the false discovery rate. The algorithm stops when any p-value is greater than a predefined significance level α, and in that case, the clusters from the previous step are returned.

Algorithm 2: Optimal number of clusters using FCM

Input: unitless rainstorms U; maximum number of clusters to test n_{max}; significance level $\alpha = 0.05$
1   compute the row-wise cumulative values U' of U
2   **for** $i \leftarrow 2$ *to* n_{max} *and all p-values* $< \alpha$ **do**
3   apply FCM on U' for $c = i$ and compute cluster centers C;
4   for all pairs in C obtain the Kolmogorov–Smirnov two sample test, p-values;
5   adjust the obtained p-values using Benjamini and Hochberg method;
Result: optimal number of clusters c_{opt} and clustering results

---

2.7. Projecting Data Using Non-Linear Mapping

The non-linear method that was used to visualize the rainstorm data in a two-dimension scatterplot was the generalized U-Matrix [59]. This method uses the results from a dimensional reduction method, such as principal components analysis, or a non-linear method, such as t stochastic neighbor embedding (t-SNE, [60]), in conjunction with emergent self-organizing maps (ESOM [61]). This step is a necessity due to the Johnson–Lindenstrauss lemma [62] stating that the "low-dimensional similarities do not represent high-dimensional distances coercively" [63]. In other words, specific measures are taken to avoid the common mistake of assuming that, when the projected points are similar to each other after dimensional reduction, such is the case in the actual high-dimensional space. The generalized U-Matrix after its calculation by ESOM is visualized by a three-dimensional space called "topographic map with hypsometric tints" or colors on surface that represent elevation ranges [64]. The topographic map has no actual real borders and is toroidal, which means that the left border is connected to the right one and the top to the bottom.

3. Results and Discussion

3.1. Rainstorms Extraction

Empirically, a minimum set of 50 values per month and station were used in Algorithm 1 and some basic statistics; results about CD are presented in Table 1. Due to the dry summer period in Greece, these monthly CD values could be computed only in a small set of stations. A seasonal sinusoidal model (Figure 3) was developed to use in Greece:

$$f(CD) = \theta_1 \sin\left(\frac{2\pi}{12}m\right) + \theta_2 \sin\left(\frac{4\pi}{12}m\right) + \theta_3 \cos\left(\frac{2\pi}{12}m\right) + \theta_4 \cos\left(\frac{4\pi}{12}m\right) \tag{10}$$

where m is the month, and $\theta_1, \ldots, \theta_4$ parameters that were estimated using linear regression.

Using the CD sinusoidal model, a set of 174,883 rainstorms were extracted from the dataset. With the exception of summer months and September, the model gives values around 6 h, close to the value selected empirically by Huff [12] and in rainfall erosivity calculations [65]. A subset of 26,678 rainstorms that met the criterion of minimum depth were preprocessed and used to compile the U matrix.

Table 1. Average statistical properties of monthly **CD** values for the stations. SD is an abbreviation for standard deviation and CV for coefficient of variation (the ratio of the standard deviation to the mean) and h for hours.

CD (h)	Min	Mean	Median	Max	SD	Skew	Kurtosis	CV
January	2	5.4	5	13	1.6	1.40	4.77	0.19
February	2	5.0	5	10	1.4	0.90	1.39	0.16
March	2	5.9	5	12	1.8	0.95	0.82	0.20
April	4	6.3	6	10	1.4	0.62	−0.20	0.16
May	4	6.8	6	12	1.9	0.91	0.49	0.24
June	4	8.2	8	13	2.1	0.15	−0.50	0.34
July	5	9.3	9	13	2.0	−0.17	−0.01	0.58
August	5	7.8	8	11	2.1	0.12	−1.73	0.70
September	6	9.1	9.5	11	1.6	−0.52	−1.03	0.45
October	2	7.4	7	13	1.9	0.36	0.78	0.25
November	2	6.7	6	11	1.6	0.25	0.17	0.19
December	2	5.2	5	11	1.4	0.85	2.00	0.16

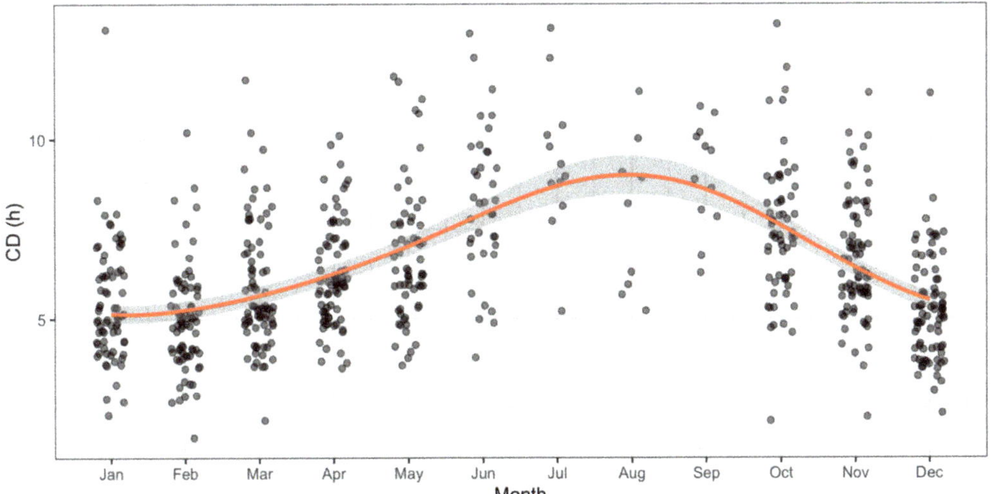

Figure 3. Red represents the monthly sinusoidal model of CD (Equation (11)). The grey area corresponds to the standard error of the model using linear regression. The grey dots represent the monthly CD values per station, as computed from Algorithm 1.

3.2. Clustering Tedency

Due to computation issues, a random sample of 10% was used to compute the value of the Hopkins index with H = 0.886, so the null hypothesis of random data could safely be rejected. This result indicated that there was a physical meaning in the categorization of the rainstorms based on their internal structure. The VAT method (Figure 4) created an image matrix, where the clusters are indicated as at least four darker blocks along the diagonal. Given the results from these two methods, there is strong evidence that the dataset contains meaningful clusters and the next step can follow, to apply the proposed method and select the optimal number of clusters.

Figure 4. Image matrix created by the VAT method, in order to visually inspect the clustering tendency of data. Distances (dissimilarities) among unitless rainstorms are unitless. The rainstorms that belong to the same cluster are displayed in successive order.

3.3. Clustering Results and Visualization

After the application of Algorithm 2, the optimal number of clusters are proposed to be four, a suitable number comparing it with VAT results. In Figure 5, the matrix of unitless rainstorms **U** is depicted, with its rows, the rainstorms, reordered by the cluster they belong to. This representation shows that the rainstorms belonging to clusters number one and four are more similar. The latter statement is strengthened by Figure 6 that illustrates the distribution of the membership of every unitless rainstorm that belongs to a cluster (that value must be >0.25 to be classified to a cluster). In the same figure, the mean values of the memberships for each one of the clusters are {0.67, 0.60, 0.59, 0.67}. The latter values are not normally distributed, and cluster one and four are left-skewed meaning less equivocality, in contrast to clusters two and three that are right-skewed.

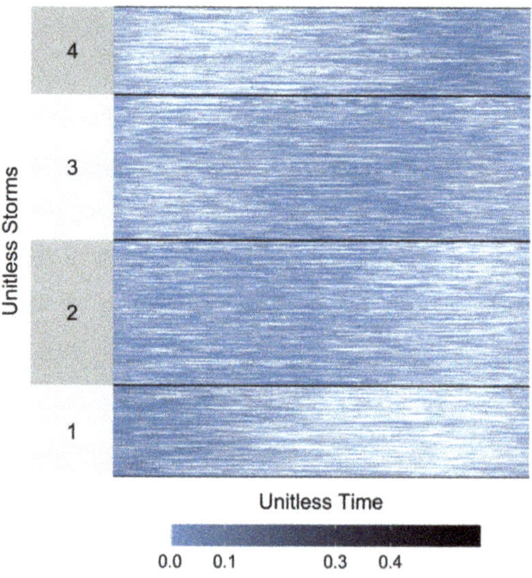

Figure 5. Unitless rainstorms values re-order by clustering results. The names of the clusters were used in order to match with Huff's classification. White symbolizes zero precipitation depth, and different blues depict the presence of unitless precipitation.

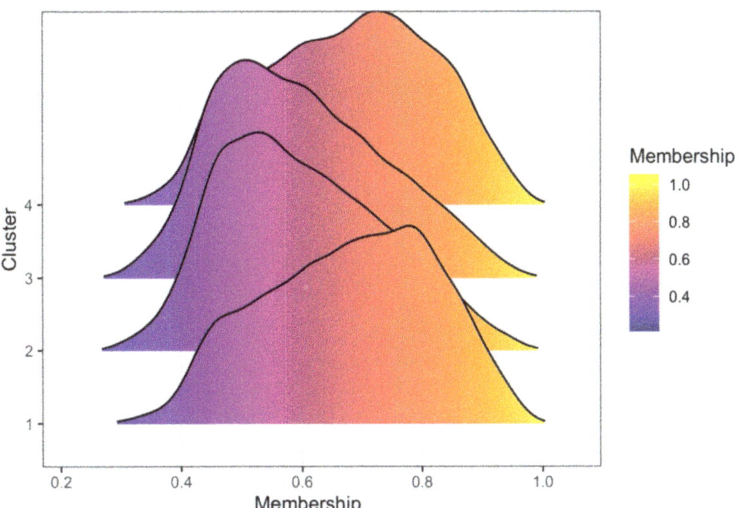

Figure 6. Distribution of the membership values from fuzzy c-means (FCM) of every unitless rainstorm that belongs to a cluster.

From Figure 5, initially, it seems that the selection of Huff about the grouping in four clusters has physical meaning, in contrast to criticism about being artificial [66].

The average values related to precipitation depth, duration and intensity per cluster are presented in Table 2. Cluster one has on average shorter duration rainstorms, with lower precipitation depth, but higher mean intensity. Clusters two and three have both rainstorms with similar statistics, with larger duration and precipitation depth than one and four. Cluster four has analogous statistics to one with the exception of intensity, which is lower.

Table 2. Average values of occurrence of clusters, duration and precipitation depth intensity of clusters' rainstorms.

	Cluster			
	1	**2**	**3**	**4**
Cluster Ratio (%)	19.5	30.86	30.84	18.79
Duration (h)	12.5	16	16.5	14.5
Precipitation depth (mm)	20.7	23.2	23.7	21.9
Intensity (mm/h)	1.82	1.66	1.66	1.71

Cluster number one has notably higher variance and maximum occurrence during spring and summer months that match convective activity over Greece (Figure 7). Clusters two and three have distributions similar to each other and inverse to the first one, related to the prevailing winter weather systems, having their minima during summer. Cluster four has a more uniform distribution with the exception of winter months, when it has its maximum. The latter cluster, with its relatively higher intensity values and, also, higher ratio of occurrence during the wetter winter months in Greece, can be utilized in hydrologic design in the construction of design storms.

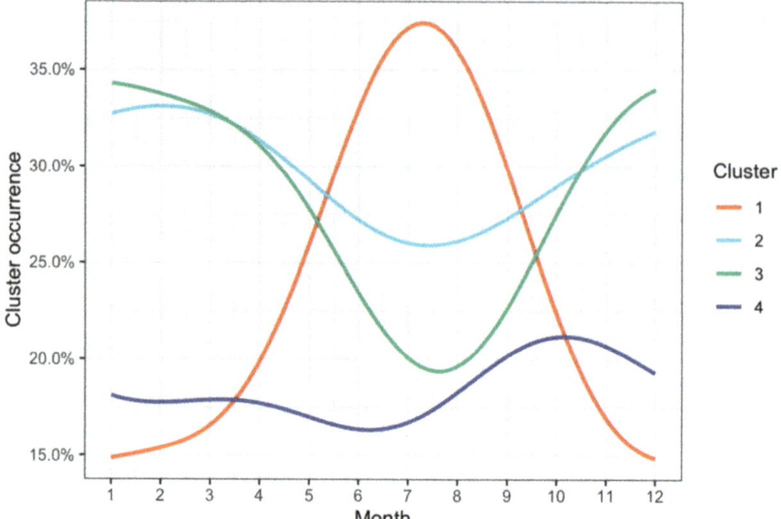

Figure 7. Seasonal variability of clusters' monthly occurrence.

In order to examine if elevation affects the distribution of rainstorms in the clusters, density plots were compiled (Figure 8) for each cluster. The distribution of the elevations of the stations related to rainstorms appears to be almost identical, despite the fact that altimetry is closely connected to the precipitation regime in Greece. This result about elevation is in accordance with the reported one for Huff's curves by Loukas et al. in Canada [15].

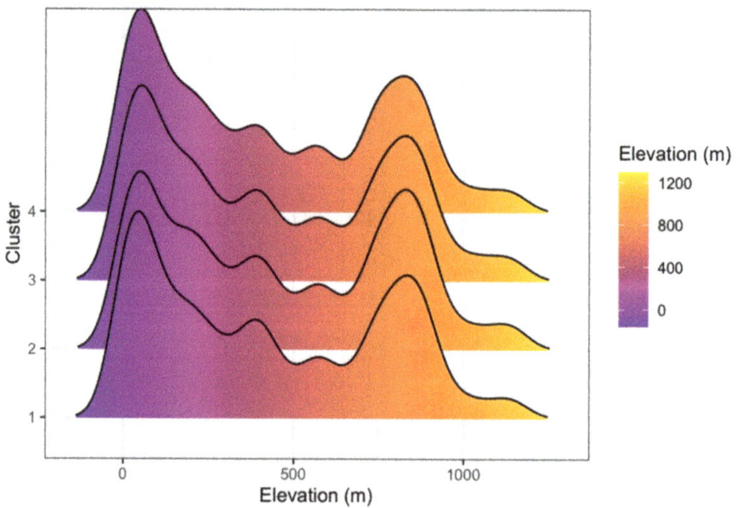

Figure 8. Distribution of elevation values from FCM of every unitless rainstorm given its cluster.

In order to examine the temporal patterns in every station using the results from FCM, correlation matrices were computed using Pearson's r coefficient [67]:

$$r = \frac{1}{n-1} \sum_{i=1}^{n} \left(\frac{x_i - \overline{x}}{s_x} \right) \left(\frac{y_i - \overline{y}}{s_y} \right) \tag{11}$$

where x_i is the average values of unitless rainstorms coming from a station x and cluster c, y_i, similarly, is the average values of unitless rainstorms coming from a station y and the same cluster c, $n = 100$ (the ordinates of unitless rainstorms), s_x and s_y are the standard deviations of x_i and y_i and, finally, $\overline{x}$ and $\overline{y}$ the mean values, respectively. The patterns per station and given cluster showed very high similarity with r $\geq$ 0.974 and on average for all values r $=$ 0.999. These results indicate that despite the rich topography and the variability of micro-climates in Greece, the clusters have regional stability in long distances.

In the sequel (Figure 9), the cumulative values of the centers of the four clusters are compared to the ones calculated using Huff's method of classification that is based on the quartile with the highest intensity. The two methods produce different results, while fuzzy clustering creates unitless cumulative curves that do not overlap. The four Huff curves were also tested, for all possible pairs, whether or not are drawn from the same distribution using the Kolmogorov–Smirnov test and three pairs failed to reject that hypothesis for both significance levels $a = 5\%$ and $a = 10\%$.

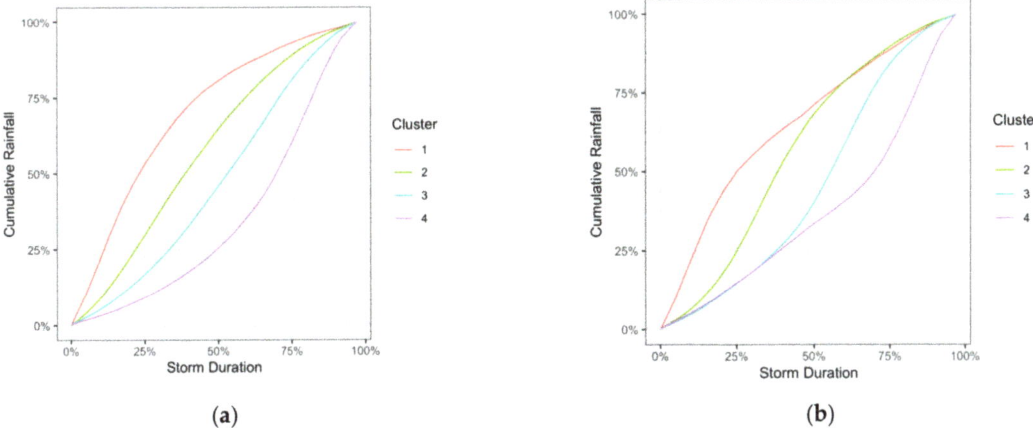

(a)   (b)

Figure 9. (a) Unitless cumulative values using FCM. **(b)** Unitless cumulative curves using Huff's classification.

Finally, the topographic maps utilizing the generalized U-matrix using ESOM, as described in Section 2.7, using both FCM and Huff's classification results were produced (Figure 10). In both maps, the unitless rainstorms with membership >80% are used to remove equivocation from the data. The visualization indicates the presence of four clusters and that Huff's method misclassifies the data. The topographic map produced valleys with blue and green color (indicating smaller distances of the points) that are separated by ridges of brown and white color (representing larger distances).

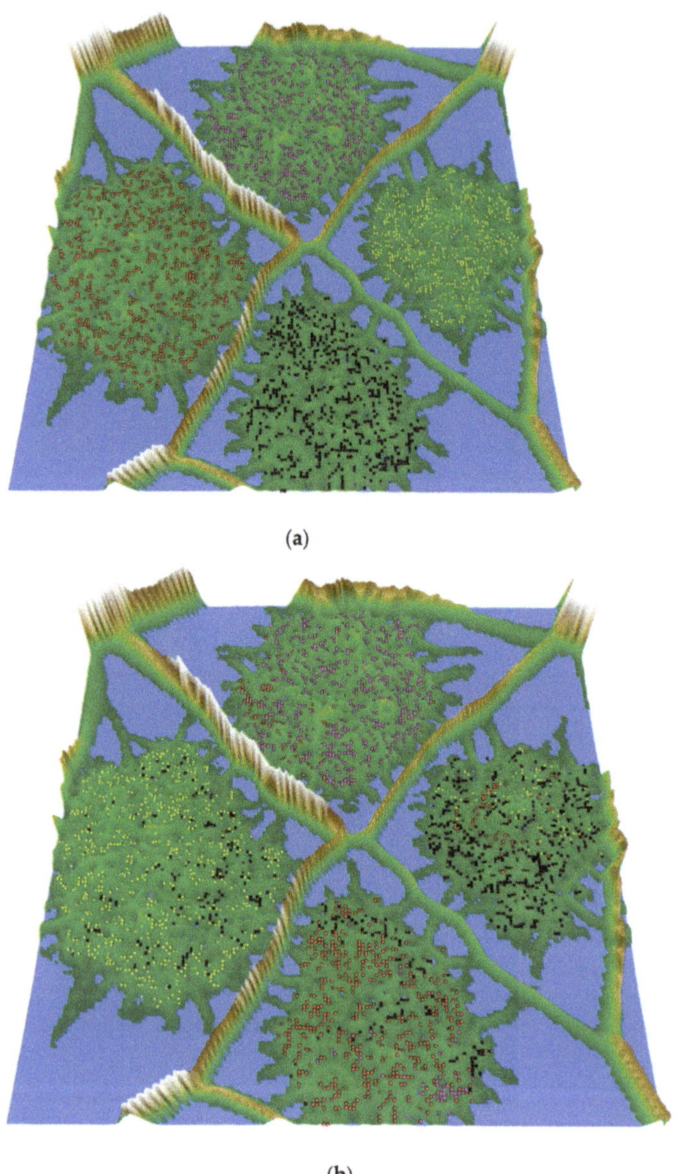

Figure 10. Topographic maps with hypsometric tints using t stochastic neighbor embedding (t-SNE) and the generalized U-matrix. The topology of the map is toroidal (the top and bottom and as well the right and left are joined). Each colored sphere represents a rainstorm with its different classification, as it has already computed. (**a**) Results using FCM: the rainstorms that belong to the same cluster are separated by mountain ranges. (**b**) Results using Huff's classification: the rainstorms of different classification are mixed, despite the visual separation of mountain ridges in the map.

4. Conclusions

In this paper, the timeseries of rainfall data were processed in order to detect patterns. The data, coming from the Greek National Bank of Hydrological and Meteorological

Information, cover the totality of the country's regions. In the various stages of the project, novel methods were presented and implemented. First, extraction of rainstorms was executed via a stochastic method. Then, the timeseries were properly transformed to unitless form. The data were tested for clustering tendency prior to the application of clustering algorithms. An iterative form of fuzzy clustering was presented that does not assume a priori knowledge of the number of clusters. It consists of a repeated execution of fuzzy c-means clustering in combination with statistical testing for the choice of the relevant number of clusters. Finally, a verification of the clustering results and a comparison to the widely used Huff's method was attained through topographic maps. The overall approach extends previous rainfall pattern recognition efforts of the authors and of the literature in the sense of the unsupervised learning and in the direction of fuzzy analysis in rainfall and in any follow-up study of floods or water management problems.

Author Contributions: Conceptualization, methodology and software K.V.; writing—original draft preparation, K.V. and E.S.; writing—review and editing, E.S. All authors have read and agreed to the published version of the manuscript.

Funding: This research received no external funding.

Institutional Review Board Statement: Not applicable.

Informed Consent Statement: Not Applicable.

Data Availability Statement: Precipitation data are available from the Greek National Bank of Hydrological and Meteorological Information.

Acknowledgments: The data importing, analysis and presentation were done using the open source R language for statistical computing and graphics [68] using the packages: hydroscoper [43], hyetor [69], e1071 [70], FCPS [71], GeneralizedUmatrix [63], factoextra [72] and ggplot2 [73].

Conflicts of Interest: The authors declare no conflict of interest.

References

1. Haan, C.T.; Allen, D.M.; Street, J.O. A Markov Chain Model of Daily Rainfall. *Water Resour. Res.* **1976**, *12*, 443–449. [CrossRef]
2. Gao, C.; Xu, Y.-P.; Zhu, Q.; Bai, Z.; Liu, L. Stochastic Generation of Daily Rainfall Events: A Single-Site Rainfall Model with Copula-Based Joint Simulation of Rainfall Characteristics and Classification and Simulation of Rainfall Patterns. *J. Hydrol.* **2018**, *564*, 41–58. [CrossRef]
3. Urdiales, D.; Meza, F.; Gironás, J.; Gilabert, H. Improving Stochastic Modelling of Daily Rainfall Using the ENSO Index: Model Development and Application in Chile. *Water* **2018**, *10*, 145. [CrossRef]
4. Onof, C.; Wang, L.-P. Modelling Rainfall with a Bartlett–Lewis Process: New Developments. *Hydrol. Earth Syst. Sci.* **2020**, *24*, 2791–2815. [CrossRef]
5. Rodriguez-Iturbe, I.; De Power, B.F.; Valdes, J.B. Rectangular Pulses Point Process Models for Rainfall: Analysis of Empirical Data. *J. Geophys. Res. Atmos.* **1987**, *92*, 9645–9656. [CrossRef]
6. Burton, A.; Fowler, H.J.; Blenkinsop, S.; Kilsby, C.G. Downscaling Transient Climate Change Using a Neyman–Scott Rectangular Pulses Stochastic Rainfall Model. *J. Hydrol.* **2010**, *381*, 18–32. [CrossRef]
7. Onof, C.; Chandler, R.E.; Kakou, A.; Northrop, P.; Wheater, H.S.; Isham, V. Rainfall Modelling Using Poisson-Cluster Processes: A Review of Developments. *Stoch. Environ. Res. Risk Assess.* **2000**, *14*, 384–411. [CrossRef]
8. Rodriguez-Iturbe, I.; Cox, D.R.; Isham, V. Some Models for Rainfall Based on Stochastic Point Processes. *Proc. R. Soc. Lond. Math. Phys. Sci.* **1987**, *410*, 269–288.
9. Rodriguez-Iturbe, I.; Cox, D.R.; Isham, V. A Point Process Model for Rainfall: Further Developments. *Proc. R. Soc. Lond. Math. Phys. Sci.* **1988**, *417*, 283–298.
10. Brigandì, G.; Aronica, G.T. Generation of Sub-Hourly Rainfall Events through a Point Stochastic Rainfall Model. *Geosciences* **2019**, *9*, 226. [CrossRef]
11. Vandenberghe, S.; Verhoest, N.E.C.; Buyse, E.; De Baets, B. A Stochastic Design Rainfall Generator Based on Copulas and Mass Curves. *Hydrol. Earth Syst. Sci.* **2010**, *14*, 2429–2442. [CrossRef]
12. Huff, F.A. Time Distribution of Rainfall in Heavy Storms. *Water Resour. Res.* **1967**, *3*, 1007–1019. [CrossRef]
13. Bonta, J. Development and Utility of Huff Curves for Disaggregating Precipitation Amounts. *Appl. Eng. Agric.* **2004**, *20*, 641. [CrossRef]
14. Yin, S.; Xie, Y.; Nearing, M.A.; Guo, W.; Zhu, Z. Intra-Storm Temporal Patterns of Rainfall in China Using Huff Curves. *Trans. ASABE* **2016**, *59*, 1619–1632.

15. Loukas, A.; Quick, M.C. Spatial and Temporal Distribution of Storm Precipitation in Southwestern British Columbia. *J. Hydrol.* **1996**, *174*, 37–56. [CrossRef]
16. Dunkerley, D. Identifying Individual Rain Events from Pluviograph Records: A Review with Analysis of Data from an Australian Dryland Site. *Hydrol. Process. Int. J.* **2008**, *22*, 5024–5036. [CrossRef]
17. Yu, R.; Xu, Y.; Zhou, T.; Li, J. Relation between Rainfall Duration and Diurnal Variation in the Warm Season Precipitation over Central Eastern China. *Geophys. Res. Lett.* **2007**, *34*. [CrossRef]
18. USDA-ARS. *Science Documentation: Revised Universal Soil Loss Equation, Version 2 (RUSLE 2)*; USDA-Agricultural Research Service: Washington, DC, USA, 2013.
19. Restrepo-Posada, P.J.; Eagleson, P.S. Identification of Independent Rainstorms. *J. Hydrol.* **1982**, *55*, 303–319. [CrossRef]
20. Wang, W.; Yin, S.; Xie, Y.; Nearing, M.A.; Wang, W.; Yin, S.; Xie, Y.; Nearing, M.A. Minimum Inter-Event Times for Rainfall in the Eastern Monsoon Region of China. *Trans. ASABE* **2019**, *62*, 9–18. [CrossRef]
21. Vantas, K.; Sidiropoulos, E.; Loukas, A. Robustness Spatiotemporal Clustering and Trend Detection of Rainfall Erosivity Density in Greece. *Water* **2019**, *11*, 1050. [CrossRef]
22. Hartigan, J.A. *Clustering Algorithms*; John Wiley & Sons: New York, NY, USA, 1975.
23. Sheikholeslami, G.; Chatterjee, S.; Zhang, A. Wavecluster: A Multi-Resolution Clustering Approach for Very Large Spatial Databases. In Proceedings of the 24th International Conference on Very Large Databases (VLDB), New York, NY, USA, 24–27 August 1998; Volume 98, pp. 428–439.
24. Nayak, J.; Naik, B.; Behera, H. Fuzzy C-Means (FCM) Clustering Algorithm: A Decade Review from 2000 to 2014. *Comput. Intell. Data Min. Vol.* **2015**, *2*, 133–149.
25. Theodoridis, S.; Koutroumbas, K. *Pattern Recognition*; Academic Press: Burlington, MA, USA, 2009; ISBN 978-1-59749-272-0.
26. Milligan, G.W.; Cooper, M.C. An Examination of Procedures for Determining the Number of Clusters in a Data Set. *Psychometrika* **1985**, *50*, 159–179. [CrossRef]
27. Charrad, M.; Ghazzali, N.; Boiteau, V.; Niknafs, A. NbClust: An R Package for Determining the Relevant Number of Clusters in a Data Set. *J. Stat. Softw.* **2014**, *61*, 1–36. [CrossRef]
28. Dikbas, F.; Firat, M.; Koc, A.C.; Gungor, M. Classification of Precipitation Series Using Fuzzy Cluster Method. *Int. J. Climatol.* **2012**, *32*, 1596–1603. [CrossRef]
29. Zeybekoğlu, U.; Keskin, A.Ü. Defining Rainfall Intensity Clusters in Turkey by Using the Fuzzy C-Means Algorithm. *Geofizika* **2020**, *37*, 181–195. [CrossRef]
30. Hsu, K.-C.; Li, S.-T. Clustering Spatial–Temporal Precipitation Data Using Wavelet Transform and Self-Organizing Map Neural Network. *Adv. Water Resour.* **2010**, *33*, 190–200. [CrossRef]
31. Lana, X.; Rodríguez-Solà, R.; Martínez, M.; Casas-Castillo, M.; Serra, C.; Burgueño, A. Characterization of Standardized Heavy Rainfall Profiles for Barcelona City: Clustering, Rain Amounts and Intensity Peaks. *Theor. Appl. Climatol.* **2020**, *142*, 255–268. [CrossRef]
32. Nojumuddin, N.S.; Yusof, F.; Yusop, Z. Identification of Rainfall Patterns in Johor. *Appl. Math. Sci.* **2015**, *9*, 1869–1888. [CrossRef]
33. Bonta, J.V. Stochastic Simulation of Storm Occurence, Depth, Duration and within-Storm Intensities. *Trans. ASAE* **2004**, *47*, 1573–1584. [CrossRef]
34. Wu, S.-J.; Yang, J.-C.; Tung, Y.-K. Identification and Stochastic Generation of Representative Rainfall Temporal Patterns in Hong Kong Territory. *Stoch. Environ. Res. Risk Assess.* **2006**, *20*, 171–183. [CrossRef]
35. Williams-Sether, T. *Empirical, Dimensionless, Cumulative-Rainfall Hyetographs Developed from 1959–1986 Storm Data for Selected Small Watersheds in Texas*; US Geological Survey: Reston, VA, USA, 2004.
36. Azli, M.; Rao, A.R. Development of Huff Curves for Peninsular Malaysia. *J. Hydrol.* **2010**, *388*, 77–84. [CrossRef]
37. Amponsah, A.O.; Daraio, J.A.; Khan, A.A. Implications of Climatic Variations in Temporal Precipitation Patterns for the Development of Design Storms in Newfoundland and Labrador. *Can. J. Civ. Eng.* **2019**, *46*, 1128–1141. [CrossRef]
38. Zeimetz, F.; Schaefli, B.; Artigue, G.; Hernández, J.G.; Schleiss, A.J. Swiss Rainfall Mass Curves and Their Influence on Extreme Flood Simulation. *Water Resour. Manag.* **2018**, *32*, 2625–2638. [CrossRef]
39. Bezak, N.; Šraj, M.; Rusjan, S.; Mikoš, M. Impact of the Rainfall Duration and Temporal Rainfall Distribution Defined Using the Huff Curves on the Hydraulic Flood Modelling Results. *Geosciences* **2018**, *8*, 69. [CrossRef]
40. Jiang, P.; Yu, Z.; Gautam, M.R.; Yuan, F.; Acharya, K. Changes of Storm Properties in the United States: Observations and Multimodel Ensemble Projections. *Glob. Planet. Chang.* **2016**, *142*, 41–52. [CrossRef]
41. Vantas, K.; Sidiropoulos, E.; Vafeiadis, M. Optimal Temporal Distribution Curves for the Classification of Heavy Precipitation Using Hierarchical Clustering on Principal Components. *Glob. NEST J.* **2019**, *21*, 530–538.
42. Vantas, K.; Sidiropoulos, E.; Vafeiadis, M. A Data Driven Approach for the Temporal Classification of Heavy Rainfall Using Self-Organizing Maps. In Proceedings of the EGU General Assembly 2019, Vienna, Austria, 7–12 April 2019; Volume 21, p. 1.
43. Vantas, K. Hydroscoper: R Interface to the Greek National Data Bank for Hydrological and Meteorological Information. *J. Open Source Softw.* **2018**, *3*, 625. [CrossRef]
44. Babu, G.J.; Rao, C.R. Goodness-of-Fit Tests When Parameters Are Estimated. *Sankhyā Indian J. Stat.* **2004**, *66*, 63–74.
45. Vantas, K.; Sidiropoulos, E.; Vafeiadis, M. Rainfall Temporal Distribution in Thrace by Means of an Unsupervised Machine Learning Method. In Proceedings of the Protection and Restoration of the Environment XIV, Thessaloniki, Greece, 3–6 July 2018; Volume 1, pp. 555–564.

46. Bezdek, J.C.; Hathaway, R.J. VAT: A Tool for Visual Assessment of (Cluster) Tendency. In Proceedings of the 2002 International Joint Conference on Neural Networks, IJCNN'02 (Cat. No.02CH37290). Honolulu, HI, USA, 12–17 May 2002; pp. 2225–2230.
47. Hopkins, B.; Skellam, J.G. A New Method for Determining the Type of Distribution of Plant Individuals. *Ann. Bot.* **1954**, *18*, 213–227. [CrossRef]
48. Husson, F.; Lê, S.; Pagès, J. *Exploratory Multivariate Analysis by Example Using R*, 2nd ed.; Chapman and Hall/CRC: New York, NY, USA, 2017; ISBN 978-0-429-22543-7.
49. Dunn, J.C. A Fuzzy Relative of the ISODATA Process and Its Use in Detecting Compact Well-Separated Clusters. *J. Cybern.* **1973**, *3*, 32–57. [CrossRef]
50. Bezdek, J.C. *Pattern Recognition with Fuzzy Objective Function Algorithms*; Springer: Boston, MA, USA, 1981; ISBN 978-1-4757-0452-5.
51. Huang, M.; Xia, Z.; Wang, H.; Zeng, Q.; Wang, Q. The Range of the Value for the Fuzzifier of the Fuzzy C-Means Algorithm. *Pattern Recognit. Lett.* **2012**, *33*, 2280–2284. [CrossRef]
52. Oppel, H.; Fischer, S. A New Unsupervised Learning Method to Assess Clusters of Temporal Distribution of Rainfall and Their Coherence with Flood Types. *Water Resour. Res.* **2020**, *56*. [CrossRef]
53. Lin, G.-F.; Wu, M.-C. A SOM-Based Approach to Estimating Design Hyetographs of Ungauged Sites. *J. Hydrol.* **2007**, *339*, 216–226. [CrossRef]
54. Ester, M.; Kriegel, H.-P.; Xu, X. A Density-Based Algorithm for Discovering Clusters in Large Spatial Databases with Noise. In *KDD'96*; AAAI Press: Palo Alto, CA, USA, 1996.
55. Conover, W.J. *Practical Nonparametric Statistics*, 3rd ed.; John Wiley & Sons: New York, NY, USA, 1980.
56. Bonta, J.V.; Rao, A.R. Regionalization of Storm Hyetographs. *JAWRA J. Am. Water Resour. Assoc.* **1989**, *25*, 211–217. [CrossRef]
57. Bonta, J.V.; Shahalam, A. Cumulative Storm Rainfall Distributions: Comparison of Huff Curves. *J. Hydrol.* **2003**, *42*, 65–74.
58. Benjamini, Y.; Hochberg, Y. Controlling the False Discovery Rate: A Practical and Powerful Approach to Multiple Testing. *J. R. Stat. Soc. Ser. B Methodol.* **1995**, *57*, 289–300. [CrossRef]
59. Ultsch, A. *U*-Matrix: A Tool to Visualize Clusters in High Dimensional Data*; Technical Report, Nr. 36; University of Marburg, Department of Computer Science: Marburg, Germany, 2003.
60. Van der Maaten, L.; Hinton, G. Visualizing Data Using T-SNE. *J. Mach. Learn. Res.* **2008**, *9*, 2579–2605.
61. Ultsch, A.; Mörchen, F. *ESOM-Maps: Tools for Clustering, Visualization, and Classification with Emergent SOM*; Technical Report, Nr. 46; University of Marburg, Department of Computer Science: Marburg, Germany, 2003.
62. Dasgupta, S.; Gupta, A. An Elementary Proof of a Theorem of Johnson and Lindenstrauss. *Random Struct. Algorithms* **2003**, *22*, 60–65. [CrossRef]
63. Thrun, M.C.; Ultsch, A. Uncovering High-Dimensional Structures of Projections from Dimensionality Reduction Methods. *MethodsX* **2020**, *7*, 101093. [CrossRef]
64. Thrun, M.C.; Lerch, F. Visualization and 3D Printing of Multivariate Data of Biomarkers. In Proceedings of the 24th Conference on Computer Graphics, Visualization and Computer Vision, Plzen, Czech Republic, 6 March 2016.
65. Vantas, K.; Sidiropoulos, E.; Evangelides, C. Rainfall erosivity and its estimation: Conventional and machine learning methods. In *Soil Erosion-Rainfall Erosivity and Risk Assessment*; IntechOpen: London, UK, 2019.
66. Koutsoyiannis, D. A Stochastic Disaggregation Method for Design Storm and Flood Synthesis. *J. Hydrol.* **1994**, *156*, 193–225. [CrossRef]
67. Hensel, D.R.; Hirsch, R.M. *Statistical Methods in Water Resources*; Book 4, Hydrologic Analysis and Interpretation; U.S. Geological Survey: Reston, VA, USA, 2002.
68. R Core Team. *R: A Language and Environment for Statistical Computing*; R Core Team: Vienna, Austria, 2021.
69. Vantas, K. Hyetor: R Package to Analyze Fixed Interval Precipitation Time Series. Available online: https://github.com/kvantas/hyetor (accessed on 20 March 2021).
70. Meyer, D.; Dimitriadou, E.; Hornik, K.; Weingessel, A.; Leisch, F. E1071: Misc Functions of the Department of Statistics, Probability Theory Group (Formerly: E1071). TU Wien. Available online: https://cran.r-project.org/web/packages/e1071 (accessed on 20 March 2021).
71. Thrun, M.C.; Stier, Q. Fundamental Clustering Algorithms Suite. *SoftwareX* **2021**, *13*, 100642. [CrossRef]
72. Kassambara, A.; Mundt, F. Factoextra: Extract and Visualize the Results of Multivariate Data Analyses. Available online: https://cran.r-project.org/web/packages/factoextra/ (accessed on 20 March 2021).
73. Wickham, H. *Ggplot2: Elegant Graphics for Data Analysis*; Springer: New York, NY, USA, 2016.

Article

Rainfall-Runoff Modeling Using the HEC-HMS Model for the Al-Adhaim River Catchment, Northern Iraq

Ahmed Naseh Ahmed Hamdan [1], Suhad Almuktar [2,3] and Miklas Scholz [3,4,5,*]

1 Department of Civil Engineering, College of Engineering, The University of Basra, Al Basra 61004, Iraq; ahmed.hamdan@uobasrah.edu.iq
2 Department of Architectural Engineering, Faculty of Engineering, The University of Basra, Al Basra 61004, Iraq; suhad.almuktar@tvrl.lth.se or suhad.abdulameer@uobasrah.edu.iq
3 Division of Water Resources Engineering, Faculty of Engineering, Lund University, P.O. Box 118, 221 00 Lund, Sweden
4 Department of Civil Engineering Science, School of Civil Engineering and the Built Environment, University of Johannesburg, Kingsway Campus, P.O. Box 524, Aukland Park, Johannesburg 2006, South Africa
5 Department of Town Planning, Engineering Networks and Systems, South Ural State University (National Research University), 76, Lenin Prospekt, 454080 Chelyabinsk, Russia
* Correspondence: miklas.scholz@tvrl.lth.se; Tel.: +46-(0)-462228920

Citation: Hamdan, A.N.A.; Almuktar, S.; Scholz, M. Rainfall-Runoff Modeling Using the HEC-HMS Model for the Al-Adhaim River Catchment, Northern Iraq. *Hydrology* **2021**, *8*, 58. https://doi.org/10.3390/hydrology8020058

Academic Editor: Andrea Petroselli

Received: 26 February 2021
Accepted: 24 March 2021
Published: 26 March 2021

Abstract: It has become necessary to estimate the quantities of runoff by knowing the amount of rainfall to calculate the required quantities of water storage in reservoirs and to determine the likelihood of flooding. The present study deals with the development of a hydrological model named Hydrologic Engineering Center (HEC-HMS), which uses Digital Elevation Models (DEM). This hydrological model was used by means of the Geospatial Hydrologic Modeling Extension (HEC-GeoHMS) and Geographical Information Systems (GIS) to identify the discharge of the Al-Adhaim River catchment and embankment dam in Iraq by simulated rainfall-runoff processes. The meteorological models were developed within the HEC-HMS from the recorded daily rainfall data for the hydrological years 2015 to 2018. The control specifications were defined for the specified period and one day time step. The Soil Conservation Service-Curve number (SCS-CN), SCS Unit Hydrograph and Muskingum methods were used for loss, transformation and routing calculations, respectively. The model was simulated for two years for calibration and one year for verification of the daily rainfall values. The results showed that both observed and simulated hydrographs were highly correlated. The model's performance was evaluated by using a coefficient of determination of 90% for calibration and verification. The dam's discharge for the considered period was successfully simulated but slightly overestimated. The results indicated that the model is suitable for hydrological simulations in the Al-Adhaim river catchment.

Keywords: hydrologic model; remote sensing; HEC-HMS; Al-Adhaim River; rainfall/runoff

1. Introduction

Surface water from the Tigris, Euphrates and Shatt Al-Arab rivers is the main water resource in Iraq. The water flow of these rivers varied from year to year depending on the incoming water flows coming from outside of Iraq and the annual rainfall intensity during the studied period [1].

The main factors affecting water shortage challenges in Iraq are the dams located outside of Iraq, in the Tigris and the Euphrates Rivers and their tributaries, climate change, and the mismanagement of Iraq's water resources [2]. It has been reported that the growing water consumption in Turkey and Syria could lead to the complete drying-up of the Tigris and Euphrates rivers by the year 2040 [1].

The rivers of Iraq are also affected by the annual and seasonal precipitations. A year with heavy precipitations may lead to major floods and disasters, while a year with low precipitation amounts can lead to droughts damaging soil and endangering crops [3]. Excessive precipitations may also lead to runoffs, which occur when water that is discharged exceeds river, lake or artificial reservoir capacity. Often rainfall-runoff models are used for modeling or predicting possible floods as well as rivers and lakes' water levels during different boundary conditions [4].

Hydrologic Engineering Center (HEC-HMS) is a hydrologic modeling software developed by the US Army Corps of Engineers of the Hydrologic Engineering Center (HEC), which contains an integrated tool for modeling hydrologic processes of dendritic watershed systems. This model consists of several components for processing rainfall loss, direct runoffs and routing. The HEC-HMS model has been widely used, for example, in many hydrological studies because of its simplicity and capability to be used in common methods [5].

The Geospatial Hydrologic Modeling Extension (HEC-GeoHMS) is a public-domain software package for use with Geographical Information Systems (GIS), GeoHMS ArcView and Spatial Analysis to develop several hydrological modeling inputs. After analyzing the information of the Digital Elevation Model (DEM), HEC-GeoHMS transforms the drainage paths and watershed boundaries into a hydrologic data structure that represents the watershed response to rainfall [6]. An important feature of the HEC-GeoHMS model is the assignment of the curve numbers (CN), which are related to the land use/land cover and soil type of the Al-Adhaim catchment.

Several researchers have used the HEC-HMS hydrological model to represent flow by simulated rainfall-runoff processes. In this regard, a study conducted by Oleyiblo et al. [7] on the Misai and Wan'an catchments in China used the HEC-HMS model for flood forecasting. The results were calibrated and verified using historical observed precipitation data, which showed satisfactory results. Saeedrashed et al. [8] used computational hydrological and hydraulic modeling systems designed by using the interface method, which links GIS with the modeling systems (HEC-HMS and HEC-RAS). They conducted a floodplain analysis of the Greater Zab River. Their model calibration and verification processes showed satisfactory results. Martin et al. [9] used HEC-GeoHMS in the Arc-Map environment to predict floods by hydraulic modeling. They obtained flood hazard maps by exporting the HEC-RAS model output results to Arc- Map, where they were processed to identify the flood-expanded areas. The results from the flood hazard maps showed that the areas more prone to floods were located in the river middle reach.

HEC-HMS has also been used by Tassew et al. [10] to conduct a rainfall-runoff simulation of the Lake Tana Basin of the Gilgel Abay catchment in the upper Blue Nile basin in Ethiopia by using six extreme daily time-series events. The Nash Sutcliffe Efficiency (NSE) accounts for the comparison between the simulated and observed hydrographs. The coefficient of determination R^2 was 0.93 indicating that the model was appropriate for hydrological simulations. HEC-HMS and GIS were successfully used to simulate the rainfall-runoff process in the Karun river basin in Iran [11], as well as the inflow designed floods and dam breach hydraulic analyses of four reservoirs in the Black Hills, South Dakota [6]. Other examples include the modeling of the watershed in Ahvaz, Iran [12,13], the simulation of the runoff process for the adjoining areas of the Lagos Island and Eti-Osa Local Government Areas and the assessment of the influence of different levels of watershed spatial discretization on the HEC-HMS model's performance for the Uberaba River Basin region [14].

Generally, it is important to mention that in Iraq, and especially in the area under assessment, very few studies have used such a modeling approach. Large input data have been prepared for the model, including maps of DEM, land use/land cover, soil type and curve number, in addition to the rainfall data and observed discharge values. Furthermore, the Al-Adhaim catchment characteristics have been created and prepared using HEC-Geo HMS, which were subsequently exported to HEC-HMS. This study will provide good support for other researchers to continue with such studies for adjacent catchments as well as other catchments, providing water resources management control in Iraq.

The main objective of this study is to develop the HEC-HMS hydrological model using remotely sensed data such as DEM. The hydrological model was used in combination with HEC-GeoHMS and GIS to identify the flow by simulating the rainfall-runoff processes for the Al-Adhaim catchment. Model input parameters were calculated for the HEC-HMS model. The rapid increase of water demand in Iraq, as well as the predicted water storage problems due to various factors (mainly climate change), can be assessed by hydrological modeling, which can help decision-makers to take preemptive actions such as storing water in the dams, depending on rainfall-runoff predictions. Furthermore, the calibrated parameters of this model can be used for future hydrological studies in this and adjacent catchments.

2. Methodology

The Al Adhaim hydrologic model was created by means of HEC-GeoHMS and particularly by using DEM of the studied areas, obtained from the United States Geological Survey (USGS) website. HEC-GeoHMS creates an HMS input file, such as the catchment of the area under study, a stream network, sub-basin boundaries, and different hydrological parameters in an Arc map environment via a series of steps. An HEC-HMS model for the Al-Adhaim catchment was developed, and simulations were run with daily rainfall recorded data.

2.1. Description of Study Area

The studied area (Al-Adhaim catchment) is located in the northeast of Iraq and comprises a dam with a reservoir downstream of the river (Figure 1). There is also a discharge gauge downstream of the dam. The Al-Adhaim river is the main source of freshwater for the Kirkuk Province. The length of the river is close to 330 km. The river joins the Tigris River at about 15 km to the south of the Balad district. The river drains an area of 13,000 km^2 and then runs almost dry from June to October. Floods take place from November to May each year, as they are mainly fed by rainfall. The total summation of the rainfall for the Al-Adhaim catchment for the recent years equals to 489, 915, 415 and 623 mm for the hydrological years 2014 to 2018 in that order. Note that it was not raining during the months of June, July, August and September [15], while the temperature ranged from 2 °C to 48 °C. Al-Adhaim can be classified as an arid catchment with 71% of its surface being covered by forests and 29% by agricultural lands [2].

Figure 1. Layout of Al-Adhaim Dam (Arc-GIS by authors).

Al-Adhaim dam is located downstream of the river, within the administrative borders of the Diyala Province (34°32′ N and 44°30′ E, about 135 km northeast of the capital Baghdad, 65 km southeast of Kirkuk Province, and 70 km upstream from Al-Adhaim's confluence with the Tigris River). The main purposes of this dam are flood control, irrigation and obtaining hydropower energy (Figure 1).

The Al-Adhaim dam is a 3.8 km long and 150 m high earth-filled embankment dam. Its width is about 10 m at its highest point. It was built in 1999, and consists of a spillway intake, a bottom outlet gate shaft, and a power intake (Figure 1). Its storage capacity, live storage and spillway discharge are about 1.5×10^9 m^3, 1×10^9 m^3 and 1150 m^3, respectively. The crest of the dam, which is 15 m long, follows an ogee spillway shape profile according to the United States Bureau of Reclamation standard. It has an elevation of 131.54 m and is designed to pass a probable maximum flood discharge of 1150 m^3/s [15].

The bottom outlet gate shaft is used for water release for irrigation purposes. The structure consists of an intake structure, which was used for releasing water through a closed conduit, which gradually transitions from a squared to a circular section. The intake is divided into two openings with the transition from a square section of 6×6 m^2 to a 6 m diameter circular one, which starts the circular tunnel section. The gate shaft accommodates two gates: the upstream one (the guard gate) and the downstream one (the regulating gate).

The power intake was used for releasing water for irrigation and power generation purposes. This outlet includes a 50-m high reinforced concrete shaft at the beginning of the tunnel, which includes two emergency gates. The inlet is a 6.5×16 m^2 streamlined rectangular opening with a shape gradually changing from a 4.5×4.5 m^2 square section to 4.5 m diameter circular section.

The reservoir's water surface area is 270 km^2 at an elevation of 143 m, which is the maximum water elevation, and 122 km^2 at an elevation of 130 m. The reservoir volume is 3750 million cubic meters (MCM). The highest operating elevation is 135 m, corresponding to an area of 170 km^2 and a volume of 2150 MCM (Table 1), and the lowest operating elevation is 118 m, corresponding to 52 km^2 and a capacity of 450 MCM. The water elevation and corresponding area and volume of the reservoir at the beginning of the hydrological year are shown in Table 2 [15,16]. The gates' opening coefficients were used for calibration, because the discharge gate was downstream of the dam. The initial elevation of the water in the reservoir at the beginning of the hydrological year was also provided by the Central Statistical Organization of Iraq [17].

Table 1. Water elevation and corresponding area and volume of the reservoir [15,16].

Elevation [1] (m)	Area (km^2)	Volume (MCM [2])
100	3	70
115	41	310
118	52	450
120	60	520
125	85	980
130	122	1400
135	170	2150
140	233	3130
143	270	3750

[1] Elevation: meters above sea level; [2] MCM: million cubic meters.

Table 2. Elevations and the corresponding water volumes at the start of the hydrological year [17].

Date	Elevation [1] (m)	Capacity (BCM [2])
1/10/2015	120.71	0.60
1/10/2016	115.22	0.31
1/10/2017	113.94	0.27
1/10/2018	117.20	0.42

[1] Elevation: meters above sea level; [2] BCM: Billion Cubic Meters.

2.2. Daily Rainfall Data

The rainfall season begins in October and ends in April for the majority of the Iraqi territory. The average rainfall increases from southwest to northeast due to topographical effects. The recorded daily rainfall data were downloaded from the Power Data Access Viewer website. Figure 2 shows the maximum rainfall amount for the last ten years for the studied area that occurred exactly on 17 November 2015, which approached 39 mm [18].

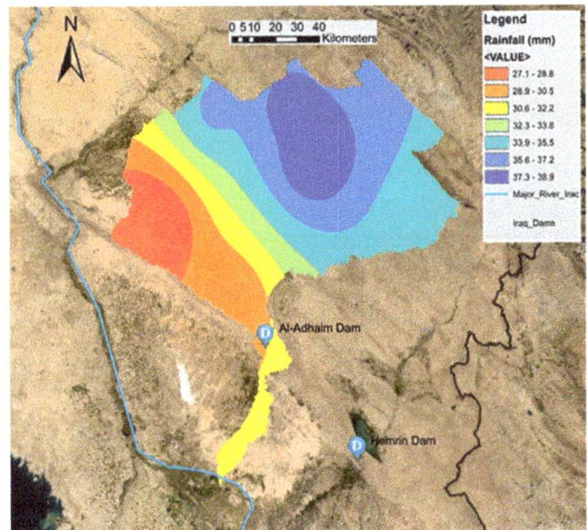

Figure 2. Distribution of the maximum rainfall in the study area, which occurred on 17 November 2015 (Power Data Access Viewer website; https://power.larc.nasa.gov/data-access-viewer, (accessed on 26 March 2021)).

2.3. HEC-GeoHMS

HEC-GeoHMS is working under the GIS environment, which is a geospatial hydrology tool allowing users to determine sub-basin streams as an input for the HEC-HMS hydrological model [17]. In this study, Arc Map 10.5 and HEC-GeoHMS were used.

2.4. HEC-HMS Project Set-Up

The HEC-HMS software was applied to convert the rainfall data into direct flow taking into account the topography and the surface characteristics of the modelled location (e.g., length of the reach). The software also considers routing, loss, and flow transformation in the runoff computation. The data were checked, and the HEC-HMS model was developed, with the creation of several HEC-HMS parameters and the basin and meteorological model files.

Model Input Parameters

The parameters required for running the HEC-HMS model are listed in Table 3.

Table 3. The hydrological model named Hydrologic Engineering Center (HEC-HMS) catchment model parameters for Al-Adhaim.

No.	Model	Method	Parameters Required (Unit)
1	Loss Rate Parameter	SCS Curve Number	Initial abstraction (mm), Curve Number and Impervious area (%)
2	Runoff Transform	SCS Unit Hydrograph	Lag time (min)
3	Routing Method Constants	Muskingum	Travel time (K) and dimensionless weight (X)

3. Results and Discussion

The temporal variation of the flow at the outlet of the catchment has been assessed by considering the model's response for three years of separate precipitation. The basin was divided into nine sub-basins based on the river network. Runoff from the sub-basins was estimated by using the SCS-CN, SCS Unit Hydrograph and Muskingum methods for loss, transformation, and routing calculations, respectively. Some hydrological parameters were obtained by the calibration process that performed by comparing the simulated flow with the observed flow data measured by using a discharge gauge, which is located near to the outlet of the catchment.

3.1. Digital Elevation Model (DEM)

The DEM is an essential input to define the topography of the catchment on the basis of semi-DEM shown in Figure 3. Data were downloaded from the United States Geological Survey (USGS) website [19] and modified for the study area using Arc-map by the authors.

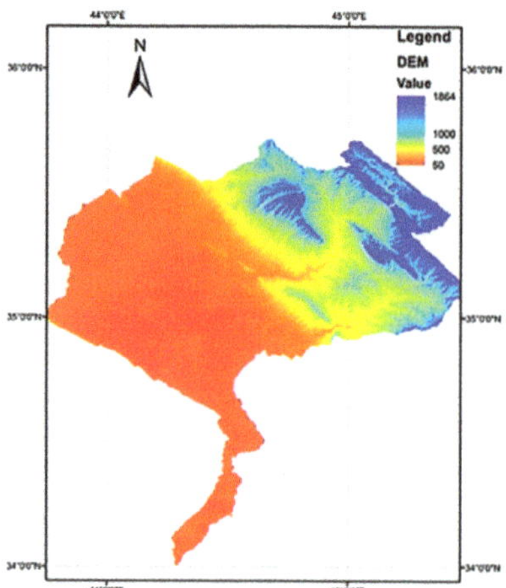

Figure 3. Digital elevation model for Al-Adhaim (downloaded from United States Geological Survey (USGS) [19] and modified by the authors).

3.2. HEC-GeoHMS Development

Terrain pre-processing and model development using HEC-GeoHMS is shown in Figure 4. In terrain pre-processing, the DEM sinks were filled, flow direction and flow accumulation were estimated, and catchments were separated. The catchment boundaries were drawn and stored as different shapefiles. Then, a suitable outlet point was selected at the outlet point located at 44°17′33.937″ E and 34°0′14.846″ N. After those basins were merged, the longest flow path and basin centroid were determined. After processing, the developed model became ready to export to HEC-HMS software. Extra processing was with HEC-GeoHMS and includes the estimation of hydrologic parameters, such as the curve numbers.

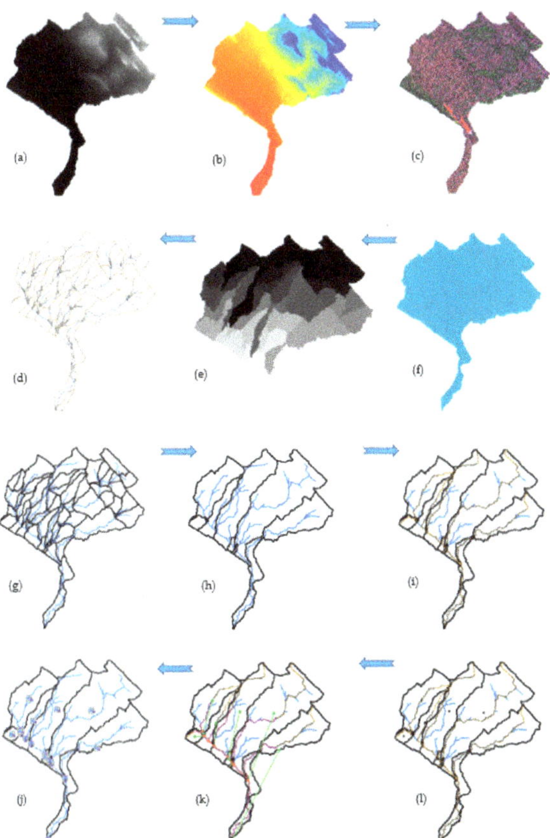

Figure 4. Pre-processing and model development: (**a**) raw digital elevation model; (**b**) fill sinks; (**c**) low directions; (**d**) catchment polygon; (**e**) basin raster; (**f**) flow accumulation; (**g**) generated project; (**h**) basin merge; (**i**) longest flow path; (**j**) Hydrologic Engineering Center (HEC-HMS) legend; (**k**) HEC-HMS schematic; and (**l**) basin centroid.

3.2.1. Soil Map and Land Use/Cover Shape Files

Procedures to get the soil types for the Al-Adhaim catchment include downloading of the raster file of the global soil type from the website http://www.fao.org/soils-portal/ data-hub/soil-maps-and-databases/harmonized-world-soil-database-v12/en/, (accessed on 26 March 2021). Data were then exported to Arc map to clip the study area from the global map of soil types and convert it from the raster file to the shape file. A symbology has been performed afterwards. Finally, the created shape file has been merged with the land use/land cover type file to calculate the curve number.

The soil upper layer in most of the areas of the Al-Adhaim catchment is considered as homogeneous, except in the middle region of the country, which is covered by mounds. The soil consists of mosaic clay or loam within a surface layer consisting mainly of silt, clay and silt loam. The sub-surface layer is mainly made of clay loam or clay (Figure 5). Land use (or land cover) is used as an input for a catchment model as it can affect surface erosion and water runoff [20].

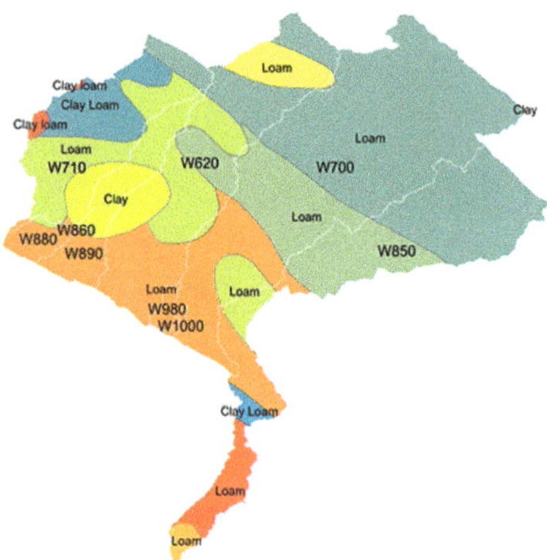

Figure 5. Soil map of the Al-Adhaim catchment (adapted and modified by the authors).

Procedures to get the land use/cover for the Al-Adhaim catchment include downloading of the raster file of the global soil type from the website http://due.esrin.esa.int/page_globcover.php, (accessed on 26 March 2021). Data were then exported to Arc map to clip the study area from the global map of soil types and convert it from the raster file to the shape file. A symbology has been performed afterwards. Finally, a shape file has been created by converting the raster to polygons to get a file that can be merged with the soil type file to calculate the curve number. The land use in the Al-Adhaim catchment is either Ever Green Forest, Dwarf Scrup or Open Sea (Figure 6).

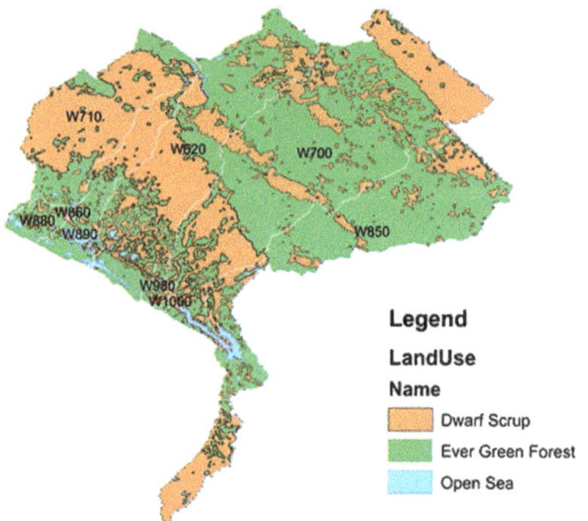

Figure 6. Land use classification of the Al-Adhaim catchment (adapted and modified by the authors).

3.2.2. Curve Number

Soil map and land use datasets were used to generate the Curve Number (CN) grid file, which is required to build the HEC-HMS model. CN values were used to determine the stream/sub-basin characteristics and to estimate the hydrological parameters used in the model [10]; Figure 7. CN values range from 30 corresponding to permeable soils with high infiltration rates to approximately 100 corresponding to water bodies.

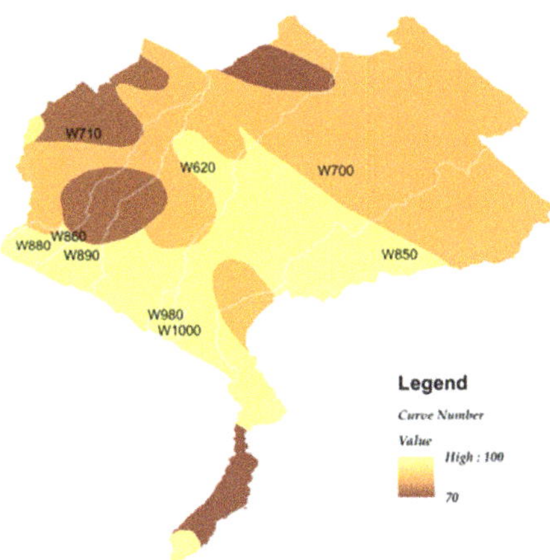

Figure 7. Curve Number calculations for the Al-Adhaim catchment (adapted and modified by authors).

3.3. Parameters Estimation

3.3.1. Loss Model—Soil Conservation Service Curve Number

The loss models in HEC-HMS were calculated by subtracting the volume of water that was intercepted, infiltrated, stored, evaporated or transpired to the rainfall water volume [10]. We used the Soil Conservation Service Curve Number loss (SCS curve number) method to calculate the direct runoff from a design rainfall.

For the loss model, the SCS-CN has two parameters: the curve number (CN) and the initial abstraction (Ia). The default initial abstraction ratio was equal to 0.2 but then varied after the model calibration. The CN is a function of land use and soil type estimated by using the HEC-GeoHMS toolkit of Arc Map 10.5. The percentage of imperviousness for each sub-basin was assumed to be 0% (the entire catchment was assumed to be completely pervious). The CN values for each sub-basin were calculated by using formula (1) [21,22].

$$CN = \frac{\sum A_i CN_i}{\sum A_i} \tag{1}$$

where A_i is the area (km^2) of the sub-basin and CN_i is the corresponding curve number. Ia (mm) is obtained by multiplying the loss coefficient by the potential abstraction S (mm). The potential abstraction is a function of CN and calculated by using the formula (2) [22].

$$S = \frac{25,400}{CN} - 254 \tag{2}$$

3.3.2. Transform Model—Soil Conservation Service Unit Hydrograph Method

The transform prediction models in HEC-HMS simulate the process of excess rainfall direct runoff in the catchment and transform the rainfall excess in point runoff [10]. During the analysis of the study data, the SCS Unit Hydrograph model was used to transform the excess rainfall into runoff.

The basin lag time parameter values have been calculated during data processing by means of the HEC GeoHMS application and stored in the attributes' table of the sub-basin data layer. Basin lag times were initially calculated in hours for the sub-basins by using Equation (3) and were then converted to minutes when used with HEC-HMS.

$$Lag = \frac{(S+1)^{0.7} L^{0.8}}{1900 * Y^{0.5}} \tag{3}$$

where S = maximum retention (mm) as defined by Equation (2), lag = basin lag time (hour), L= hydraulic length of the catchment (longest flow path) (feet) and Y = basin slope (%). Table 4 shows the Loss and Transform Model parameter value estimations.

Table 4. The estimation of Loss and Transform Model parameter values.

Sub-Basin	Basin Area (km^2)	Basin Slope (%)	Curve Number (CN)	Potential Abstraction (mm)	Ia (mm)	Basin Lag (Hours)
W880	193.5333	37.00	87	1.49	7.59	5.3959
W890	708.2559	29.87	80	2.50	12.70	6.1618
W710	1635.3000	14.20	78	2.82	14.33	9.0085
W860	8.15850	35.00	95	0.53	2.67	5.4074
W620	2490.1000	18.24	75	3.33	16.93	8.0493
W980	81.4450	47.00	100	0.00	0.00	4.5998
W1000	24.7302	49.00	100	0.00	0.00	4.50491
W700	4485.1000	21.04	79	2.66	13.50	7.37090
W850	2718.1000	38.06	70	4.29	21.77	5.70099

3.3.3. Routing—Muskingum Method

As the flood runoff moves through the channel reach, it weakens because of the channel storage effects. The routing model available in the HEC-HMS software for this scenario was the Muskingum method [23].

The Muskingum method is a common lumped flow routing technique. In this model, a calibration for two parameters, X and K, was required. X is a dimensionless weight, which is a constant coefficient that varies between 0 and 0.5, where X is a factor representing the relative influence of flow on storage levels. It can be assumed that the value equals 0.1 as an initial value of the calibration parameters, which was corrected during the calibration process. K is the parameter having a unit of time and value ranging from 1 to 5 h. It is related to the delay between discharge peaks [24]. K is estimated using Equation (4).

$$K = \frac{L}{V_w} \tag{4}$$

where V_w is the flood wave velocity, which can be taken as 1.5 times the average velocity, and L is the reach length. The average velocity was obtained from the stream gauging sites. The value of K was used also in the calibration process within short limits based on Equation (4) until the simulated hydrographs approached the observed ones.

The rainfall runoff processes of the dendritic catchment systems were simulated by using the hydrological modeling system of the HEC-HMS software. After considering the pre-processing in the HEC-GeoHMS, the model was imported to the HEC-HMS software as a basin file. Figure 8 shows the basin model file. HEC-HMS input data are important to run the rainfall-runoff modeling.

The calculated parameters, such as loss parameters (curve number, initial abstraction and percentage of imperviousness), transform parameters (lag time) and routing parameters (k and x), were added to the sub-basins and the reaches either manually or automatically from the GIS attribute tables. The precipitation, temperature, evaporation and discharge gauge data were added as time series using the time series data manager, while the elevation-area table was added as paired data.

Three files were created for rainfall data input in the meteorological folder, corresponding to the hydrological year intervals 2015–2016, 2016–2017 and 2017–2018. For the control run, daily rainfall was started on 1 October at 00:00 and ended on 30 September at 00:00. The selected time interval for the hydrograph was of one day for the three corresponding hydrological years.

Figure 8. Hydrologic modeling system of the Al-Adhaim catchment by HEC-HMS.

3.4. Model Calibration

The model is calibrated by using the daily rainfall data from the hydrological year intervals 2016–2017 and 2015–2016. Manual calibration was applied to estimate the values of the different parameters. The optimal values of the Muskingum Model parameters (K, X) were obtained by comparing the observed and simulated flows, while the parameters of the Loss Model and the Transform Model were calculated as explained in Sections 3.3.1 and 3.3.2. The dam data and the data needed for running the HEC-HMS models such as the spillway level, the gates opening area and the center elevation were obtained from the Central Statistical Organization of Iraq [17].

3.5. Comparison of the Simulated and Observed Hydrograph and Validation of Model

The simulated and observed hydrographs for the calibration period intervals 2015–2016 and 2016–2017 are shown in Figure 9a,b and Figure 10a,b for both the simulated and observed hydrographs exhibiting nearly similar trends and shapes. However, the peak discharge of the simulated hydrographs is greater than the observed ones. The results of the model in this study showed an acceptable fit between the simulated values and observations. The trend and shape of the hydrograph seems compatible. The R^2 values for the calibration periods 2015–2016 and 2016–2017 were 0.90 for the outlet section.

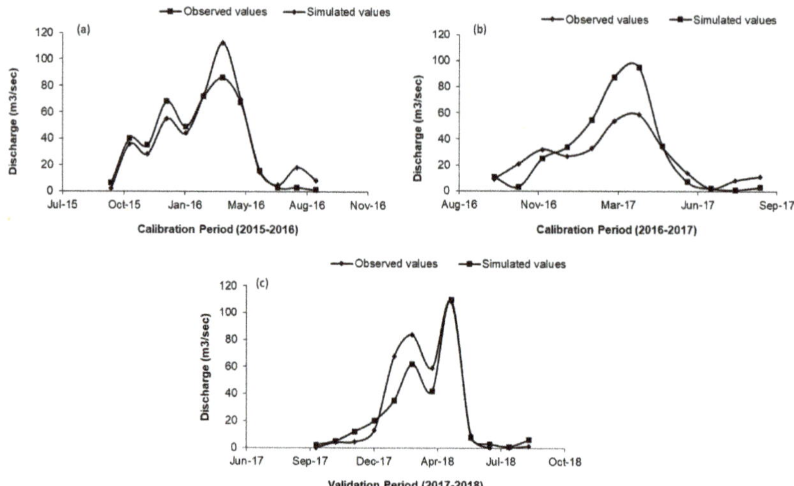

Figure 9. Observed and simulated discharge values for the dam outlet (a) calibration of 2015–2016; (b) calibration of 2016–2017; and (c) validation of 2017–2018.

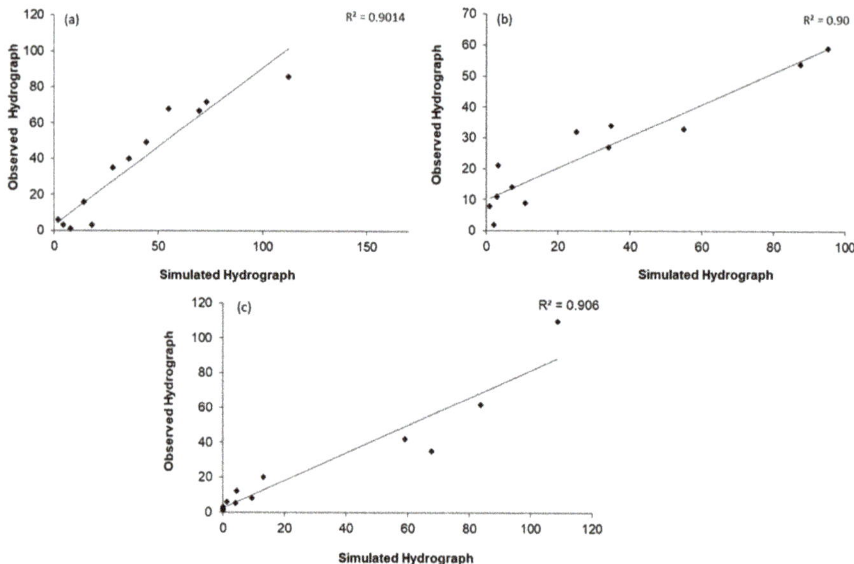

Figure 10. Regression scatter plot for the (a) calibration period for the dam outlet for the years 2015–2016; (b) calibration period for the dam outlet for the years 2016–2017; and (c) validation period for the dam outlet for the years 2017–2018.

The model was run for one year of daily rainfall data for validation purposes. The runoff was simulated by using the hydrological year interval 2017–2018 in the validation model. The model calibration parameters were applied to the validation model. The simulated and observed hydrograph and regression scatter plot for the outlet section for the validation period of the years 2017 and 2018 is presented in Figures 9c and 10c. As observed in the calibration plots, the observed and simulated hydrographs are almost identical except for the peak discharge, which is higher for the simulated graph. The R^2 for the validation period of 2017–2018 is 0.906 for the outlet section.

A similar result for R^2 of 0.9 was obtained by Oleyiblo et al. [7] on Misai and Wan'an catchments in China using HEC-HMS. Tassew et al. [10] did a comparison of the observed and simulated hydrographs and the performance of the model was as follows: NSE = 0.884 and the R^2 = 0.925. It follows that the model is suitable for hydrological simulations in the Gilgel Abay Catchment. Barbosa [14] used seven different methods to investigate the performance of the HEC-HMS model: MAE, RMSE, RSR, NSE, PBIAS, R2, and KGE. The researcher concluded that the HEC-HMS model represents the hydrological processes of the basin under investigation efficiently. The results suggest that the subdivision of a catchment does not result in the improvement of the HEC-HMS model's performance without significant differences in physiographic characteristics; for example, the values of R^2 ranged between 0.72 for two sub-basins to 0.73 for 32 sub-basins. From the above, it can be concluded that the model performs considerable well, and the simulation can be judged to be satisfactory.

During modeling using HEC-HMS, it was noticed that the main parameters which affect runoff quantities were the curve number and then initial abstraction. However, lag time and percentage of impervious area were less affected by the runoff results.

3.6. Reservoir Modelling

According to the Department of Environment Statistics [17], the observed annual volumes discharged from the dam concerning its outlet were 1.15, 0.81 and 0.81 billion cubic meters (BCM), while the simulated volumes were 1.22, 0.93 and 0.909 BCM for the hydrological year intervals 2015–2016, 2016–2017 and 2017–2018, respectively (Figure 11a). Additionally, the simulated values for the average annual discharge flow were 51.6, 29.9 and 29.3 m^3/s, while the observed values [17] for the outlet were 36.42, 25.83 and 25.58 m^3/s for the hydrological year intervals 2015–2016, 2016–2017 and 2017–2018, respectively, as shown in Figure 11b. The figures indicate a slightly overestimated discharge flow.

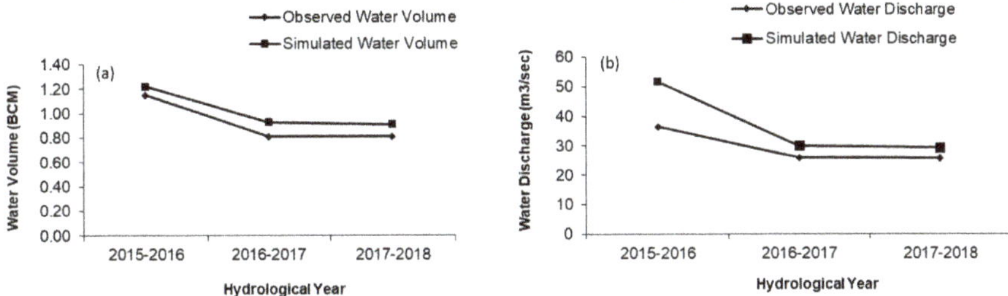

Figure 11. Observed and simulated hydrographs corresponding to (**a**) the annual water volume discharged; and (**b**) the annual discharge flow.

Table 5 shows the results of the simulation run for the dam storage area. The results indicate that the storage and the pool elevation at the beginning of the hydrological years were compatible with Table 2 for all years assessed in this study. The results showed that the years 2015–2016 can be considered as wet ones, with peak storage of 0.86 BCM and an elevation of 124.6 m. However, for this period the storage capacity was just under the maximum limits, accounting for approximately 1.5 BCM at the elevation of 131.54 m, which is when the spillway comes into operation.

Table 5. Results of a simulation run for the dam storage area.

Year	Storage (BCM)/Elevation (m) at the Start of the Hydrological Year	Peak Inflow (m^3/s)/Date	Peak Discharge (m^3/s)/Date	Inflow Volume (BCM)	Peak Storage (BCM)/Elevation (m)
2015–2016	0.60/120.71	742.4/ 12 April 2016	133.3/ 16 April 2016	1.320	0.860/124.6
2016–2017	0.31/115.20	396.9/ 24 March 2017	142.1/ 26 March 2017	0.895	0.438/118.4
2017–2018	0.27/113.94	625.1/ 18 February 2018	142.4/ 26 February 2018	0.953	0.438/118.4

BCM, billion cubic meters.

Figure 12a–c show the Al-Adhaim reservoir simulations for the year intervals 2015–2016, 2016–2017 and 2017–2018. These figures show that during the summer season when there was almost no rain, no flow occurred, and the pool elevation approached its minimum. This is due to the fact that the main water source for the Al-Adhaim river is rainfall. In contrast, during the period of precipitation, the storage capacity increased, reaching peak inflow values of 742.4, 396.9 and 625.1 m^3/s corresponding to maximum rainfall values of 39.91, 29.14 and 33.41 mm for the year intervals 2015–2016, 2016–2017 and 2017–2018, respectively.

The model used in our research is useful in predicting runoff volumes and flooding in the area of interest. Figure 12 shows that the peaks of the estimated discharges for the hydrological years from 2015 to 2018 occurred between March and May, which is considered to be the time in which severe flooding takes place in the area. Moreover, it can be noted that during the March to May period, runoff depth and volume increased, and therefore, special attention should be dedicated to dam outlet management during this period in the coming years.

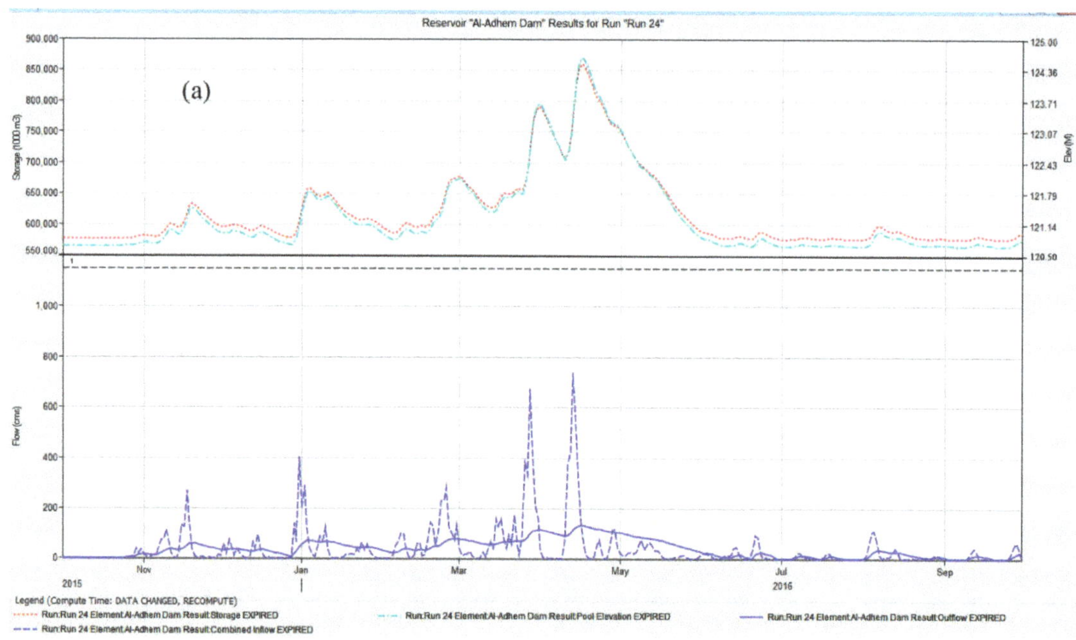

Figure 12. *Cont.*

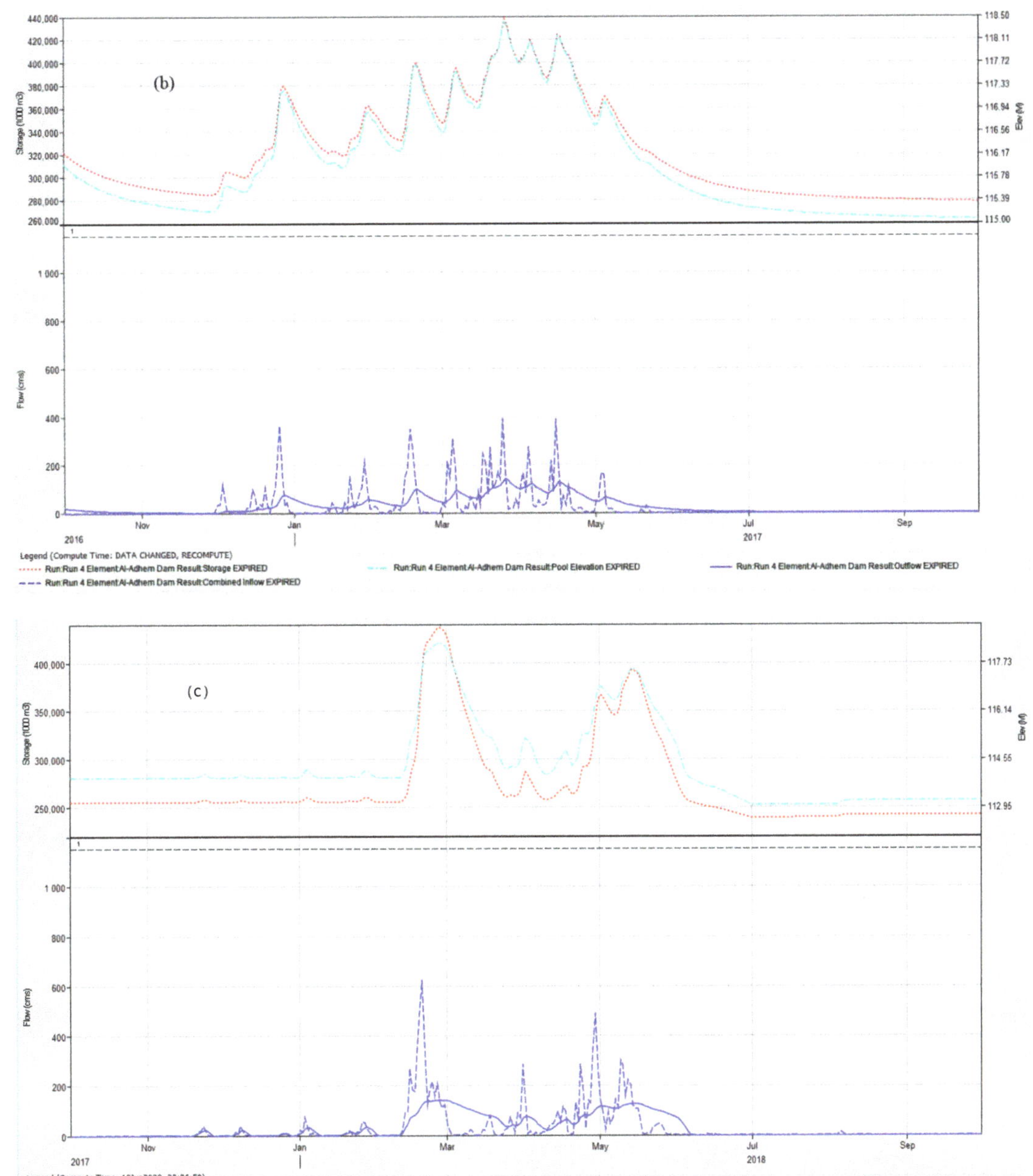

Figure 12. The simulation run for the Al-Adhaim reservoir concerning the intervals (**a**) 2015–2016; (**b**) 2016–2017; and (**c**) 2017–2018.

4. Conclusions and Recommendations

The HEC-HMS hydrologic model was used in combination with the HEC-GeoHMS and GIS to identify flow by simulated rainfall-runoff processes. The outlet flow discharge of the Al Adhaim Dam has been simulated and compared with the flow obtained from the discharge gauge located downstream of the dam by calibration of the parameters for two hydrological years and verification for one hydrological year. The results show a good agreement between observed and simulated flows, and R^2 was 0.9 for both calibration and verification. The correlation between simulation and observation was good, but the total volume of discharge storage for these years was slightly overestimated. The conclusions from the analysis are listed below:

1. The HEC-HMS model can be used to obtain satisfactory simulated hydrological models and is a valuable tool for the management of dam storage by forecasting rainfall amounts.
2. The simulation results of runoff discharge peaks are slightly different compared with the observed data.
3. In the summer season with almost no precipitation, there was no flow and the pool elevation approached minimum limits. On the contrary, during the period of precipitation, the storage capacity approached the peak inflow of 742.4 m^3/s for the years 2015–2016, which corresponds to maximum daily rainfall of 39.91 mm.
4. The area of interest does not have an available discharge station other than the one located near the outlet. Discharge stations could provide real observed discharge data that can be used to validate the modeling results. Therefore, the provision of an upstream discharge station is vital.
5. The development of serious water policy and planning strategies in accordance with the results obtained from this study could reduce the probability of floods and may help in the management and control of the dam outlet.
6. During modeling using HEC-HMS, it was noticed that the main parameters which affect runoff quantities were the curve number and then initial abstraction.
7. HEC-HMS model results were good for flood forecasting concerning the Al-Adhaim catchment. Data can be exported to simulate a 2-dimensional flood inundation map using a hydraulic model such as HEC-RAS. They can also be used for forecasting the rainfall using a suitable program to predict flooding for long-time periods.

Author Contributions: Conceptualization, A.N.A.H., S.A., and M.S.; methodology, S.A.; software, A.N.A.H.; validation, S.A., A.N.A.H. and M.S.; formal analysis, A.N.A.H.; investigation, S.A.; resources, A.N.A.H.; data curation, S.A. and A.N.A.H.; writing—original draft preparation, A.N.A.H.; writing—review and editing, S.A. and M.S.; visualization, S.A., A.N.A.H. and M.S. All authors have read and agreed to the published version of the manuscript.

Funding: This research received no external funding.

Institutional Review Board Statement: Not applicable.

Informed Consent Statement: Not applicable.

Data Availability Statement: Not applicable.

Conflicts of Interest: The authors declare no conflict of interest.

References

1. Al-Ansari, N. Management of water resources in Iraq: Perspectives and prognoses. *Engineering* **2013**, *5*, 667–684. [CrossRef]
2. Abbas, N.; Wasimia, S.A.; Al-Ansari, N. Assessment of climate change impacts on water resources of Al-Adhaim, Iraq using SWAT model. *Engineering* **2016**, *8*, 716–732. [CrossRef]
3. Rahi, K.A.; Al-Madhhachi, A.-S.T.; Al-Hussaini, S.N. Assessment of Surface Water Resources of Eastern Iraq. *Hydrology* **2019**, *6*, 57. [CrossRef]
4. Jia, Y.; Zhao, H.; Niu, C.; Jiang, Y.; Gan, H.; Xing, Z.; Zhao, X.; Zhao, Z. A WebGIS-based system for rainfall-runoff prediction and real-time water resources assessment for Beijing. *Comput. Geosci.* **2009**, *35*, 1517–1528. [CrossRef]
5. Halwatura, D.; Najim, M. Application of the HEC-HMS model for runoff simulation in a tropical catchment. *Environ. Modell. Softw.* **2013**, *46*, 155–162. [CrossRef]
6. Hoogestraat, G.K. *Flood hydrology and Dam-Breach Hydraulic Analyses of Four Reservoirs in the Black Hills, South Dakota*; U.S. Geological Survey Scientific Investigations Report 2011–5011; U.S. Geological Survey: Reston, VA, USA, 2011; p. 37.

7. Oleyiblo, J.O.; Li, Z.-j. Application of HEC-HMS for flood forecasting in Misai and Wan'an catchments in China. *Water Sci. Eng.* **2010**, *3*, 14–22.
8. Saeedrashed, Y.S. Hydrologic and Hydraulic Modelling of the Greater Zab River-Basin for an Effective Management of Water Resources in the Kurdistan Region of Iraq Using DEM and Raster Images. In *Environmental Remote Sensing and GIS in Iraq*; Springer: Cham, Switzerland, 2020; pp. 415–446.
9. Martin, O.; Rugumayo, A.; Ovcharovichova, J. Application of HEC HMS/RAS and GIS tools in flood modeling: A case study for river Sironko–Uganda. *J. Eng. Des. Technol.* **2012**, *1*, 19–31.
10. Tassew, B.G.; Belete, M.A.; Miegel, K. Application of HEC-HMS model for flow simulation in the Lake Tana basin: The case of Gilgel Abay catchment, upper Blue Nile basin, Ethiopia. *Hydrology* **2019**, *6*, 21. [CrossRef]
11. Tahmasbinejad, H.; Feyzolahpour, M.; Mumipour, M.; Zakerhoseini, F. Rainfall-runoff Simulation and Modeling of Karon River Using HEC-RAS and HEC-HMS Models, Izeh District, Iran. *J. Appl. Sci.* **2012**, *12*, 1900–1908. [CrossRef]
12. Radmanesh, F.; Hemat, J.P.; Behnia, A.; Khond, A.; Mohamad, B.A. Calibration and assessment of HEC-HMS model in Roodzard watershed. In Proceedings of the 17th International Conference of River Engineering, University of Shahid Chamran, Ahva, Iran, 26–28 February 2020.
13. Olayinka, D.; Irivbogbe, H. Estimation of Hydrological Outputs using HEC-HMS and GIS. *N. J. Environ. Sci. Technol.* **2017**, *1*, 390–402. [CrossRef]
14. Barbosa, J.; Fernandes, A.; Lima, A. The influence of spatial discretization on HEC-HMS modelling: A case study. *Int. J. Hydrol.* **2019**, *3*, 442–449.
15. Salman, A.H.; Habib, Z.H.; Rahman, A.A. Environmental impacts of water dams in Iraq. *J. Geog. Univ. Kufa* **2014**, *20*, 329–364.
16. Jasim, S.M. The natural constituents of the Great Dam Lake and its impact on developing tourism demand. *J. Admin. Econ.* **2008**, *71*, 207–231.
17. The Central Statistical Organization of Iraq/Environment Statistics/Bagdad. Available online: http://cosit.gov.iq/en/ (accessed on 5 August 2020).
18. UN-ESCWA. Inventory of shared water resources in Western Asia. In *United Nations Economic and Social Commission for Western Asia*; Federal Institute for Geosciences and Natural Resources: Beirut, Lebanon, 2013.
19. USGS. Available online: https://earthexplorer.usgs.gov (accessed on 5 August 2020).
20. Mhaina, A.S. Modeling suspended sediment load using SWAT model in data scarce area-Iraq (Al-Adhaim Watershed as a Case Study). Master's Thesis, University of Technology, Baghdad, Iraq, 2017.
21. FAO. *State of Food and Agriculture 2012: Investing in Agriculture for a Better Future*; FAO: Rome, Italy, 2012.
22. Manoj, N.; Kurian, C.; Sudheer, K. Development of a flood forecasting model using HEC-HMS. In Proceedings of the National Conference on Water Resources & Flood Management, National Conference on Water Resources & Flood Management with special reference to Flood Modelling, Sardar Vallabhbhai National Institute of Technology, Surat, India, 14–15 October 2016; pp. 14–15.
23. McCarthy, G.T. The unit hydrograph and flood routing. In Proceedings of the Conference of North Atlantic Division, US Army Corps of Engineers, New London, CT, USA, 1 January 1939; pp. 608–609.
24. Din, S.U.; Khan, N.M.; Israr, M.; Nabi, H.; Khan, M. Runoff modelling using HEC HMS for rural watershed. *Development* **2019**, *6*. [CrossRef]

Article

STORAGE (STOchastic RAinfall GEnerator): A User-Friendly Software for Generating Long and High-Resolution Rainfall Time Series

Davide Luciano De Luca [1,*] **and Andrea Petroselli** [2]

[1]   Department of Informatics, Modelling, Electronics and System Engineering, University of Calabria, 87036 Arcavacata di Rende, Italy

[2]   Department of Economics, Engineering, Society and Business Organization (DEIM), Tuscia University, 01100 Viterbo, Italy; petro@unitus.it

*   Correspondence: davide.deluca@unical.it

Abstract: The MS Excel file with VBA (Visual Basic for Application) macros named STORAGE (STOchastic RAinfall GEnerator) is introduced herein. STORAGE is a temporal stochastic simulator aiming at generating long and high-resolution rainfall time series, and it is based on the implementation of a Neymann–Scott Rectangular Pulse (NSRP) model. STORAGE is characterized by two innovative aspects. First, its calibration (i.e., the parametric estimation, on the basis of available sample data, in order to better reproduce some rainfall features of interest) is carried out by using data series (annual maxima rainfall, annual and monthly cumulative rainfall, annual number of wet days) which are usually longer than observed high-resolution series (that are mainly adopted in literature for the calibration of other stochastic simulators but are usually very short or absent for many rain gauges). Second, the seasonality is modelled using series of goniometric functions. This approach makes STORAGE strongly parsimonious with respect to the use of monthly or seasonal sets for parameters. Applications for the rain gauge network in the Calabria region (southern Italy) are presented and discussed herein. The results show a good reproduction of the rainfall features which are mainly considered for usual hydrological purposes.

Keywords: rainfall generator; stochastic processes; STORAGE; VBA macros in Excel

Citation: De Luca, D.L.; Petroselli, A. STORAGE (STOchastic RAinfall GEnerator): A User-Friendly Software for Generating Long and High-Resolution Rainfall Time Series. *Hydrology* **2021**, *8*, 76. https://doi.org/10.3390/hydrology8020076

Academic Editors: Kwok-Wing Chau

Received: 8 April 2021
Accepted: 29 April 2021
Published: 3 May 2021

Publisher's Note: MDPI stays neutral with regard to jurisdictional claims in published maps and institutional affiliations.

Software Information

- Name of software: STORAGE.xlsm
- Developers and contact information: Davide Luciano De Luca (davide.deluca@unical.it); Andrea Petroselli (petro@unitus.it)
- Year first available: 2021
- Software required: Windows 8 or later versions as Operating System (OS); Microsoft Excel 2013 or later versions
- OS settings: dot as decimal separator is mandatory. The folder "C:\NSRP\", where the output generated rainfall will be printed, must be created.
- Availability: https://sites.google.com/unical.it/storage
- Cost: free
- Program language: Visual Basic for Application (VBA) macros in MS Excel
- Program size: 6.5 MB

1. Introduction

Many hydrological applications, mainly related to small and ungauged catchments that are characterized by a short response time of runoff to rainfall, require the use of continuous rainfall time series at high resolutions [1]. However, these data series usually present a very short sample size or they are absent for many sites of interest, where only

Annual Maximum Rainfall (AMR) series are available (but they are often not so long at the finest time scale, e.g., 1 or 5 min, [2]). In this context, the use of Stochastic Rainfall Generators (SRGs) appears helpful for a more in-depth analysis of rainfall processes [3,4]. SRGs generally present a simple mathematical formulation and low computational costs, and large ensembles of long rainfall time series can be quickly obtained [5]. Moreover, an SRG can be easily used for obtaining perturbed time series [6,7] that are representative of future rainfall on hydrological scales, which are finer than the spatial and time scale investigated by Regional Climate Models (RCM). In fact, concerning this latter aspect, RCM outputs are mainly available at daily scale and are averaged over large spatial resolutions, so they require statistical downscaling or bias correction methods [8,9] for hydrological analyses. Only very recent RCM applications regarded high resolutions (hourly) and small spatial scales (e.g., [10,11]).

Specifically, the Poisson cluster models are the SRGs widely used in literature [12–33]; they include the Neyman–Scott and Bartlett–Lewis families, which provide similar performances [34]. These models can satisfactorily recreate the observed summary statistics (used for calibration) within the generated rainfall series at several fine time scales, but they usually underestimate extreme value distributions on hourly and sub-hourly scales (e.g., [35]). Thus, many variants were proposed, aimed at overcoming this problem. Unfortunately, they implied:

(i) an increasing parameterization [36–43], induced by a change of model structure and/or by estimating parameters for each month or season; this is clearly unsuitable for case studies characterized by very short samples of continuous rainfall data series at a high resolution;

(ii) the impossibility of reproduction of the proportion of dry/wet periods [44], which can be of interest for some applications.

Moreover, other kinds of SRGs, based on different stochastic engines than Poisson cluster models, were also proposed in literature [45,46].

Recent works [6,7,47] investigated the possibility to calibrate an SRG (i.e., to carry out the parametric estimation, on the basis of available sample data, in order to better reproduce some features of interest) by using only sample series at coarser time scales, which are usually longer than continuous data with a high resolution. In this framework, a modified version of the Neymann–Scott Rectangular Pulse (NSRP) model was implemented with Visual Basic for Application (VBA) macros in MS Excel, and the realized software, named STORAGE, is discussed in the present work. STORAGE is the acronym of STOchastic RAinfall GEnerator and its innovative aspects, with respect to other SRGs proposed in literature, can be summarized as follows:

1. the model calibration is carried out by using summary statistics from annual maxima rainfall (AMR), annual / monthly cumulative rainfall, and annual number of wet days, which are usually longer than continuous observed high-resolution series (mainly adopted for SRG calibration but typically very short or absent in many locations). In this way, the SRG generates 1 min or 5 min continuous rainfall series which present, at coarser resolutions, summary statistics which are comparable with those of the above-mentioned sample data;

2. the seasonality is modelled by using series of goniometric functions. This approach makes STORAGE more parsimonious with respect to the use of monthly or seasonal sets for parameters.

Concerning the latter aspect, the proposed approach is very flexible, because it is possible to model seasonality:

- by using goniometric series only for some rainfall descriptors, and by considering the other ones as invariant during the year;
- by setting the maximum number of harmonics for each selected descriptors, with the goal of having a parsimonious model.

Obviously, this methodology can be applied for any SRG proposed in literature.

The present manuscript is organized as described in the following. A brief overview of the investigated study area, i.e., the rain gauge network of the Calabria region in southern Italy, is presented in Section 2. The theoretical background of the STORAGE model and the user-friendly interface are described in Section 3. Applications are then discussed in Section 4, and the conclusions are drawn in Section 5.

2. Study Area

The investigated study area is the rain gauge network of the Calabria region (southern Italy). The employed data were downloaded from the website of the Multi Risks Centre of the Calabria region [48]. In particular, authors selected as reference the rain gauges with at least 30 years of observed data concerning AMR series with rainfall durations from 1 to 24 h. In total, time series of AMR, annual and monthly cumulative rainfall values, and annual number of wet days were analyzed for 64 stations (Figure 1). It is noteworthy that in Italy a day is classified as wet if the daily rainfall is greater than or equal to 1 mm.

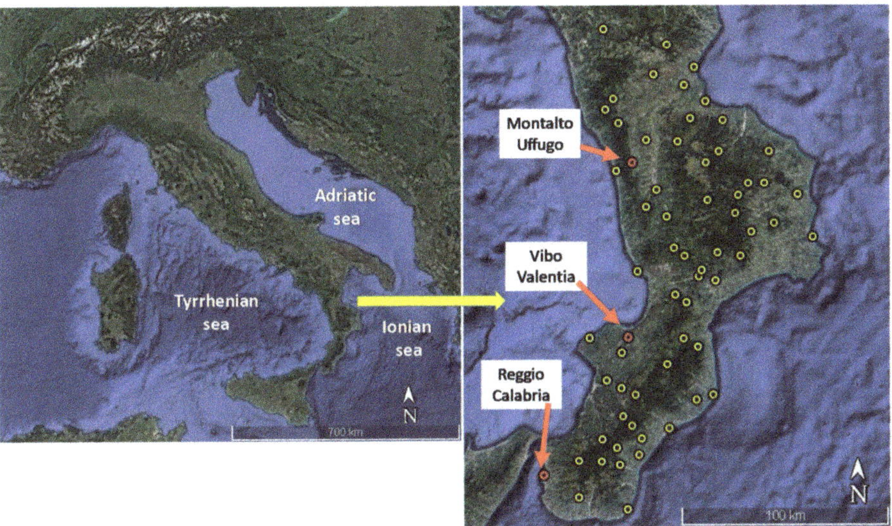

Figure 1. Location of the investigated rain gauges (yellow and red dots) in the Calabria region (southern Italy). The stations characterized by red dots (Montalto Uffugo, Reggio Calabria and Vibo Valentia) are described in detail in the present manuscript (see Section 4).

The Calabria region is located in the central part of the Mediterranean area and the total area is about 15,000 km^2; the territory is hilly in 49.2% and mountainous in 41.8% of the total area. From the collected data, the mean annual precipitation (MAP) assumes an average value of about 1150 mm, with higher values in mountainous areas and lower values in the coastal areas (particularly on the north-eastern one). As explained in [49], many rainfall events are induced by cyclones that develop close to the Alps and in the western part of the Mediterranean, and impact on the Tyrrhenian side, moving from west to east. Cyclones from North Africa and the Balkans are less frequent and mainly affect the region eastern side. In general, in the western part of Calabria there are the greatest rainfall amounts, while in the eastern part the most extreme events occur, as they are exposed to more intense cyclones [50].

3. Methods

3.1. Theoretical Overview of the Implemented Model

The basic version of the NSRP model [13,51] is suitable for stationary (i.e., without any seasonality and trend) continuous rainfall processes. In such model, five quantities, which are considered as random variables, hence following assigned probability distributions, play a crucial role. In detail, the five quantities are (see also Figure 2):

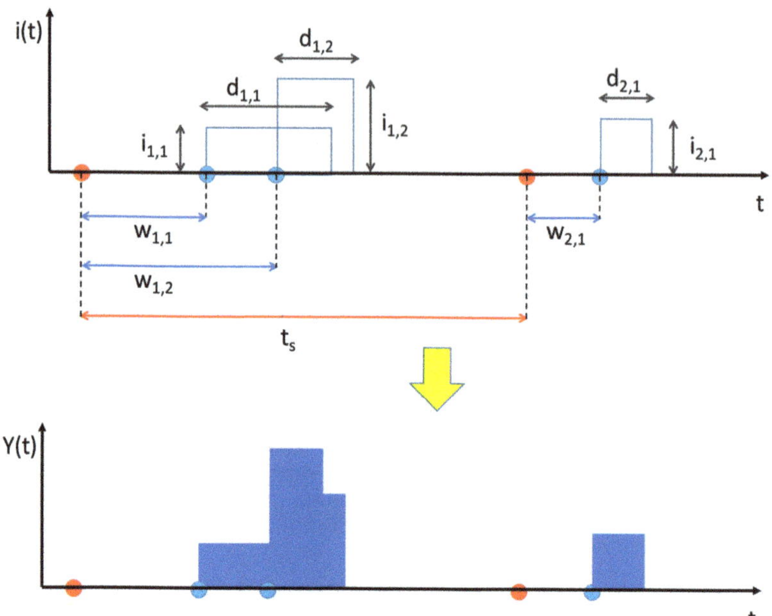

Figure 2. Representation of the Neyman–Scott Rectangular Pulses (NSRP) stochastic process for at-site rainfall modeling. In the upper part of the Figure, 2 storm occurrences (red dots) with an inter-arrival t_s, 2 bursts for the first storm and 1 burst for the second storm, are represented. The corresponding waiting times, intensities and duration are also indicated. Then, in the lower part of the Figure, the total precipitation intensity at time t can be calculated as the sum of all the intensities from the active bursts at time t.

- the inter-arrival time, T_s, between the origins of two consecutive storms, which is assumed to be an exponential random variable. Consequently, the probability $P[T_s \leq t_s]$ to have a new storm origin after a waiting time $T_s \leq t_s$ from the previous one can be calculated as:

$$P_{T_S}(t_S) = P[T_s \leq t_s] = 1 - e^{-\lambda \cdot t_S} \tag{1}$$

  where $1/\lambda$ represents the mean value for the inter-arrival times, i.e., $E[T_s] = 1/\lambda$;
- the number M of rain cells (also indicated as bursts or pulses) inside a specific storm, which is set in this work as a geometric random variable with a mean value $E[M] = \theta$;
- the waiting time W between a specific burst origin and the origin of the associated storm, which follows an exponential distribution:

$$P_W(w) = P[W \leq w] = 1 - e^{-\beta_W \cdot w} \tag{2}$$

  with $E[W] = 1/\beta_W$;
- the intensity I and the duration D of a specific burst, having a rectangular shape, belonging to a storm. Both I and D are assumed as exponentially distributed, with

parameters β_I and β_D, respectively, and mean values $E[I] = 1/\beta_I$, $E[D] = 1/\beta_D$, so that:

$$P_I(i) = P[I \leq i] = 1 - e^{-\beta_I \cdot i} \tag{3}$$

$$P_D(d) = P[D \leq d] = 1 - e^{-\beta_D \cdot d} \tag{4}$$

By considering all these five mentioned quantities, the total precipitation intensity $Y(t)$ at time t is then calculated as the sum of all the intensities from the active bursts at time t (see also Figure 2), and the rainfall height $R_j^{(\tau)}$, aggregated on the temporal τ resolution and related to the time interval j with extremes $(j-1)\tau$ and $j\tau$, is:

$$R_j^{(\tau)} = \int_{(j-1)\cdot\tau}^{j\cdot\tau} Y(t) \cdot dt \tag{5}$$

An SRG model such as NSRP is usually calibrated by minimizing an Objective Function (OF), which is defined as the sum of residuals (normalized or not) concerning the considered (by user) statistical properties of the observed data at chosen time resolutions and their theoretical expressions. The statistical properties are typically referred to high-resolution continuous time series (e.g., 1 or 5-min rainfall time series): mean, variance, and k-lag autocorrelation for $R_j^{(\tau)}$, at several values of τ can be mentioned as examples.

A first crucial aspect of the NSRP model is represented by the seasonality modelling of the rainfall process, for which monthly or seasonal parameter sets are usually considered, i.e., by carrying out a specific calibration for each considered month or season. This procedure clearly implies an increase in the number of the parameters to be estimated, and then a reduced ratio data/parameters.

In this context, another important aspect emerges, i.e., continuous high-resolution data sets are typically very short (in general no more than 15–20 years) or absent in many locations, and then a calibration with these data sets could not be suitable for a robust estimation of parameters.

To overcome these critical issues, a modified version of NSRP was implemented in STORAGE software, which is discussed in this work. STORAGE represents the implementation of the framework presented in [6,7], and its innovation regards the following features:

- In order to reproduce the seasonality of the rainfall process, goniometric series are adopted (Section 3.1.1). In doing so, the model is more parsimonious, with respect to the use of monthly or seasonal sets for parameters. Moreover, this approach is very flexible, because it is possible to model seasonality:
  - by using goniometric series only for some rainfall descriptors, and by considering the other ones as invariant during the year;
  - by setting the maximum number of harmonics for each selected descriptors, with the goal of having a parsimonious model.
- Moreover, model calibration is carried out by using data series, such as AMR, annual and monthly rainfall, and annual number of wet days series (Section 3.1.2), which are usually longer than continuous observed high-resolution series.

Obviously, like for other SRGs proposed in literature, a transient version can be implemented [6,7] in order to obtain perturbed synthetic series, which are representative of future hypothesized rainfall scenarios on spatial and temporal hydrological scales. However, in this work we describe only the implementation in STORAGE software of the cycle-stationary process (i.e., without temporal trends).

3.1.1. Seasonality Modelling with Goniometric Series

Focusing on the five NSRP summary statistics:

1. $1/\lambda$: mean value for the inter-arrival times between two consecutive storms;
2. θ: mean value for the number of rain cells (or bursts) for each storm;

3. $1/\beta_W$: mean value for the waiting time between a specific rain cell and the associated storm;
4. $1/\beta_I$: mean value for intensity of the cells with a rectangular shape;
5. $1/\beta_D$: mean value for duration of the cells with a rectangular shape.

The adoption of different sets for each month would imply the estimation of 60 parameters. Alternatively, it is possible to use goniometric series for the seasonal variation of an investigated quantity p:

$$p(t) = p_0 + \sum_{n=1}^{K} A_n \cdot \cos\left(\frac{2\pi \cdot n}{T_y} \cdot t + \phi_n\right) \tag{6}$$

where $p(t)$ is the summary statistic along the time t (expressed in min); p_0 is the mean value of $p(t)$ in the whole year; K is the maximum number of goniometric functions (also named as harmonics) to be considered; n is the n-th harmonic function; T_y is total number of minutes in the whole year (here considered with 365 days); A_n corresponds to the amplitude for the n-th harmonic function; ϕ_n corresponds to the phase shift for the n-th harmonic function.

Adoption of Equation (6) implies the estimation of $1 + 2K$ parameters for each summary statistic, i.e., the annual mean value and the K couples regarding amplitude and phase shift for the harmonics.

Under the assumption that the seasonal variation regards all the five summary statistics, the proposed SRG is characterized by: 15 parameters if $K = 1$ for all, 25 parameters if $K = 2$ for all, 35 parameters if $K = 3$ for all and so on. Obviously, K can be also different from a summary statistic to the other.

For the selected case study (described in Section 2), STORAGE software was organized with the following assumptions:

(a) The quantities $1/\lambda, \theta, 1/\beta_I$ and $1/\beta_D$ present a seasonal variation. Specifically, $K = 2$ is used for $1/\lambda$ (according to [52]):

$$\frac{1}{\lambda(t)} = \frac{1}{\lambda_0} + A_{1,\lambda} \cdot \cos\left(\frac{2\pi}{T_y} \cdot t + \phi_{1,\lambda}\right) + A_{2,\lambda} \cdot \cos\left(\frac{4\pi}{T_y} \cdot t + \phi_{2,\lambda}\right) \tag{7}$$

where $\frac{1}{\lambda_0}$ represents the mean annual value without any seasonal variation; $A_{1,\lambda} = \frac{1}{\lambda_0} - \left(\frac{1}{\lambda}\right)_{min}$, and $\left(\frac{1}{\lambda}\right)_{min}$ is equal to the smallest value for mean inter-arrival times between two consecutive storms; $A_{2,\lambda} = \xi \cdot A_{1,\lambda}$; $\phi_{1,\lambda}$ and $\phi_{2,\lambda}$ are the phase shifts for the two adopted harmonics.

(b) As regards $\theta, 1/\beta_I$ and $1/\beta_D$, we adopted $K = 1$:

$$\theta(t) = \theta_0 + A_{1,\theta} \cdot \cos\left(\frac{2\pi}{T_y} \cdot t + \phi_{1,\theta}\right) \tag{8}$$

$$\frac{1}{\beta_I(t)} = \frac{1}{\beta_{I,0}} + A_{1,\beta_I} \cdot \cos\left(\frac{2\pi}{T_y} \cdot t + \phi_{1,\beta_I}\right) \tag{9}$$

$$\frac{1}{\beta_D(t)} = \frac{1}{\beta_{D,0}} + A_{1,\beta_D} \cdot \cos\left(\frac{2\pi}{T_y} \cdot t + \phi_{1,\beta_D}\right) \tag{10}$$

where

- $\theta_0, \frac{1}{\beta_{I,0}}$ and $\frac{1}{\beta_{D,0}}$ are the mean annual values without any seasonal variation;
- $A_{1,\theta} = \theta_0 - \theta_{min}$, and θ_{min} is the smallest value for the mean number of cells for each storm;
- $A_{1,\beta_I} = \frac{1}{\beta_{I,0}} - \left(\frac{1}{\beta_I}\right)_{min}$, and $\left(\frac{1}{\beta_I}\right)_{min}$ is the smallest value for the mean intensity of a rain cell. We considered $\left(\frac{1}{\beta_I}\right)_{min} = \chi \cdot \frac{1}{\beta_{I,0}}$ with $0.5 \leq \chi < 1$.

- $A_{1,\beta_D} = \frac{1}{\beta_{D,0}} - \left(\frac{1}{\beta_D}\right)_{\min}$, and $\left(\frac{1}{\beta_D}\right)_{\min}$ is the smallest value for the mean duration of a rain cell. We considered $\left(\frac{1}{\beta_D}\right)_{\min} = \eta \cdot \frac{1}{\beta_{D,0}}$ with $0.5 \leq \eta < 1$.

(c) $\phi_{1,\theta} = 0$, $\phi_{1,\beta_D} = 0$ and $\phi_{1,\beta_I} = \pi$, in order to obtain $\theta(t) = \theta_{\min}$ and $\frac{1}{\beta_D(t)} = \left(\frac{1}{\beta_D}\right)_{\min}$ in summer months and $\frac{1}{\beta_I(t)} = \left(\frac{1}{\beta_I}\right)_{\min}$ during the winter.

These assumptions are compatible with the climatology of the Calabria region. In this part of Italy, the summer period is characterized by a lower average number of rain events with respect to the winter season. Moreover, the summer season usually presents rain events with higher intensities and shorter durations, compared with winter months, due to convective phenomena [53]. No seasonal variation (i.e., $K = 0$) is assumed for $1/\beta_W$.

Overall, calibration requires the estimation of twelve parameters: $1/\lambda_0$, $(1/\lambda)_{\min}$, ξ, $\phi_{1,\lambda}$, $\phi_{2,\lambda}$, θ_0, $\theta_{\min}$, $1/\beta_W$, $1/\beta_{I,0}$, $1/\beta_{D,0}$, χ e η.

Obviously, as also reported in Section 4, future developments of STORAGE will allow to consider a more comprehensive ensemble of combinations of K for the involved parameters, together with more flexibility about the phase shifts here fixed, in order to adequately model rainfall series in other climatic areas around the world.

3.1.2. Calibration

An a priori ensemble of simulations, described below, was carried out and the results were filed into an "information reservoir" in STORAGE software, ready to be queried for a specific site of interest. In detail, all the previously mentioned twelve parameters were considered uniform random variables with assigned ranges of variation, reported in Table 1 [7,54]. Then, 50,000 parametric sets were generated with the Monte Carlo technique and, for each one, a simulation of a 200-year rainfall series with resolution of 1 min was carried out by using the same macros which were afterwards implemented in STORAGE. At the end, we filed in STORAGE software only the parametric sets for which the 200-year synthetic series presented summary statistics according to the variation ranges of those from the observed data of a wide area of interest (i.e., all the rain gauges of the Calabria region for the presented application). Specifically, we focused on the following summary statistics:

- Mean Annual Precipitation (MAP), and
- mean annual number of wet days (i.e., mean annual number of days for which the daily rainfall is greater than or equal to 1 mm), and
- parameters of Amount-Duration-Frequency (ADF) curves, related to rainfall durations from 1 to 24 h, and
- mean values for seasonal rainfall in DJF (December–January–February), MAM (March–April–May), JJA (June–July–August), and SON (September–October–November).

The results of this composite filter, constituted by a subset of 50,000 parametric sets, are illustrated in Section 4 for the Calabria region. The storage of this information further justifies the choice of the acronym STORAGE. In fact, the software allows to use, for the synthetic generation related to a single rain gauge of interest, parametric sets belonging to this pre-existing "information reservoir" (regarding a wide previously investigated area), for which the corresponding series of AMR, annual rainfall, seasonal rainfall and number of wet days are comparable with those related to the sample historical data. Obviously, this aspect considerably reduces the calculation times for the model calibration on a specific site of interest, with respect to a usual calibration procedure that is carried out without any a priori indication about possible model outcomes. It is clear that this "information reservoir" can be continuously updated when other areas are investigated as case studies. Moreover, refinement algorithms will be implemented in future versions of STORAGE, in order to enhance the performance of calibration for a specific rain gauge.

Table 1. Ranges of variation for parameters in STORAGE, according to [7,54].

Parameter	Min	Max
$1/\lambda_0$ (days)	5	30
θ_0 (-)	2	20
$1/\beta_W$ (h)	5	24
$1/\beta_{I,0}$ (mm/h)	5	20
$1/\beta_{D,0}$ (h)	0.1	0.6
$(1/\lambda)_{min}$ (days)	0.5	5
θ_{min} (-)	1	2
$\phi_{1,\lambda}$ (rad)	0	$\pi/2$
χ (-)	0.5	1
η (-)	0.5	1
ξ (-)	0	1
$\phi_{2,\lambda}$ (rad)	0	2π

3.2. The User-Friendly Interface of STORAGE

When a user executes STORAGE, after having enabled the VBA macros, the **Main** worksheet will appear as in Figure 3. Two different procedures are allowed for the generation of a synthetic rainfall time series, and each one is associated with a specific command button:

- **RUN with parameter values chosen by the user;**
- **PARAMETER ESTIMATION AND RUN.**

Moreover, in the Main worksheet, the user can manually stop the elaborations in progress with the related command button (**manual STOP to elaborations**).

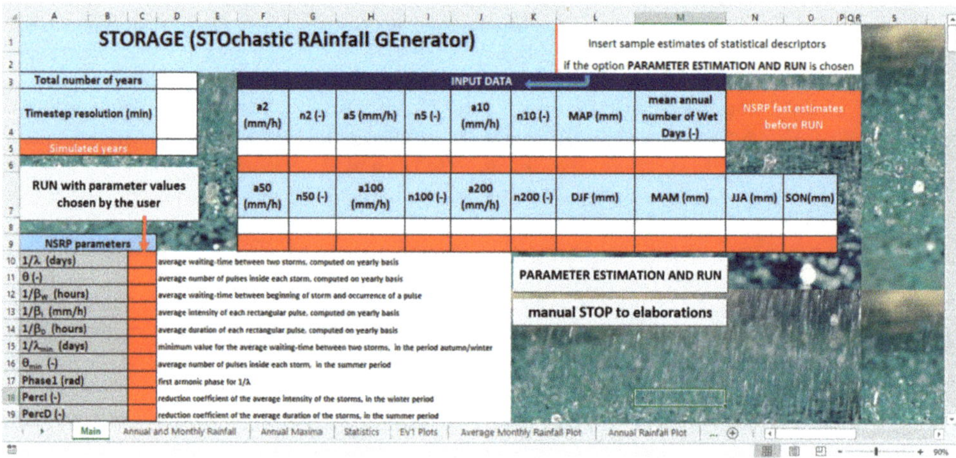

Figure 3. Interface of the Main worksheet of STORAGE after enabling the VBA macros' content.

In addition to the **Main** worksheet, STORAGE contains the following worksheets:

1. **Annual and Monthly Rainfall**, in which the generated rainfall values, aggregated at monthly and annual timescale, as well as the annual number of wet days, will be printed (for each generated year);
2. **Annual Maxima**, where the values for AMR series will be printed for rainfall durations equal to 5, 15, 30, 60 min, 3, 6, 12, 24 h, and 1 day;
3. **Statistics**, in which the mean and standard deviation values will be calculated and printed for all the quantities reported in the previous points 1 and 2;
4. **EV1 Plots**, in which the frequency distributions of all the previously listed AMR series will be represented on EV1 (Extreme Values type 1) probabilistic plots;

5.   **Average Monthly Rainfall Plot**, which contains the histogram of the average monthly rainfall values related to the generated rainfall series;
6.   **Annual Rainfall Plot**, where the annual cumulative rainfall series is represented.

Concerning the **Annual Maxima** worksheet, the series from 60 min to 24 h are estimated by considering the continuous series with a time step of 1 h. This choice is justified by the fact that many observed AMR series around the world were extracted, until 20–30 years ago, by using 1-h continuous data, while data with resolutions lower than 1 h are available only from 1990 or later [55]. Consequently, the comparison among synthetic and observed AMR series should be preferred by using this setting.

3.2.1. Data Input

For both previously mentioned procedures of time series generation, it is necessary to insert the following input information before starting the elaborations:

- the number of years to be generated (Cell D3). The maximum allowed is 500 years;
- the time resolution, expressed in minutes (Cell D4). The software allows for resolutions of 1, 5, 10, 15, 20, 30 and 60 min.

If the option **RUN with parameter values chosen by the user** is selected, then the user has to fill all the cells from C10 to C22 (Figure 3).

On the contrary, if **PARAMETER ESTIMATION AND RUN** is chosen, then the user has to insert the following input data, which are sample estimates from the observed series of the investigated case study:

- The values of parameters for Amount–Duration–Frequency (ADF) curves, expressed as a power function:

$$h_T(d) = a_T d^{n_T} \tag{11}$$

  where d is the rainfall duration (hours) ranging from 1 to 24 h, T is the return period (years), $h_T(d)$ is the d-AMR associated with T, and a_T and n_T are ADF parameters. In detail, the values for a_T and n_T, associated with specific T values, are requested:

  - concerning a_T, the cells to be filled are F5 (T = 2 years), H5 (T = 5 years), J5 (T = 10 years), F8 (T = 50 years), H8 (T = 100 years) and J8 (T = 200 years);
  - concerning n_T, the cells to be filled are G5 (T = 2 years), I5 (T = 5 years), K5 (T = 10 years), G8 (T = 50 years), I8 (T = 100 years) and K8 (T = 200 years). If the size of the sample AMR series for the investigated case study is limited (less than 20 years), then it is advisable to use only sample estimates from low T values (2, 5 and 10 years). For higher sample sizes, information deriving from higher return periods can also be entered.

- The values for Mean Annual Precipitation (MAP) into the cell L5, for the mean annual number of wet days into the cell M5, and for the mean cumulative seasonal precipitation, associated with December–January–February (DJF), March–April–May (MAM), June–July–August (JJA) and September–October–November (SON), into the cells L8, M8, N8 and O8, respectively. Moreover, also in this case, it not necessary to fill all the listed cells. The VBA macro will run the model calibration on the basis of the available information. Concerning the cell M5, strictly related to the wet day proportion, it should be remarked that the trivial rainfall (of which amount is less than the capacity of the tipping bucket of the rain gauges) could highly distort the result of the calibration in some cases, and so not filling this cell could avoid this possibility.

An example of Data Input is shown in Figure 4, if the option **PARAMETER ESTIMATION AND RUN** is selected by the user.

3.2.2. Synthetic Generation of Rainfall Time Series at a High Resolution

After completing the Data Input step, it is possible to run one of the two generation procedures. In the following pages, attention is focused on the **PARAMETRIC ESTIMATION AND RUN** button (Figure 4), which further allows for different generation alternatives. The table and graphic outputs, associated to **RUN with parameter values chosen by the user** button (Section 3.3), are similar.

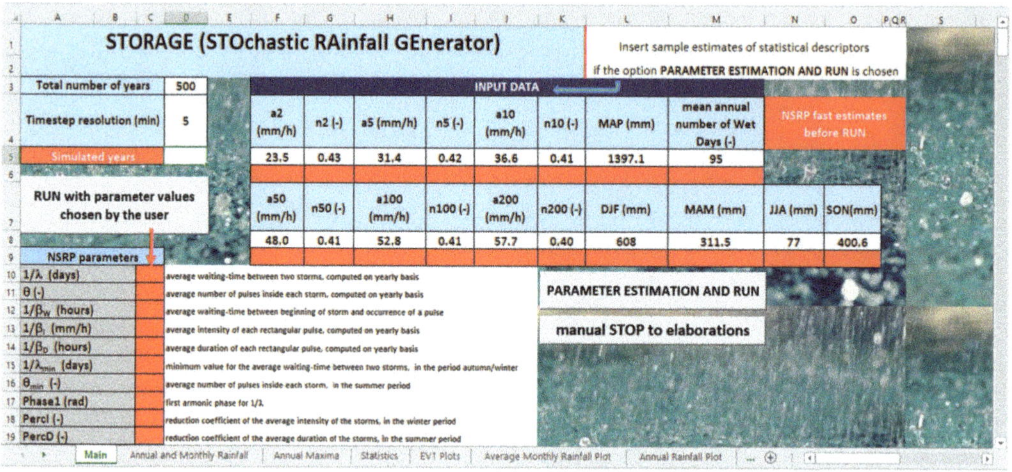

Figure 4. Example of Data Input step in the **Main** worksheet.

It must be highlighted that, in a worksheet hidden for the user, the results deriving from the use of about 3500 parametric sets, in terms of a_T and n_T for the ADF, MAP and the mean annual number of wet days, and mean cumulative seasonal rainfalls (DJF, MAM, JJA, SON), are stored. In detail (see also Sections 3.1.2 and 4), for each single parametric set, 200 years of precipitation were synthetically generated.

By clicking on the **PARAMETRIC ESTIMATION AND RUN** button, the userform shown in Figure 5 is displayed; from the combobox at the top (Figure 6) it is possible to select the statistical descriptors to be reproduced, i.e.,:

1. only the parameters a_T and n_T of the ADF curves;
2. only MAP and the mean value for annual number of wet days (NumWetDays);
3. a_T, n_T, MAP and NumWetDays;
4. a_T, n_T, MAP, NumWetDays and the mean cumulative seasonal rainfalls (DJF, MAM, JJA, SON).

After the choice of the descriptors to be reproduced (for example, a_T, n_T, MAP, NumWetDays, DJF, MAM, JJA, and SON, as in Figure 6), it is possible to click on the **PARAMETRIC ESTIMATION** button. The software will display by default, in the cell range C10:C22, the parametric set (indicated with ID SET 1) which is characterized, among the 3500 used offline, by the best value (i.e., the lowest value) of the evaluated Objective Function (OF) (Figure 7), in percentage terms, as:

$$OF = \begin{cases} OF\ a\_n & Option\ 1 \\ OF\ MAP\_NumWetDays & Option\ 2 \\ OF\ a\_n + OF\ MAP\_NumWetDays & Option\ 3 \\ OF\ a\_n + OF\ MAP\_NumWetDays + OF\ Seasons & Option\ 4 \end{cases} \tag{12}$$

in which:

$$\bullet\ OF\ a\_n = \sum_{i=1}^{K_a} \frac{\left| a_i - a_i^* \right|}{a_i} + \sum_{j=1}^{K_n} \frac{\left| n_j - n_j^* \right|}{n_j} \tag{13}$$

where:

- a_i is the i-th value (i = 1, ... K_a) of parameter **a** for an ADF curve of an assigned T, inserted by the user into an input cell, while a_i^* is the correspondent NSRP value. K_a is the number of return periods T which are considered by the user for parameter **a**.

- n_j is the j-th value (j = 1, ... K_n) of parameter **n** for an ADF curve of an assigned T, inserted by the user into an input cell, while n_j^* is the correspondent NSRP value. K_n is the number of return periods T which are considered by the user for parameter **n**.

$$OF\ MAP\_NumWetDays = \frac{|MAP-MAP^*|}{MAP} + \frac{|NumWetDays-NumWetDays^*|}{NumWetDays} \tag{14}$$

where MAP and $NumWetDays$ are the sample values which are inserted by the user, while MAP^* e $NumWetDays^*$ are the correspondent NSRP values.

$$OF\ Seasons = \frac{|DJF-DJF^*|}{DJF} + \frac{|MAM-MAM^*|}{MAM} + \frac{|JJA-JJA^*|}{JJA} + \frac{|SON-SON^*|}{SON} \tag{15}$$

where DJF, MAM, JJA and SON are the sample values which are inserted by the user, while DJF^*, MAM^*, JJA^* and SON^* are the correspondent NSRP values.

Whatever option is selected in the combobox, STORAGE will provide the correspondent values for all the OFs (Equations (13)–(15)) for a specific parameter set.

Moreover, by using the spin button (Figure 8), it is possible to adopt other parameter sets for simulation, which are sorted (by STORAGE in the hidden worksheet) on the basis of the values related to the selected OF.

After the choice for parametric set, the user can click on the **RUN** button for carrying out the generation of a synthetic rainfall series.

During the run, the user can control the progress of generation by analyzing the several worksheets in STORAGE.xlsm. As examples, the cell D5 in **Main** (Figure 9) and the histogram for Annual Rainfall (Figure 10) can be checked.

A message box will appear when simulation is completed. Then, the final results can be analyzed in the several tables and plots of STORAGE (Figure 11), while the whole synthetic rainfall series at the selected high resolution (cell D4 in Main), will be printed in "C:\NSRP\RainSim.txt".

As explained in the following sections, STORAGE also allows for rainfall generation with multisets approaches, as an alternative way to the run with a single parametric set.

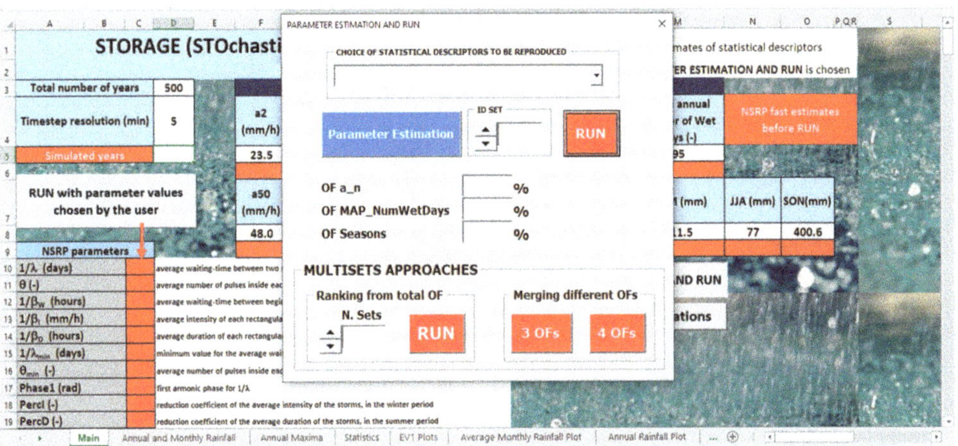

Figure 5. Userform where a user can carry out calibration and select several run options.

Figure 6. Example of procedure for calibration, by using the combobox at the top of the userform.

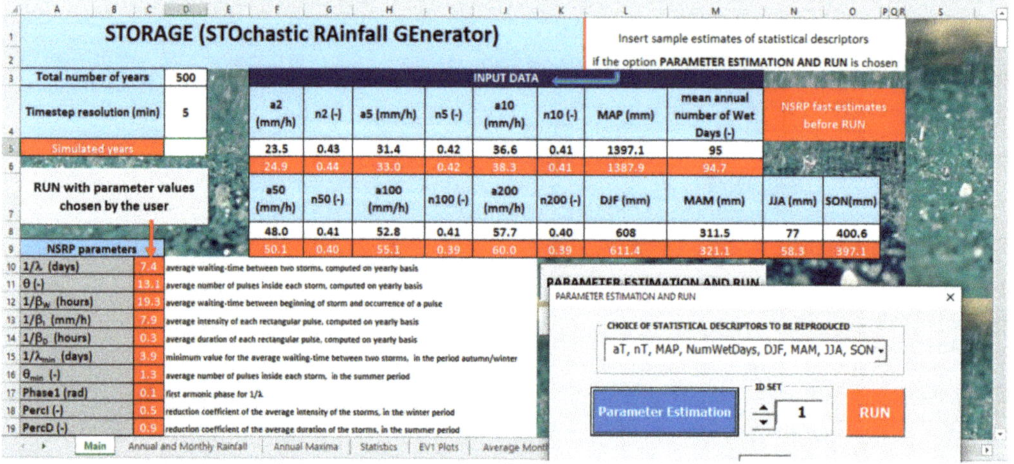

Figure 7. Example of calibration results in the Main worksheet.

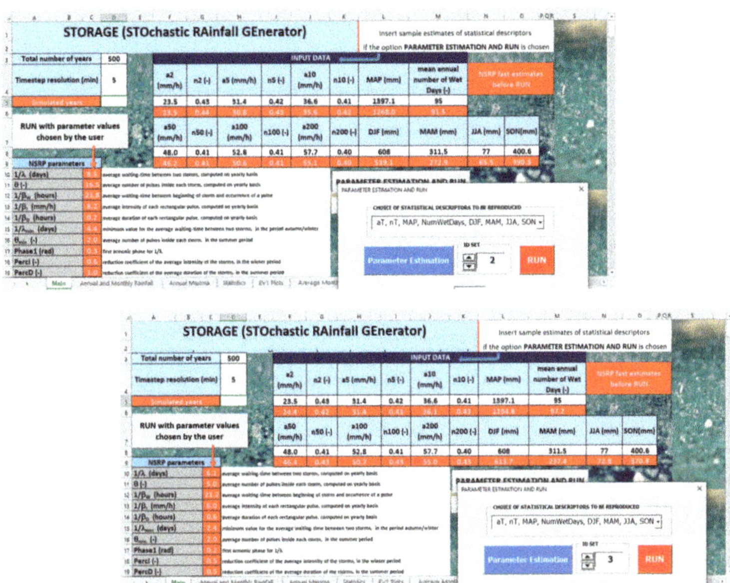

Figure 8. Example of calibration by using the spin button, which allows for using other parameter sets for simulation.

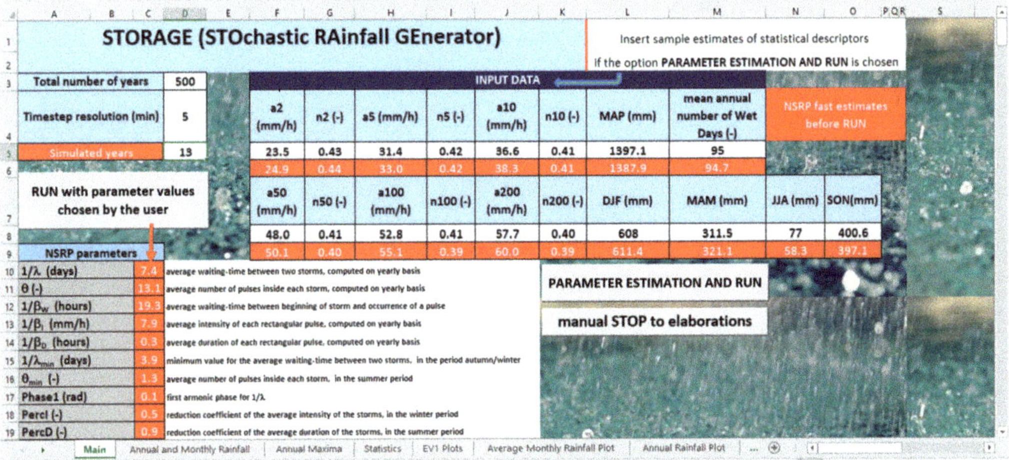

Figure 9. During the run, the user can control the progress of generation by checking the number of simulated years in the cell D5 in the **Main** worksheet.

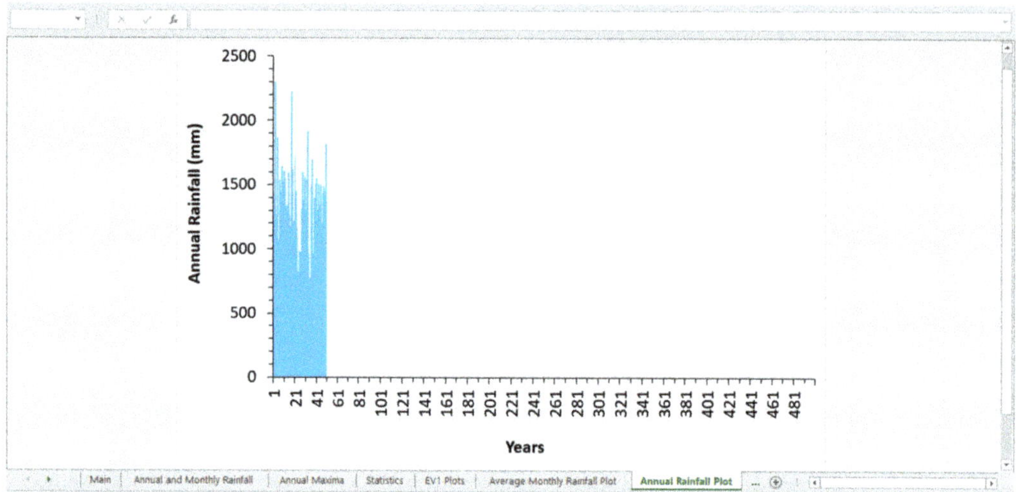

Figure 10. During the run, the user can control the progress of generation by analyzing the histogram for Annual Rainfall.

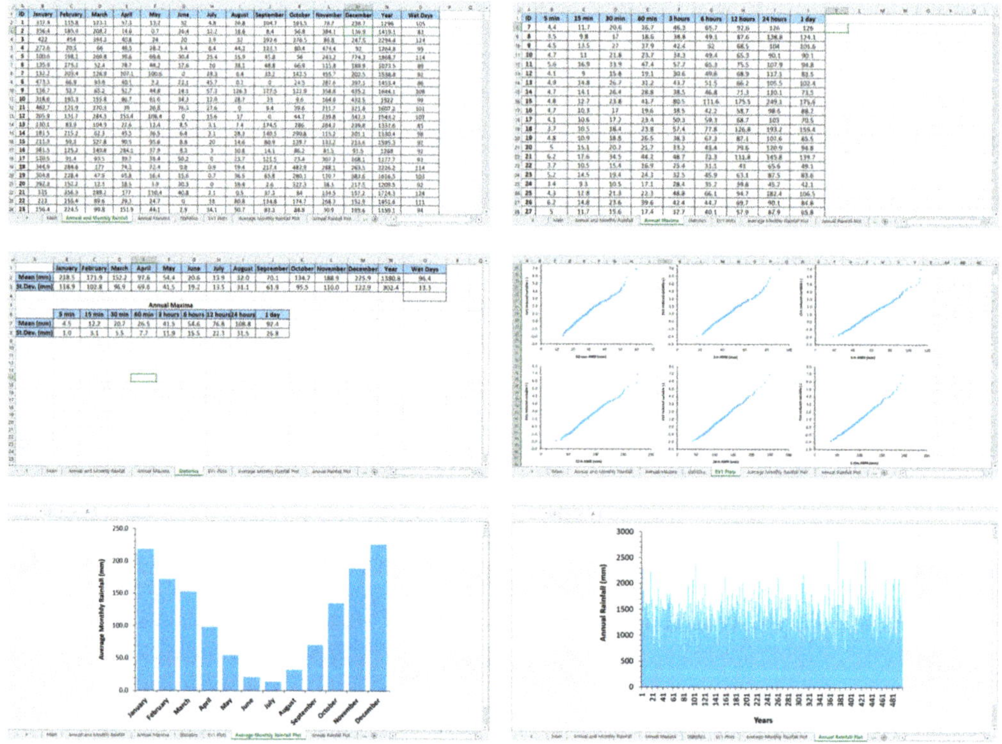

Figure 11. Examples of visualization in the different worksheets when the simulations are completed.

3.2.3. Multisets Approaches

Focusing on Option 4 of Equation (12), different parametric sets can be characterized by very similar OF values among them, but some sets could better reconstruct ADF curves, while other ones could best fit MAP and NumWetDays, and so on.

In this context, if the fourth option of Equation (12) (i.e., a_T, n_T, MAP, NumWetDays, DJF, MAM, JJA, SON) is chosen as an ensemble of statistical descriptors to be reproduced, the user can take advantages from several parametric sets by selecting one of these two options concerning multisets approaches (Figure 12):

- **Ranking from total OF;**
- **Merging different OFs**, which is further subdivided in **3 OFs** and **4 OFs.**

The proposed multisets approaches are based on the concept of equifinality [56], which means that "different parametric sets within a chosen model structure may be behavioural or acceptable in reproducing the observed behaviour of that system".

Ranking from Total OF

It is possible to select S parametric sets (sorted with increasing values of Option 4 in Equation (12)) by using the spin button of Figure 12.

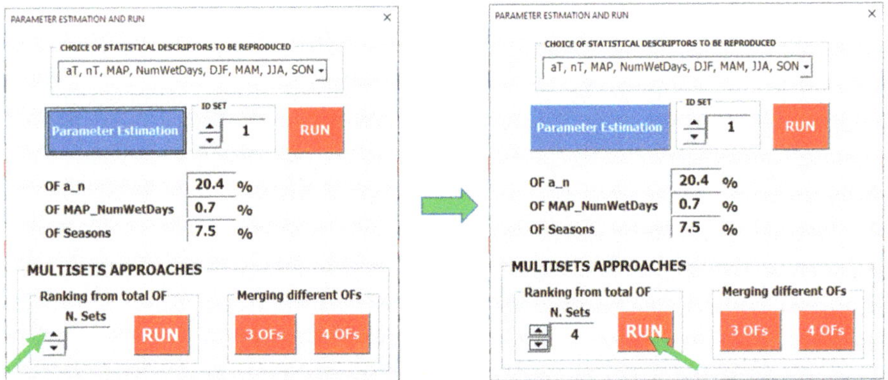

Figure 12. Example of multisets approach regarding the option "Ranking from total OF".

Automatically, STORAGE will assign (to a specific set) a frequency of use which is inversely proportional to its overall OF value (Option 4 in Equation (12)). In detail, let f_i be the frequency of use for the i-th parametric set ($i = 1, \ldots, S$) and OF_i its corresponding value of OF; f_i is computed as:

$$f_i = \frac{\frac{1}{OF_i}}{\sum\limits_{i=1}^{S} \frac{1}{OF_i}} \tag{16}$$

with, obviously, $\sum\limits_{i=1}^{S} f_i = 1$.

Then, considering the total number N of years to simulate (input data in cell D3 in the Main worksheet, Figure 4), the number $f_i \cdot N$ of years will be generated with the i-th parametric set.

It should be highlighted that:

- if a multisets approach is selected, a user should consider at most $S = 4$ and a large value for N (we suggest $N = 500$ years), in order to have a significant number of years for each set (with N = 500 years and S = 4, there are on average 125 years which are simulated with each set);

- in a context, such as in this case, of stationary/cycle-stationary process (i.e., without any climatic trend), it is not necessary to generate a large number L of N-year synthetic series (in which each i-th set should regard $f_i \cdot L$ series), but it is sufficient to consider the generation of only one year, which is repeated $L = N$ times. This is allowed by the ergodicity property of a stationary process [57], which means that the statistics from a long temporal N-year series are equal to the statistics from one year (generated N times).

After clicking on the **RUN** command button (Figure 12), the user is able to check the progress of the rainfall generation, similarly to the procedure with only one parameter set (Figures 9–11). It is clear that this approach can be well used for a more comprehensive sensitivity analysis (i.e., not only related for the first ranked parametric sets) in further upgraded versions of STORAGE software.

Merging Different OFs

This approach can be carried out in two options:

- 3 OFs;
- 4 OFs.

In the first case, from the worksheet (hidden for the user) where the information of the offline generations with about 3500 parametric sets is stored, the VBA code selects the three parametric sets with the lowest values for, respectively, Equations (13)–(15). Then, STORAGE will assign to each selected set a frequency f_i, evaluated by considering Option 4 of Equation (12) as OF_i in Equation (16).

In the second case (4 OFs), the parametric set with the lowest value of the overall OF (option 4 of Equation (12)) is also considered, together with the three above mentioned sets.

It must be highlighted that these two options are allowed by STORAGE only if all the 3 OFs of the first option are inside the first 10 positions of the ranking for OF calculated with Option 4 of Equation (12).

Also in this case, after clicking on the 3 OFs or 4 OFs buttons (Figure 13), the user is able to check the progress of the rainfall generation, similarly to the procedures with only one parameter set (Figures 9–11).

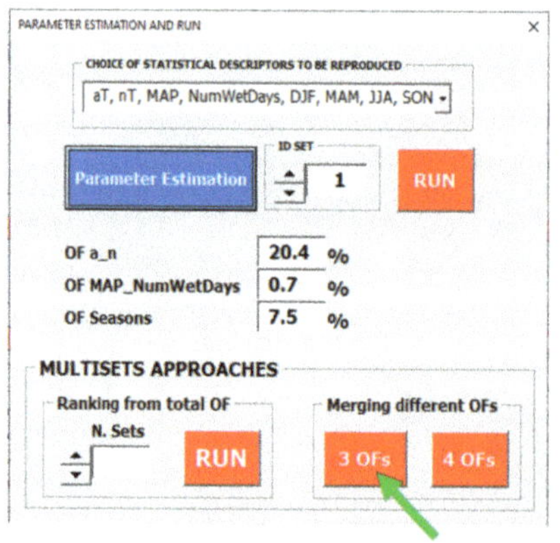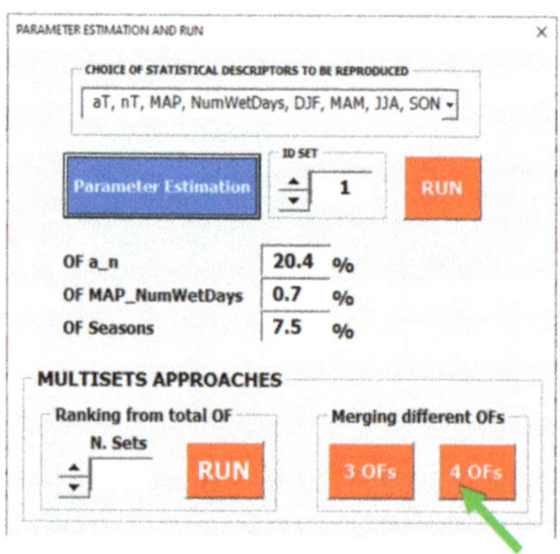

Figure 13. Use of multisets approach regarding the option "Merging different OFs".

3.3. RUN with Parameter Values Chosen by the User

This option allows for manually setting the values for the parameters in the interval C10:C21 of cells in the Main worksheet (Figure 4). Also in this case, after clicking on the corresponding command button for the run, the user is able to check the progress of the rainfall generation, similar to the previous described procedures. The ranges of variation for parameters are reported in Table 1, according to [7,54].

4. Application for Rain Gauge Network of the Calabria Region and Discussion

As regards the application for the Calabria region, we saved in STORAGE about 3500 parametric sets, for which the 200-year synthetic series presented summary statistics ranging inside specific intervals (according to the observed data in the whole region). In detail:

- concerning MAP, a value between 450 and 2500 mm;
- concerning the mean annual number of wet days, a value between 50 and 120;
- concerning the ADF curves (Equation (11)), values of **a** and **n** for T = 5 years between 20 and 65 mm/h and between 0.12 and 0.65, respectively;
- concerning the SON cumulative rainfall, a mean value inside a variation of ±50 mm with respect to the linear regression curve between observed MAP and SON of the investigated data series.

By applying this composite filter, graphical comparisons among synthetic and observed summary statistics are shown in Figures 14 and 15. From analysis of these dispersion plots, the STORAGE good reconstruction for the investigated rainfall descriptors can be assessed.

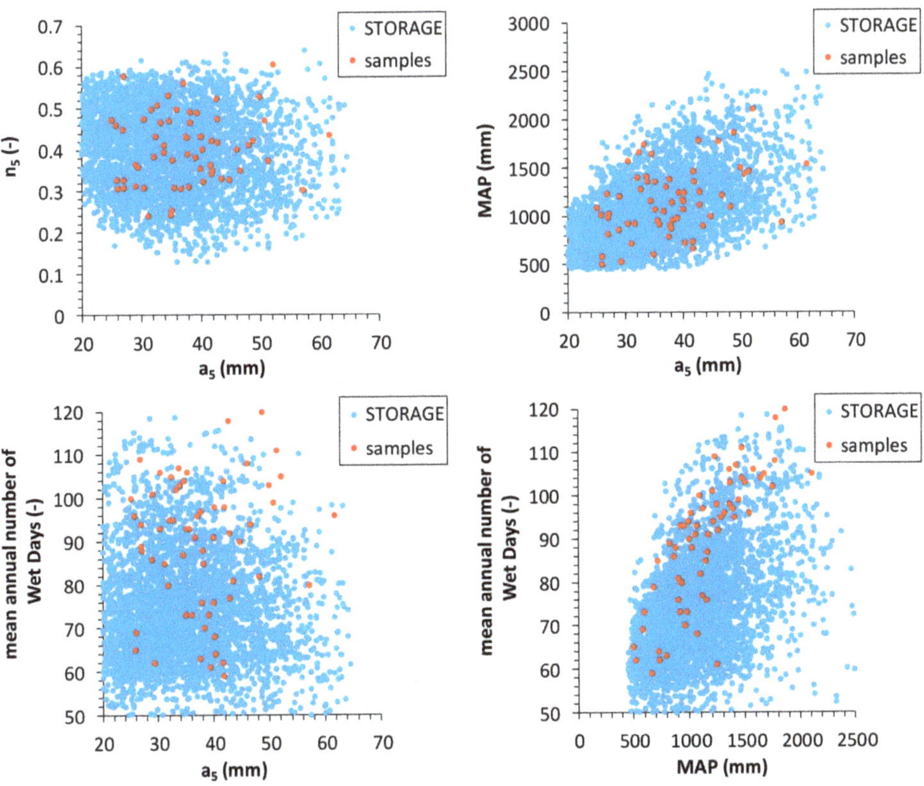

Figure 14. Comparison among synthetic and observed summary statistics.

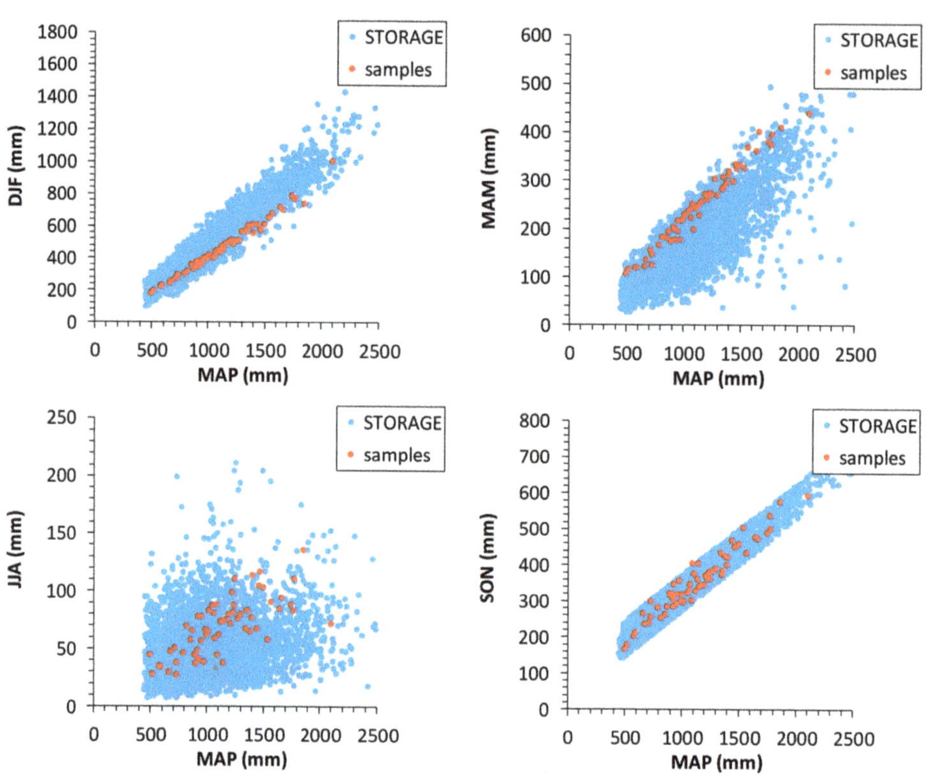

Figure 15. Comparison among synthetic and observed summary statistics.

For the sake of brevity, focusing on specific rain gauge data series, examples of STORAGE application are below described for Montalto Uffugo, Reggio Calabria and Vibo Valentia stations. Their associated sample values for the statistical descriptors are reported in Tables 2 and 3.

Table 2. Montalto Uffugo, Reggio Calabria and Vibo Valentia rain gauges: values of parameters concerning ADF curves.

Rain Gauge	Sample Size AMR Series (years)	a_2 (mm/h)	n_2 (-)	a_5 (mm/h)	n_5 (-)	a_{10} (mm/h)	n_{10} (-)
Montalto Uffugo	53	23.5	0.43	31.4	0.42	36.6	0.41
Reggio Calabria	57	25.7	0.24	35.9	0.23	42.7	0.23
Vibo Valentia	67	24.4	0.31	36.1	0.29	45.0	0.28
Rain Gauge		a_{50} (mm/h)	n_{50} (-)	a_{100} (mm/h)	n_{100} (-)	a_{200} (mm/h)	n_{200} (-)
Montalto Uffugo		48.0	0.41	52.8	0.41	57.7	0.40
Reggio Calabria		57.6	0.23	63.9	0.22	70.1	0.22
Vibo Valentia		68.2	0.27	79.0	0.27	90.1	0.26

Table 3. Montalto Uffugo, Reggio Calabria and Vibo Valentia rain gauges: sample values of mean annual and seasonal precipitation, and mean annual number of wet days.

Rain Gauge	Sample Size Daily Series (years)	MAP (mm)	Mean Annual Number of Wet Days (-)	DJF (mm)	MAM (mm)	JJA (mm)	SON (mm)
Montalto Uffugo	71	1397.1	95	608.0	311.5	77.0	400.6
Reggio Calabria	101	597.2	73	229.9	119.4	34.2	213.7
Vibo Valentia	99	949.7	93	362.2	217	77.9	292.6

For all the three stations, 500-year synthetic rainfall time series with a resolution of 5 min were generated, and we carried out model validation by analyzing the reproduction of frequency distributions for sample data of AMR, annual and seasonal rainfall, and annual number of wet days. The best STORAGE performances were obtained:

- by using the parametric set with the lowest value for the total OF (Option 4 in Equation (12)), concerning Montalto Uffugo;
- by considering the multisets approach **Ranking from total OF** for Reggio Calabria and Vibo Valentia, with S equal to 3 and 4, respectively.

For Montanto Uffugo rain gauge, STORAGE provided a 500-year synthetic rainfall time series which satisfactorily reproduces the frequency distributions of AMR sample data (see the EV1 probabilistic plots in Figure 16), with an over-estimation only for 24-h AMR series. The reproduction of the frequency distributions concerning sample series for annual rainfall, annual number of wet days, and seasonal precipitation in DJF, MAM and SON is analyzed on Gaussian plots (Figure 17): a slight underestimation is obtained only for JJA rainfall. As regards Reggio Calabria and Vibo Valentia rain gauges, the obtained results (Figures 18–21) highlighted some crucial aspects to be investigated further when future developments in STORAGE software will be carried out. In detail:

- when AMR sample data present outliers from an EV1 behaviour (Figures 18 and 20), or if extremes are underestimated, it could be useful to consider other probability distributions for cell intensity I (e.g., Weibull, Gamma or a mixture of exponential functions, [20,25,58]), and/or to use other shapes for rain cells (such as the sinusoidal one, [59]), in order to better reproduce quantiles at high values of return period T;
- though frequency distributions of annual rainfall are properly reproduced, an increase in the maximum number of harmonics for $1/\lambda$ (i.e., the mean inter-arrival time between two consecutive storms) and/or modelling seasonality also for $1/\beta_W$ (i.e., the mean waiting time between a specific burst origin and the origin of the associated storm) could improve the reconstruction of both the annual number of wet days and seasonal rainfall in some specific cases.

Starting from this latter aspect, a more in-depth investigation of the maximum number of harmonics for some quantities, and of their phase shifts, could justify the STORAGE application also in regions far from the investigated area, i.e., characterized by drier or wetter climates. This obviously means to increase the number of parametric sets to be stored in the software.

Further analyses of STORAGE performances were carried out focusing on Montalto Uffugo rain gauge, characterized by 30-year continuous time series at resolutions of 20 min. Such analyses aim to evaluate the model capacity for reproducing summary statistics of high-resolution continuous series (not used for STORAGE calibration) and to compare the STORAGE results with those from a standard NSRP (i.e., calibrated by only using continuous high-resolution data). In details:

- we calibrated a basic version of NSRP with the 1-h continuous data series (aggregated from the available 20-min one), by estimating parameters for each month (according to [14]) in order to avoid possible underestimation of extremes (as mentioned in the introduction). This version of NSRP is indicated as NSRP_v0 in the following;

- we compared STORAGE and NSRP_v0 performances, graphically and in terms of Root Mean Square Error (RMSE), as regards the modelling of:
  - mean, standard deviation and percentage of dry intervals from the continuous series at 20-min and 1-h resolutions;
  - mean values of monthly rainfall heights;
  - rainfall heights of ADF curves for return periods T = 5, 50 and 200 years.

Concerning the summary statistics of the continuous series, it is clear that NSRP_v0 provides the best performances for 1-h resolution, because this time step was used for NSRP calibration in this case. However, the obtained STORAGE results for 1-h data series can be considered acceptable for the mean and percentage of dry intervals (Table 4 and Figure 22). For a 20-min resolution, STORAGE and NSRP_v0 performances are comparable (Table 4 and Figure 23).

Table 4. Montalto Uffugo rain gauge: evaluation of STORAGE and NSRP_v0 performances. RMSE values related to the mean, standard deviation and percentage of dry intervals for the continuous 20-min and 1-h series.

RMSE	1-h Mean (mm)	1-h St.Dev. (mm)	Ratio of 1-h Dry Intervals (-)	20-min Mean (mm)	20-min St.Dev. (mm)	Ratio of 20-min Dry Intervals (-)
STORAGE	0.06	0.30	0.07	0.02	0.13	0.03
NSRP_v0	0.02	0.04	0.03	0.01	0.13	0.02

Moreover, monthly rainfall heights are very well reproduced by STORAGE, as seasonal rainfalls are used for its calibration (Sections 3.2.1 and 3.2.2), but NSRP_v0 results can be also considered good: RMSE values are 7.5 and 14.4 mm for STORAGE and NSRP_v0, respectively (see Table 5 and Figure 24).

Table 5. Montalto Uffugo rain gauge: evaluation of STORAGE and NSRP_v0 performances. RMSE values related to the mean of monthly rainfall heights and ADF curves.

RMSE	Mean of Monthly Rainfall (mm)	5-year ADF (mm)	50-year ADF (mm)	200-year ADF (mm)
STORAGE	7.5	6.0	5.5	5.6
NSRP_v0	14.4	27.6	35.5	40.1

The clear benefit of using STORAGE is highlighted by focusing on ADF curves (Table 5 and Figure 25). As expected, STORAGE provides a very good reconstruction (RMSE values are comprised between 5.5 and 6 mm) because it is calibrated with a_T and n_T of the sample ADF curves (Sections 3.2.1 and 3.2.2). On the contrary, NSRP_v0 significantly overestimates rainfall extremes in this specific case; as its parametric estimation is only based on summary statistics from high-resolution continuous series, an acceptable reproduction of ADF curves could not be guaranteed in general (such as for Montalto Uffugo rain gauge), also by using monthly or seasonal parameter sets.

This last comparison allows us to remark the most important aspect of the usefulness of STORAGE, i.e., the possibility of calibrating an SRG by only using information at coarser resolutions (AMR, MAP, and so on) and then generating continuous series which preserves sample features (often un-known for lack of data) in an acceptable way at high resolutions.

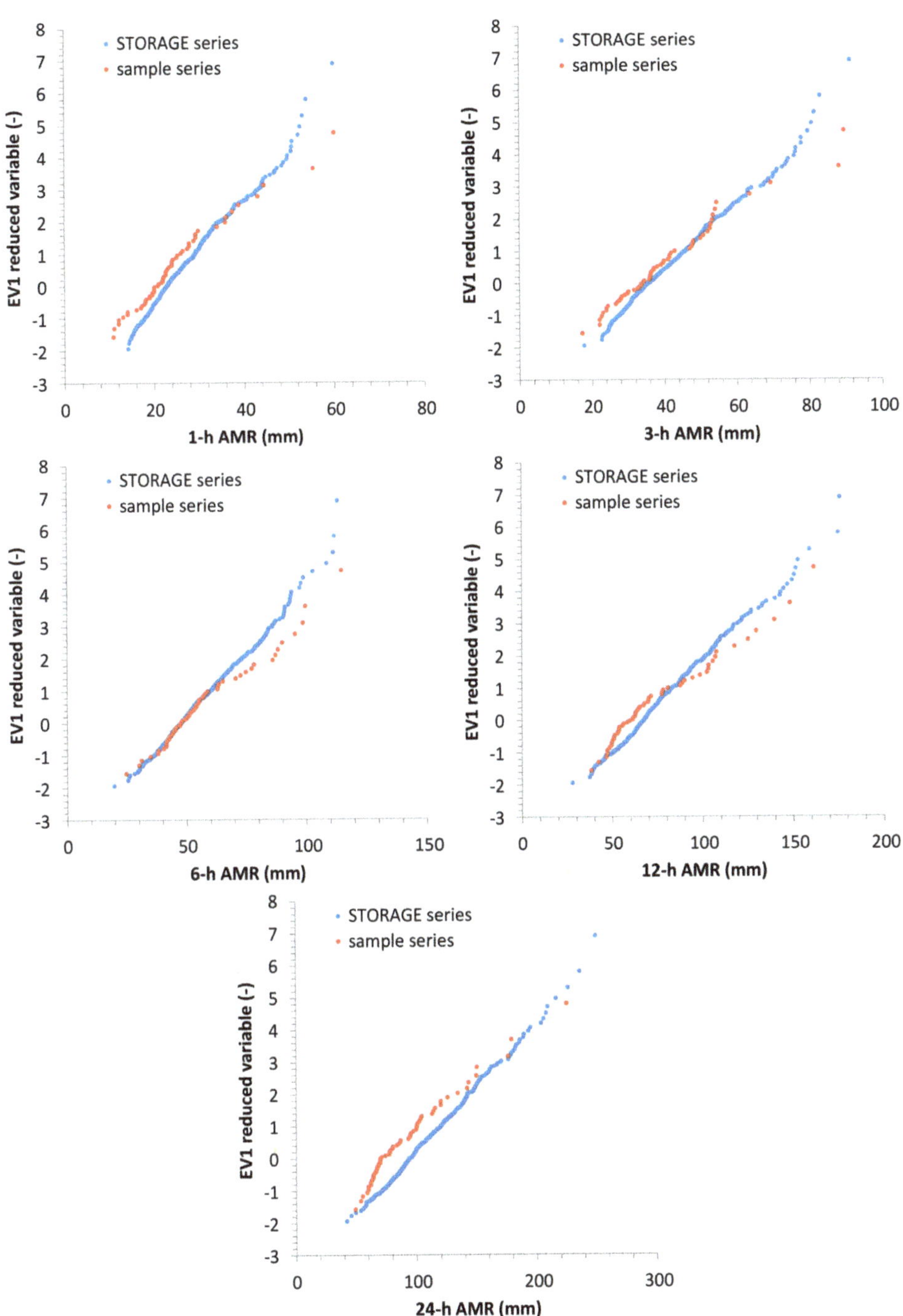

Figure 16. Montalto Uffugo rain gauge: EV1 probabilistic plots, showing the comparison among synthetic and observed AMR series.

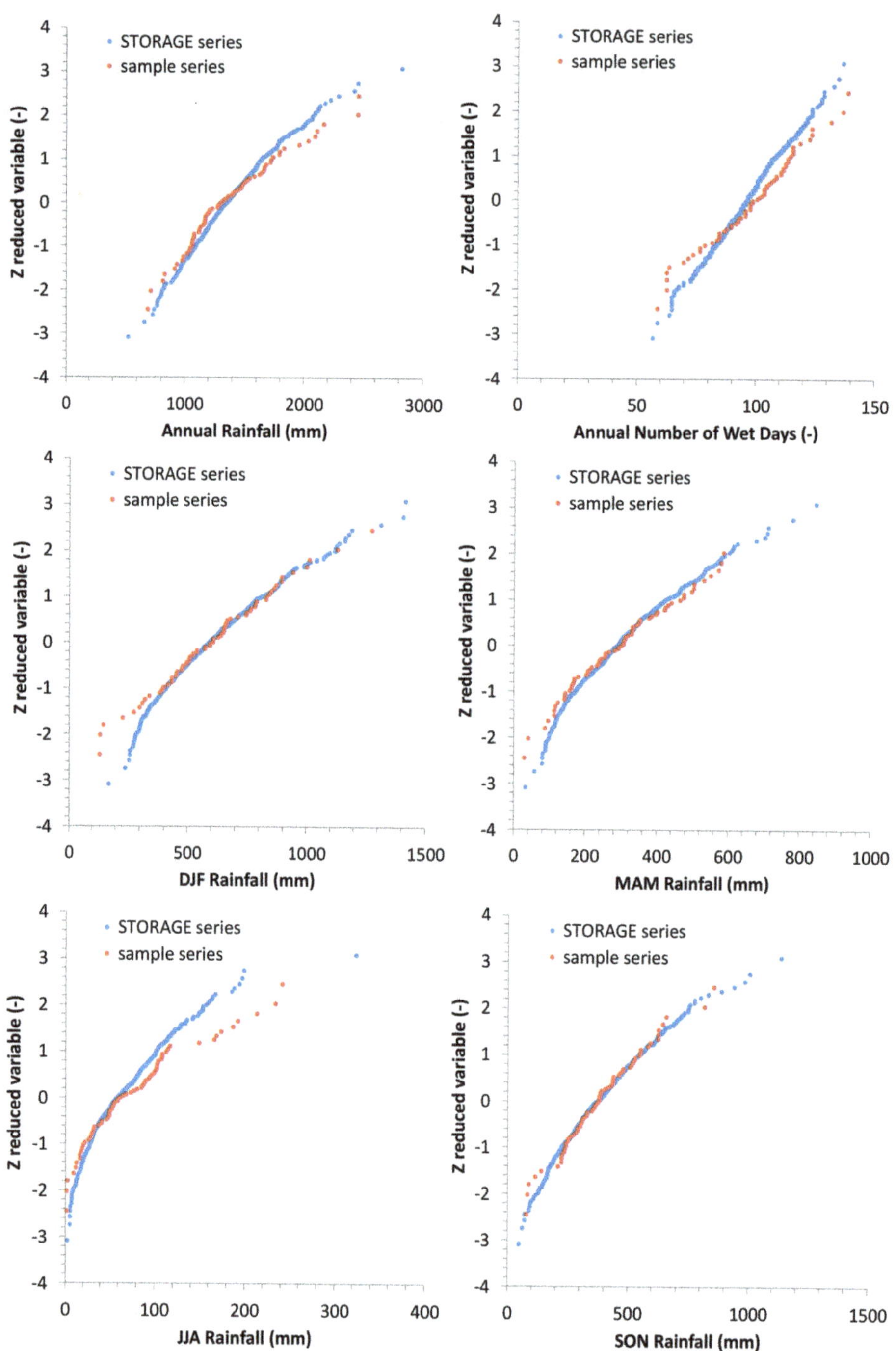

Figure 17. Montalto Uffugo rain gauge: Gaussian probabilistic plots, showing the comparison among synthetic and observed series, regarding annual and seasonal rainfall, and annual number of wet days.

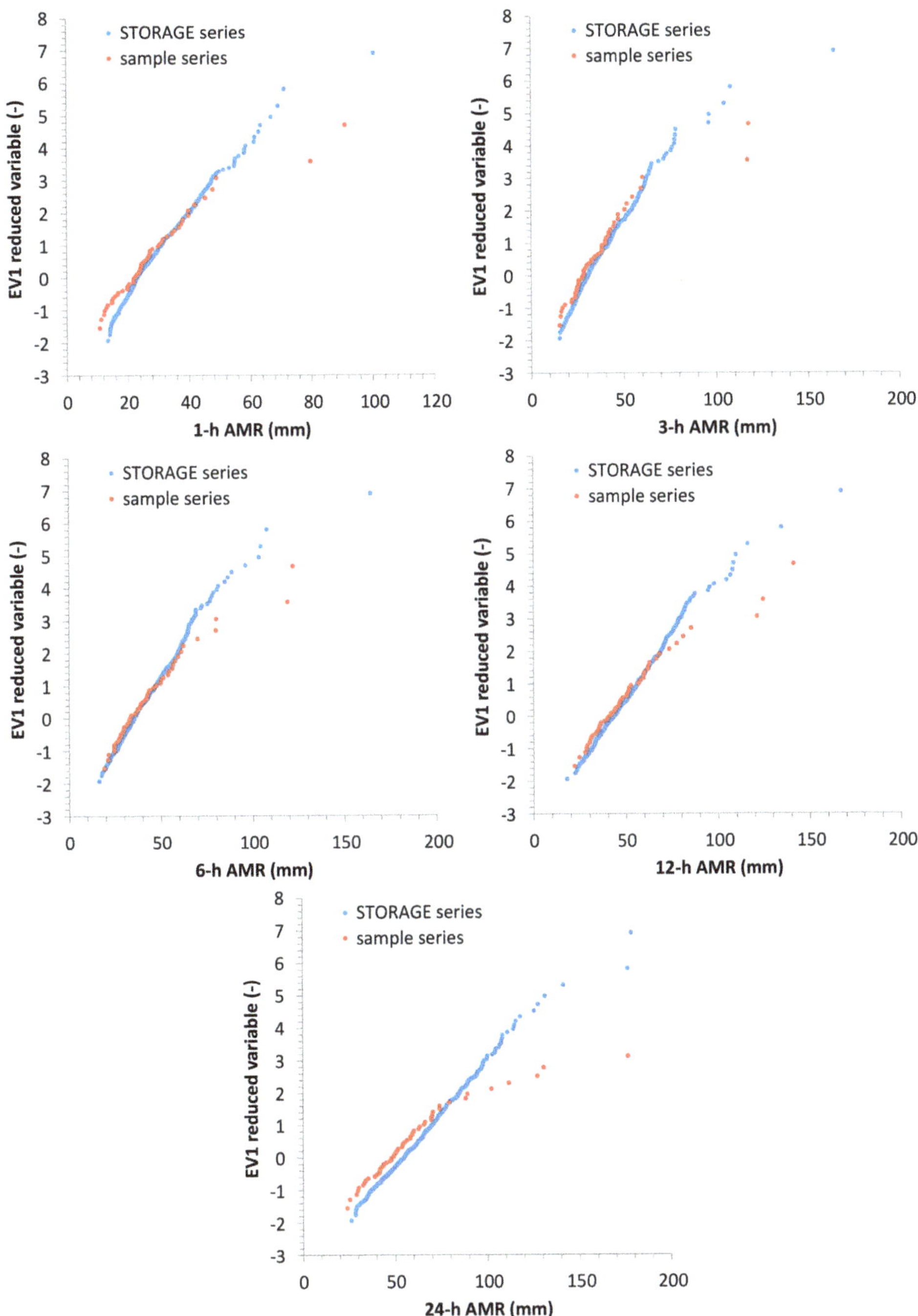

Figure 18. Reggio Calabria rain gauge: EV1 probabilistic plots, showing the comparison among synthetic and observed AMR series.

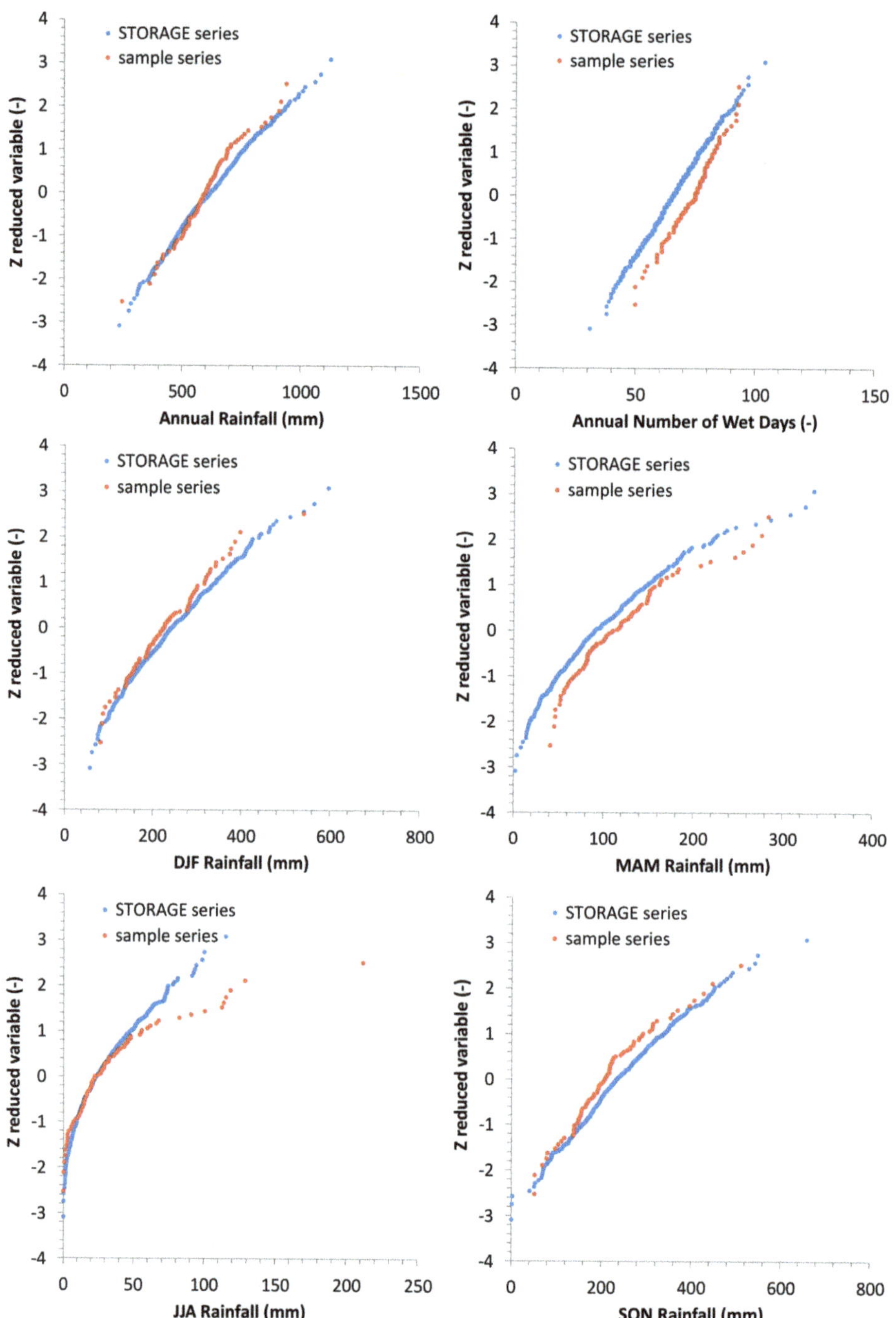

Figure 19. Reggio Calabria rain gauge: Gaussian probabilistic plots, showing the comparison among synthetic and observed series, regarding annual and seasonal rainfall, and annual number of wet days.

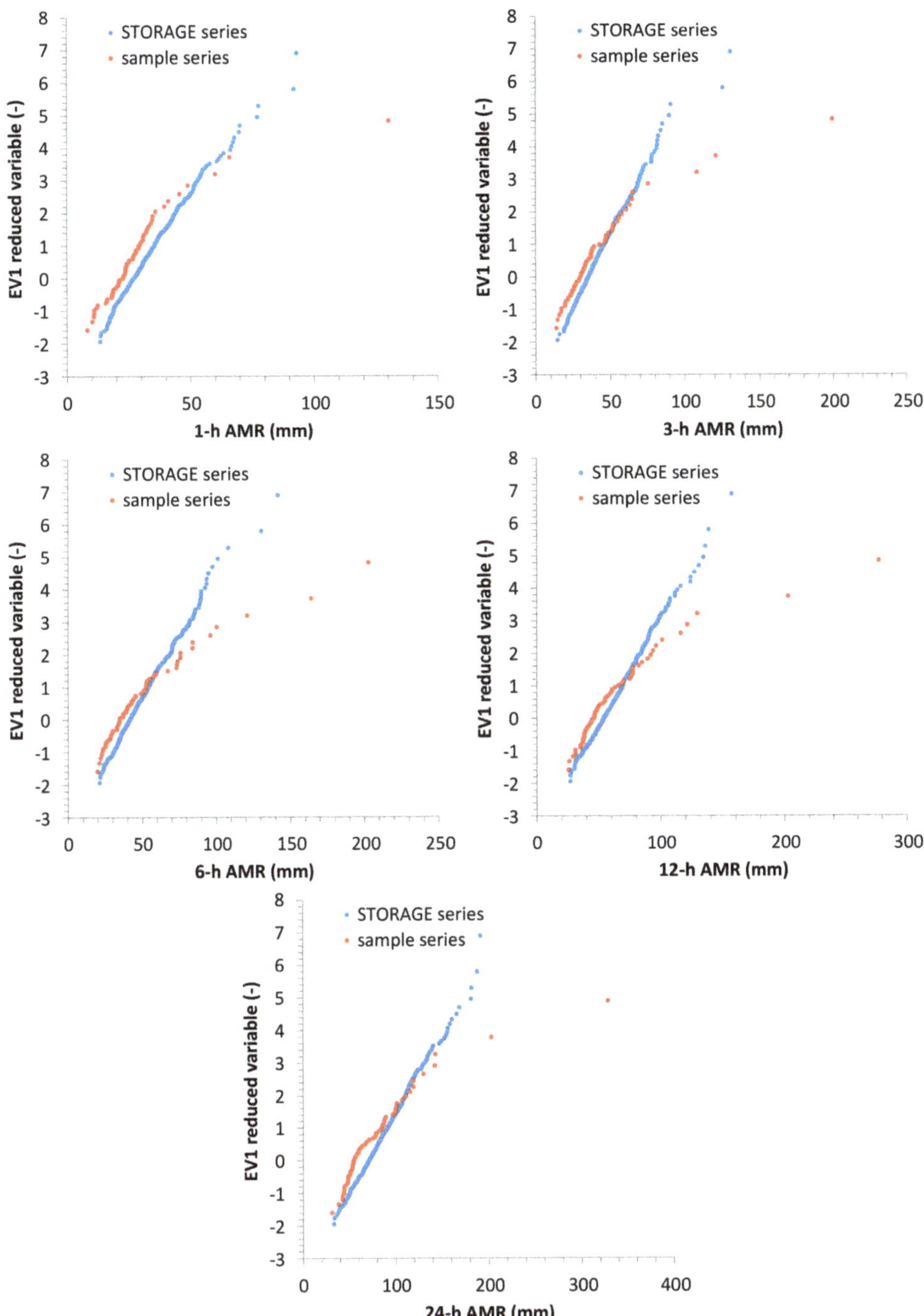

Figure 20. Vibo Valentia rain gauge: EV1 probabilistic plots, showing the comparison among synthetic and observed AMR series.

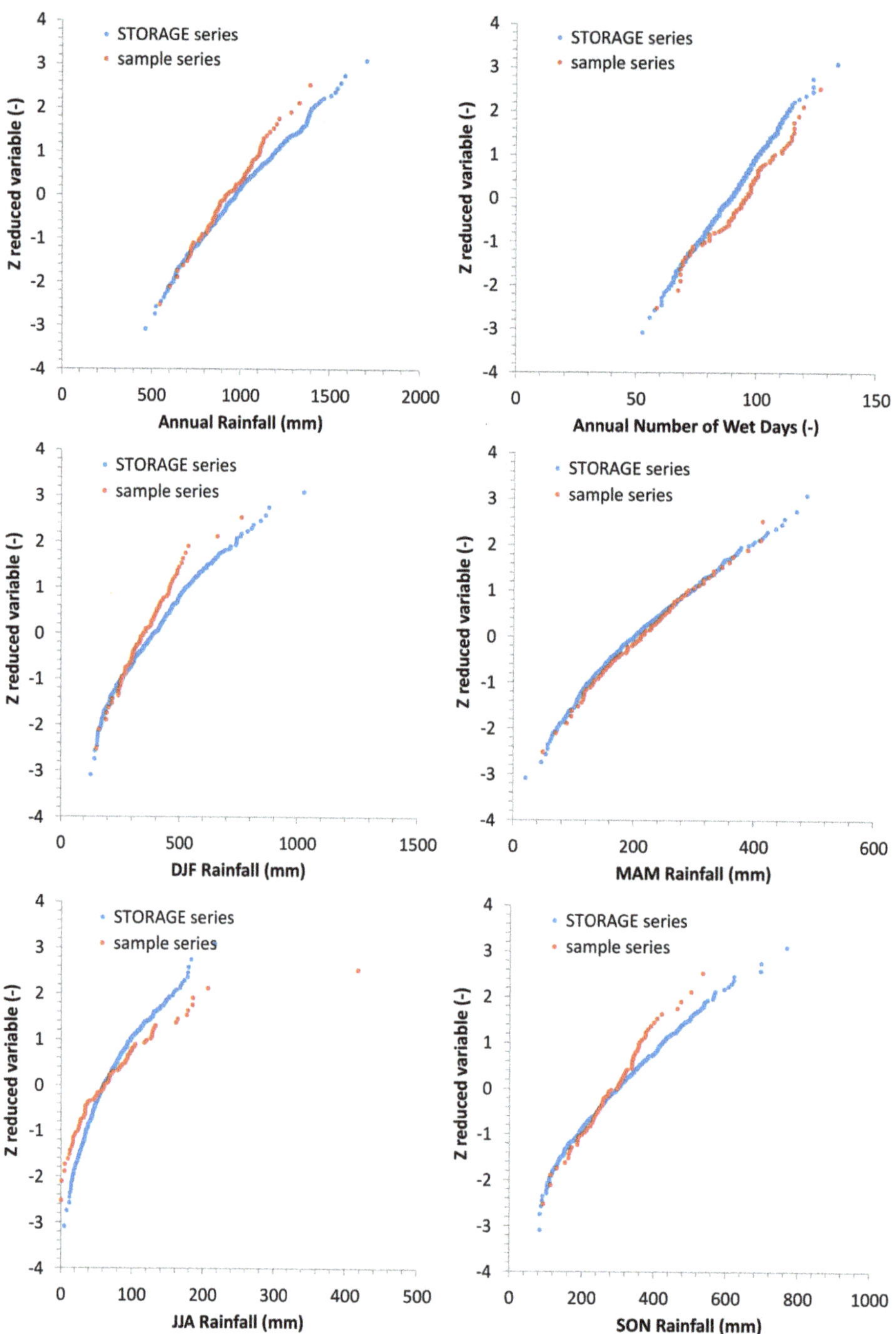

Figure 21. Vibo Valentia rain gauge: Gaussian probabilistic plots, showing the comparison among synthetic and observed series, regarding annual and seasonal rainfall, and annual number of wet days.

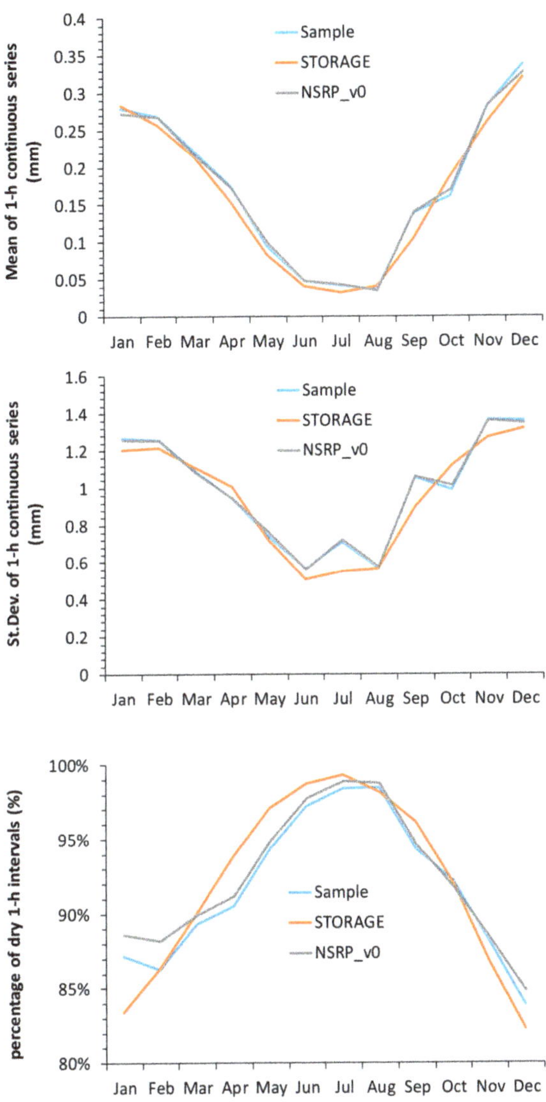

Figure 22. Montalto Uffugo rain gauge: comparison between STORAGE and NSRP_v0 performances, focusing on the mean, standard deviation and percentage of dry intervals for the continuous 1-h series.

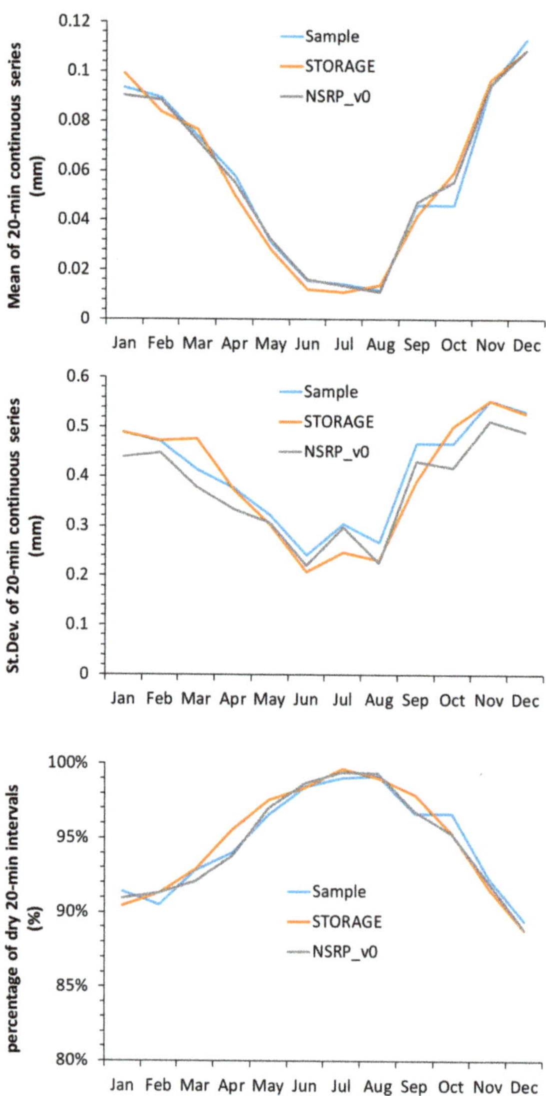

Figure 23. Montalto Uffugo rain gauge: comparison between STORAGE and NSRP_v0 performances, focusing on the mean, standard deviation and percentage of dry intervals for the continuous 20-min series.

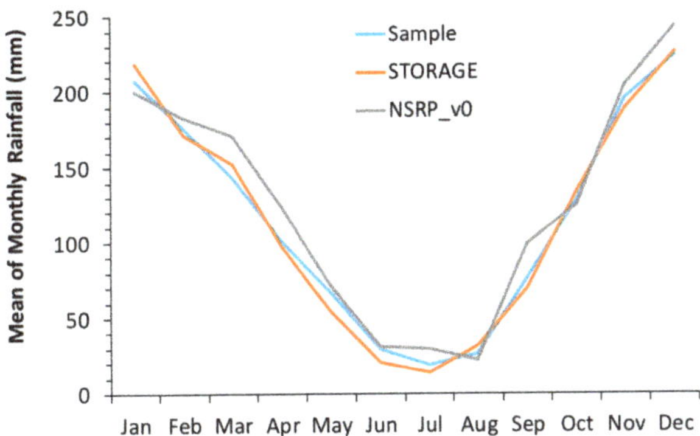

Figure 24. Montalto Uffugo rain gauge: comparison between STORAGE and NSRP_v0 performances, focusing on mean values of monthly rainfall heights.

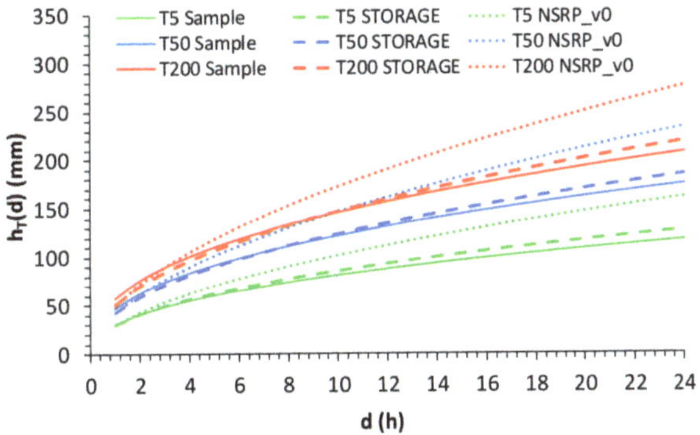

Figure 25. Montalto Uffugo rain gauge: comparison between STORAGE and NSRP_v0 performances, focusing on ADF curves.

5. Conclusions

The developed STORAGE software constitutes a very useful user-friendly tool for generating long rainfall time series at high resolutions, which could be applied as input data in many hydrological analyses, such as in the continuous rainfall-runoff modeling.

The innovative aspects of the software regard: (i) the possibility of using information, for model calibration, from observed time series which are longer than continuous data sample at high resolutions; (ii) the modelling of seasonality by adopting goniometric series, which allows for a more parsimonious approach with respect to considering monthly parametric sets (as is usually done).

The presented version of STORAGE software, available at https://sites.google.com/unical.it/storage, is currently suitable for the reproduction of rainfall series which exhibit a clear EV1 behaviour in terms of AMR and present values of annual and seasonal precipitation that are typical of the Mediterranean area.

Future developments will concern: (i) the extension of the ensemble of the parametric sets and the possibility to use other probability distributions for some rainfall features and other shapes besides the rectangular one for rain cells, in order to apply the model in other regions with different climates with respect to the investigated area; (ii) the implementation of a module for obtaining perturbed synthetic series, which can be representative of future hypothesized rainfall scenarios on spatial and temporal hydrological scales.

Moreover, the authors consider as very important the possibility of implementing in STORAGE specific modules related to soft computing methods (widely used in recent literature [60–62]), in order to provide different approaches for a specific case study. This aspect will allow to immediately compare the performances of an SRG (having a mathematical structure which is "physically-based", as it models some aspects of rainfall genesis, see Figure 2) with those from approaches such as Artificial Neural Networks (ANNs), Support Vector Regression (SVR) and Fuzzy Logic (FL), which are characterized by high nonlinearity, flexibility and data-driven learning.

Author Contributions: Conceptualization, D.L.D.L. and A.P.; methodology, D.L.D.L. and A.P.; software, D.L.D.L.; validation, D.L.D.L. and A.P.; data curation, D.L.D.L.; writing—original draft preparation, D.L.D.L. and A.P.; writing—review and editing, D.L.D.L. and A.P. All authors have read and agreed to the published version of the manuscript.

Funding: This research received no external funding.

Institutional Review Board Statement: Not applicable.

Informed Consent Statement: Not applicable.

Acknowledgments: This work was supported by the Italian Ministry for Ecological Transition (Ministero della Transizione Ecologica—Direzione Generale per la sicurezza del suolo e dell'Acqua-SUA), through the Law 5/1/2017 n. 4 "Interventi per il sostegno della formazione e della ricerca nelle scienze geologiche, per progetti di ricerca finalizzati alla previsione e alla prevenzione dei rischi geologici".

Conflicts of Interest: The authors declare no conflict of interest.

References

1. Grimaldi, S.; Nardi, F.; Piscopia, R.; Petroselli, A.; Apollonio, C. Continuous hydrologic modelling for design simulation in small and ungauged basins: A step forward and some tests for its practical use. *J. Hydrol.* **2020**, *595*, 125664. [CrossRef]
2. Młyński, D.; Wałęga, A.; Petroselli, A.; Tauro, F.; Cebulska, M. Estimating Maximum Daily Precipitation in the Upper Vistula Basin, Poland. *Atmosphere* **2019**, *10*, 43. [CrossRef]
3. Onof, C.; Chandler, R.E.; Kakou, A.; Northrop, P.; Wheater, H.S.; Isham, V. Rainfall modelling using Poisson-cluster processes: A review of developments. *Stoch. Environ. Res. Risk Assess.* **2000**, *14*, 384–411. [CrossRef]
4. Wheater, H.S.; Chandler, R.E.; Onof, C.J.; Isham, V.S.; Bellone, E.; Yang, C.; Lekkas, D.; Lourmas, G.; Segond, M.-L. Spatial-temporal rainfall modelling for flood risk estimation. *Stoch. Environ. Res. Risk Assess.* **2005**, *19*, 403–416. [CrossRef]
5. Ritschel, C.; Ulbrich, U.; Névir, P.; Rust, H.W. Precipitation extremes on multiple timescales—Bartlett–Lewis rectangular pulse model and intensity–duration–frequency curves. *Hydrol. Earth Syst. Sci.* **2017**, *21*, 6501–6517. [CrossRef]
6. De Luca, D.L.; Petroselli, A.; Galasso, L. Modelling Climate Changes with Stationary Models: Is It Possible or Is It a Paradox? In *Numerical Computations: Theory and Algorithms*; NUMTA 2019. Lecture Notes in Computer Science; Sergeyev, Y., Kvasov, D., Eds.; Springer Science and Business Media LLC: Berlin/Heidelberg, Germany, 2020; Volume 11974, pp. 84–96.
7. De Luca, D.; Petroselli, A.; Galasso, L. A Transient Stochastic Rainfall Generator for Climate Changes Analysis at Hydrological Scales in Central Italy. *Atmosphere* **2020**, *11*, 1292. [CrossRef]
8. Willems, P.; Arnbjerg-Nielsen, K.; Olsson, J.; Nguyen, V. Climate change impact assessment on urban rainfall extremes and urban drainage: Methods and shortcomings. *Atmospheric Res.* **2012**, *103*, 106–118. [CrossRef]
9. Maraun, D. Bias Correcting Climate Change Simulations—A Critical Review. *Curr. Clim. Chang. Rep.* **2016**, *2*, 211–220. [CrossRef]
10. Kendon, E.J.; Roberts, N.M.; Fowler, H.J.; Roberts, M.J.; Chan, S.C.; Senior, C.A. Heavier summer downpours with climate change revealed by weather forecast resolution model. *Nat. Clim. Chang.* **2014**, *4*, 570–576. [CrossRef]
11. Ban, N.; Schmidli, J.; Schär, C. Heavy precipitation in a changing climate: Does short-term summer precipitation increase faster? *Geophys. Res. Lett.* **2015**, *42*, 1165–1172. [CrossRef]
12. Cameron, D.; Beven, K.; Tawn, J. An evaluation of three stochastic rainfall models. *J. Hydrol.* **2000**, *228*, 130–149. [CrossRef]
13. Cowpertwait, P.S.P. Further developments of the neyman-scott clustered point process for modeling rainfall. *Water Resour. Res.* **1991**, *27*, 1431–1438. [CrossRef]
14. Cowpertwait, P.; O'Connell, P.; Metcalfe, A.; Mawdsley, J. Stochastic point process modelling of rainfall. I. Single-site fitting and validation. *J. Hydrol.* **1996**, *175*, 17–46. [CrossRef]
15. Cowpertwait, P.; Isham, V.; Onof, C. Point process models of rainfall: Developments for fine-scale structure. *Proc. R. Soc. A Math. Phys. Eng. Sci.* **2007**, *463*, 2569–2587. [CrossRef]
16. Entekhabi, D.; Rodriguez-Iturbe, I.; Eagleson, P.S. Probabilistic representation of the temporal rainfall process by a modified Neyman-Scott Rectangular Pulses Model: Parameter estimation and validation. *Water Resour. Res.* **1989**, *25*, 295–302. [CrossRef]
17. Gyasi-Agyei, Y. Identification of regional parameters of a stochastic model for rainfall disaggregation. *J. Hydrol.* **1999**, *223*, 148–163. [CrossRef]

18. Gyasi-Agyei, Y.; Willgoose, G.R. A hybrid model for point rainfall modeling. *Water Resour. Res.* **1997**, *33*, 1699–1706. [CrossRef]
19. Islam, S.; Entekhabi, D.; Bras, R.L.; Rodriguez-Iturbe, I. Parameter estimation and sensitivity analysis for the modified Bartlett-Lewis rectangular pulses model of rainfall. *J. Geophys. Res. Space Phys.* **1990**, *95*, 2093–2100. [CrossRef]
20. Kaczmarska, J.; Isham, V.; Onof, C. Point process models for fine-resolution rainfall. *Hydrol. Sci. J.* **2014**, *59*, 1972–1991. [CrossRef]
21. Khaliq, M.; Cunnane, C. Modelling point rainfall occurrences with the modified Bartlett-Lewis rectangular pulses model. *J. Hydrol.* **1996**, *180*, 109–138. [CrossRef]
22. Kim, D.; Olivera, F.; Cho, H.; Socolofsky, S.A. Regionalization of the Modified Bartlett-Lewis Rectangular Pulse Stochastic Rainfall Model. *Terr. Atmos. Ocean. Sci.* **2013**, *24*, 421–436. [CrossRef]
23. Kim, D.; Kwon, H.-H.; Lee, S.-O.; Kim, S. Regionalization of the Modified Bartlett–Lewis rectangular pulse stochastic rainfall model across the Korean Peninsula. *HydroResearch* **2016**, *11*, 123–137. [CrossRef]
24. Kim, D.; Cho, H.; Onof, C.; Choi, M. Let-It-Rain: A web application for stochastic point rainfall generation at ungaged basins and its applicability in runoff and flood modeling. *Stoch. Environ. Res. Risk Assess.* **2017**, *31*, 1023–1043. [CrossRef]
25. Kim, D.; Onof, C. A stochastic rainfall model that can reproduce important rainfall properties across the timescales from sev-eral minutes to a decade. *J. Hydrol.* **2020**, *589*, 125150. [CrossRef]
26. Kossieris, P.; Efstratiadis, A.; Koutsoyiannis, D. Coupling the strengths of optimization and simulation for calibrating Poisson cluster models. In Proceedings of the Facets of Uncertainty: 5th EGU Leonardo Conference–Hydrofractals 2013–STAHY 2013, Kos Island, Greece, 17–19 October 2013.
27. Onof, C.; Wheater, H.S. Improved fitting of the Bartlett-Lewis Rectangular Pulse Model for hourly rainfall. *Hydrol. Sci. J.* **1994**, *39*, 663–680. [CrossRef]
28. Onof, C.; Wheater, H.S. Improvements to the modelling of British rainfall using a modified Random Parameter Bartlett-Lewis Rectangular Pulse Model. *J. Hydrol.* **1994**, *157*, 177–195. [CrossRef]
29. Paschalis, A.; Molnar, P.; Fatichi, S.; Burlando, P. On temporal stochastic modeling of precipitation, nesting models across scales. *Adv. Water Resour.* **2014**, *63*, 152–166. [CrossRef]
30. Smithers, J.; Pegram, G.; Schulze, R. Design rainfall estimation in South Africa using Bartlett–Lewis rectangular pulse rainfall models. *J. Hydrol.* **2002**, *258*, 83–99. [CrossRef]
31. Velghe, T.; Troch, P.A.; De Troch, F.P.; Van De Velde, J. Evaluation of cluster-based rectangular pulses point process models for rainfall. *Water Resour. Res.* **1994**, *30*, 2847–2857. [CrossRef]
32. Verhoest, N.; Troch, P.A.; De Troch, F.P. On the applicability of Bartlett–Lewis rectangular pulses models in the modeling of design storms at a point. *J. Hydrol.* **1997**, *202*, 108–120. [CrossRef]
33. Wasko, C.; Pui, A.; Sharma, A.; Mehrotra, R.; Jeremiah, E. Representing low-frequency variability in continuous rainfall simulations: A hierarchical random Bartlett Lewis continuous rainfall generation model. *Water Resour. Res.* **2015**, *51*, 9995–10007. [CrossRef]
34. Wheater, H.S.; Isham, V.S.; Chandler, R.E.; Onof, C.J.; Stewart, E.J. *Improved Methods for National Spatial–Temporal Rainfall and Evaporation Modelling for BSM*; Department for Environment, Food and Rural Affairs (DEFRA); Flood Management Division: London, UK, 2007.
35. Verhoest, N.; Vandenberghe, S.; Cabus, P.; Onof, C.; MecaFigueras, T.; Jameleddine, S. Are stochastic point rainfall models able to preserve extreme flood statistics? *Hydrol. Process.* **2010**, *24*, 3439–3445. [CrossRef]
36. Cowpertwait, P.S.P. A generalized point process model for rainfall. *Proc. R. Soc. Lond. Ser. A Math. Phys. Sci.* **1994**, *447*, 23–37. [CrossRef]
37. Cameron, D.; Beven, K.; Tawn, J. Modelling extreme rainfalls using a modified random pulse Bartlett–Lewis stochastic rain-fall model (with uncertainty). *Adv. Water Resour.* **2000**, *24*, 203–211. [CrossRef]
38. Evin, G.; Favre, A.-C. A new rainfall model based on the Neyman-Scott process using cubic copulas. *Water Resour. Res.* **2008**, *44*, 03433. [CrossRef]
39. Koutsoyiannis, D.; Onof, C. Rainfall disaggregation using adjusting procedures on a Poisson cluster model. *J. Hydrol.* **2001**, *246*, 109–122. [CrossRef]
40. Onof, C.; Townend, J.; Kee, R. Comparison of two hourly to 5-min rainfall disaggregators. *Atmos. Res.* **2005**, *77*, 176–187. [CrossRef]
41. Onof, C.; Arnbjerg-Nielsen, K. Quantification of anticipated future changes in high resolution design rainfall for urban areas. *Atmos. Res.* **2009**, *92*, 350–363. [CrossRef]
42. Kossieris, P.; Makropoulos, C.; Onof, C.; Koutsoyiannis, D. A rainfall disaggregation scheme for sub-hourly time scales: Coupling a Bartlett-Lewis based model with adjusting procedures. *J. Hydrol.* **2018**, *556*, 980–992. [CrossRef]
43. Kim, D.; Olivera, F.; Cho, H. Effect of the inter-annual variability of rainfall statistics on stochastically generated rainfall time series: Part 1. Impact on peak and extreme rainfall values. *Stoch. Environ. Res. Risk Assess.* **2013**, *27*, 1601–1610. [CrossRef]
44. Cross, D.; Onof, C.; Winter, H.; Bernardara, P. Censored rainfall modelling for estimation of fine-scale extremes. *Hydrol. Earth Syst. Sci.* **2018**, *22*, 727–756. [CrossRef]
45. Paschalis, A.; Molnar, P.; Fatichi, S.; Burlando, P. A stochastic model for high-resolution space-time precipitation simulation. *Water Resour. Res.* **2013**, *49*, 8400–8417. [CrossRef]
46. Peleg, N.; Fatichi, S.; Paschalis, A.; Molnar, P.; Burlando, P. An advanced stochastic weather generator for simulating 2-D high-resolution climate variables. *J. Adv. Model. Earth Syst.* **2017**, *9*, 1595–1627. [CrossRef]

47. De Luca, D.L.; Galasso, L. Calibration of NSRP Models from Extreme Value Distributions. *Hydrology* **2019**, *6*, 89. [CrossRef]
48. Website of the Multi Risks Centre of Calabria Region. Available online: www.cfd.calabria.it (accessed on 8 April 2021).
49. Federico, S.; Avolio, E.; Pasqualoni, L.; De Leo, L.; Sempreviva, A.M.; Bellecci, C. Preliminary results of a 30-year daily rainfall data base in southern Italy. *Atmos. Res.* **2009**, *94*, 641–651. [CrossRef]
50. Federico, S.; Avolio, E.; Pasqualoni, L.; Bellecci, C. Atmospheric patterns for heavy rain events in Calabria. *Nat. Hazards Earth Syst. Sci.* **2008**, *8*, 1173–1186. [CrossRef]
51. Rodriguez-Iturbe, I.; Cox, D.R.; Isham, V. Some models for rainfall based on stochastic point processes. *Proc. R. Soc. London. Ser. A Math. Phys. Sci.* **1987**, *410*, 269–288. [CrossRef]
52. Sirangelo, B.; Ferrari, E.; De Luca, D.L. Occurrence analysis of daily rainfalls through non-homogeneous Poissonian processes. *Nat. Hazards Earth Syst. Sci.* **2011**, *11*, 1657–1668. [CrossRef]
53. Greco, A.; De Luca, D.L.; Avolio, E. Heavy Precipitation Systems in Calabria Region (Southern Italy): High-Resolution Observed Rainfall and Large-Scale Atmospheric Pattern Analysis. *Water* **2020**, *12*, 1468. [CrossRef]
54. Calenda, G.; Napolitano, F. Parameter estimation of Neyman–Scott processes for temporal point rainfall simulation. *J. Hydrol.* **1999**, *225*, 45–66. [CrossRef]
55. Morbidelli, R.; García-Marín, A.P.; Al Mamun, A.; Atiqur, R.M.; Ayuso-Muñoz, J.L.; Taouti, M.B.; Baranowski, P.; Bellocchi, G.; Sangüesa-Pool, C.; Bennett, B.; et al. The history of rainfall data time-resolution in a wide variety of geographical areas. *J. Hydrol.* **2020**, *590*, 125258. [CrossRef]
56. Beven, K.; Freer, J. Equifinality, data assimilation, and uncertainty estimation in mechanistic modelling of complex envi-ronmental systems using the GLUE methodology. *J. Hydrol.* **2001**, *249*, 11–29. [CrossRef]
57. Koutsoyiannis, D.; Montanari, A. Negligent killing of scientific concepts: The stationarity case. *Hydrol. Sci. J.* **2015**, *60*, 1174–1183. [CrossRef]
58. Onof, C.; Wang, L.-P. Modelling rainfall with a Bartlett–Lewis process: New developments. *Hydrol. Earth Syst. Sci.* **2020**, *24*, 2791–2815. [CrossRef]
59. Park, J.; Cross, D.; Onof, C.; Chen, Y.; Kim, D. A simple scheme to adjust Poisson cluster rectangular pulse rainfall models for improved performance at sub-hourly timescales. *J. Hydrol.* **2021**, *598*, 126296. [CrossRef]
60. Wu, C.; Chau, K. Prediction of rainfall time series using modular soft computingmethods. *Eng. Appl. Artif. Intell.* **2013**, *26*, 997–1007. [CrossRef]
61. Sattari, M.T.; Falsafian, K.; Irvem, A.; S, S.; Qasem, S.N. Potential of kernel and tree-based machine-learning models for estimating missing data of rainfall. *Eng. Appl. Comput. Fluid Mech.* **2020**, *14*, 1078–1094. [CrossRef]
62. Shiru, M.; Park, I. Comparison of Ensembles Projections of Rainfall from Four Bias Correction Methods over Nigeria. *Water* **2020**, *12*, 3044. [CrossRef]

 *hydrology*

Article

Integrative Assessment of Stormwater Infiltration Practices in Rapidly Urbanizing Cities: A Case of Lucknow City, India

Jheel Bastia [1], Binaya Kumar Mishra [2] and Pankaj Kumar [3,*]

1 Institute for the Advance Study of Sustainability, United Nations University, 5-53-70 Jingumae, Shibuya-Ku, Tokyo 150-8925, Japan; jheel.bastia89@gmail.com
2 School of Engineering, Pokhara University, Pokhara-30, Lekhnath 33700, Nepal; bkmishra@pu.edu.np
3 Natural Resources and Ecosystem Services, Institute for Global Environmental Strategies, Hayama, Kanagawa 240-0115, Japan
* Correspondence: kumar@iges.or.jp

Abstract: The lack of strategic planning in stormwater management has made rapidly urbanizing cities more vulnerable to urban water issues than in the past. Low infiltration rates, increase in peak river discharge, and recurrence of urban floods and waterlogging are clear signs of unplanned rapid urbanization. As with many other low to middle-income countries, India depends on its conventional and centralized stormwater drains for managing stormwater runoff. However, in the absence of a robust stormwater management policy governed by the state, its impact trickles down to a municipal level and the negative outcome can be clearly observed through a failure of the drainage systems. This study examines the role of onsite and decentralized stormwater infiltration facilities, as successfully adopted by some higher income countries, under physical and social variability in the context of the metropolitan city of Lucknow, India. Considering the 2030 Master Plan of Lucknow city, this study investigated the physical viability of the infiltration facilities. Gridded ModClark rainfall-runoff modeling was carried out in Kukrail river basin, an important drainage basin of Lucknow city. The HEC-HMS model, inside the watershed modeling system (WMS), was used to simulate stormwater runoff for multiple scenarios of land use and rainfall intensities. With onsite infiltration facilities as part of land use measures, the peak discharge reduced in the range of 48% to 59%. Correlation analysis and multiple regression were applied to understand the rainfall-runoff relationship. Furthermore, the stormwater runoff drastically reduced with decentralized infiltration systems.

Keywords: stormwater runoff; decentralized and onsite infiltration facilities; climate change; distributed hydrological model

Citation: Bastia, J.; Mishra, B.K.; Kumar, P. Integrative Assessment of Stormwater Infiltration Practices in Rapidly Urbanizing Cities: A Case of Lucknow City, India. *Hydrology* **2021**, *8*, 93. https://doi.org/10.3390/hydrology8020093

Academic Editors: Davide Luciano De Luca and Andrea Petroselli

Received: 30 April 2021
Accepted: 10 June 2021
Published: 12 June 2021

Publisher's Note: MDPI stays neutral with regard to jurisdictional claims in published maps and institutional affiliations.

1. Introduction

Stormwater runoff is defined as the amount of rainfall that is unable to infiltrate the ground and instead flows over the land into streams, rivers, or lakes as surface runoff. While flowing over the land surface, it picks up materials such as leaves, stones, and sediments, but also non-biodegradable plastics, trash, and other pollutants [1]. There are two key drivers which result in stormwater runoff: (i) rapid and unplanned urbanization, and (ii) climate change. The relation between urbanization and economic growth is a complex one, yet the former has been strongly associated with bringing progress to a nation [2].

Urbanization alters the nature of surface perviousness by reducing the ability of water to infiltrate soil and, thus, the majority of water runs as overland water. Stormwater runoff in urbanizing areas ranges from 10% in areas with natural land cover to 55% in areas with 75% to 100% surface imperviousness [3]. While the percent increase is relational, it is region specific and will depend upon the stormwater infiltration facilities as well as on the level of imperviousness. The higher the surface imperviousness, the more surface runoff. Unplanned urbanization, instead of enhancing natural pathways for the flow of water, replaces the surface by installing artificial drainage systems [4]. The second key driver for runoff is climate change, the manifestation of anthropogenic activities, which is characterized by extreme rainfall events. Stormwater runoff further intensifies in urban areas where natural drainages meant to carry excess stormwater are replaced by unplanned built up areas with inefficient artificial drainage systems [5].

The severity of adverse runoff impacts can vary across the physical components in an urban watershed. Urbanization and climate change result in an increased runoff rate, less groundwater recharge, a greater magnitude of river flow, and the recurrence of small to major urban floods [4,6–9].

To study the effect of stormwater runoff on downstream water quality, Mohit and Sellu [10] conducted a comparative impact study of an urban, a suburban, and a rural stream, collecting dry and wet samples from 12 sites. They found that the urban stream had the highest levels of biochemical oxygen demand (BOD), orthophosphate, total suspended sediment (TSS), and surfactant concentrations. The study highlights a strong correlation between the level of development in the watershed and impervious surface coverage with BOD. Similar results were observed in an urban catchment of the rapidly urbanizing Bhatinda city in India [11]. While the primary effects are mostly observed in the hydrological systems and urban flooding, secondary effects include vulnerability to health risks in humans as well as animals and marine or aquatic life [9,12].

Various comparative and standalone studies carried out through hydrological modeling, such as the soil and water assessment test (SWAT), and field observations emphasize that infiltration facilities have a higher efficiency than storage facilities in reducing runoff volumes [13,14]. These structures primarily mimic natural processes as they behave as a source control and decentralized stormwater absorption system to facilitate the reduction in surface runoff, minimize environmental degradation, and enhance infiltration. These are developed via a combination of sound site planning, and structural and nonstructural techniques. Some sustainable stormwater management practices include rain gardens, biofiltration swales, bioretention pits, pervious pavements, and green roofs [6,15].

Using the SWMM model, Zahmatkesh et al. [16] investigated the impact of low impact development practices, such as rainwater harvesting, porous pavement, and bioretention, in mitigating the effects of climate change on stormwater runoff in a watershed of New York City. With the implementation of low impact development practices for 2- and 50-year return period extreme precipitation events, both runoff volume and peak discharge reduced significantly. Hunt et al. [17], through field observations and hydrological modeling, reported that bioretention pits can reduce 0% to 99% of sediment and nutrient losses due to the protective vegetation layer as it acts as a natural filter. Similarly, in Monash University, Melbourne, Hatt et al. [18] observed peak flow reductions between 49% and 80% for different rain gardens. Additionally, the vegetation of rain gardens effectively removed the suspended solids and heavy metals by 90%.

Various assessment tools for evaluating the sustainability of stormwater practices have been studied. A study conducted by Zhou et al. [19] in Denmark on urban flood modeling used a hedonic valuation model for cost-benefit analysis and revealed that a conventional drainage system, with traditional storage or source control facilities, has greater efficiency in controlling and reducing the impacts of stormwater runoff. Furthermore, Zhou et al. [19] analyzed the performance of open urban drainage systems using the cost-benefit approach (CBA) and discussed the socio-economic benefits of the sustainable urban drainage system, such as mitigating flood risks and enhancing the recreational value of local neighborhoods. To validate this research finding, Zhou [5] further reviewed various assessment tools and highlighted the use of cost-benefit analysis, life cycle assessment tools, and multi-criteria analysis as measures to study economic benefits of rooftop gardens in addressing urban water management issues.

India is on the brink of an urban revolution, with an expected 600 million predicted to reside in cities by 2031 [20]. India is prone to major river flood risk and has been experiencing urban floods in recent years. Unplanned rapid urbanization is attributed as one of the main reasons behind the increase in flood risks. Indian cities need to adopt unprecedented solutions to tackle urban stormwater runoff to mitigate the unforeseeable occurrence of hydro-meteorological hazards. Studies on decentralized stormwater infiltration practices are largely lacking, especially in developing countries. Gogate et al. [15] assessed various decentralized facilities, focusing on their advantages and disadvantages in the context of India, and found varying impacts of infiltration measures with region, type of development, soil, and drainage. Andimuthu et al. [21] tested the performance of urban storm drainage networks and infiltration control measures for urban flood mitigation under a climate change scenario in Chennai city, India. Aside from the use of urban stormwater drainage networks, stormwater best management practices, such as permeable pavements, rain gardens, green roofs, street planters, rain barrels, infiltration trenches and vegetative swales, in hydrologic modeling were suggested to provide more accurate and effective flood management systems.

These drawbacks form the main rationale of this study, which is to assess the decentralized stormwater infiltration practices under different physical variability in the context of an urban area in a developing economy such as India. This study focuses on the cumulative impact of stormwater runoff at a catchment scale, when the infiltration facilities are implemented at a local scale, as it can help in reducing the numerous effects of urban floods and waterlogging in unfavorable conditions. Lucknow city is facing an unprecedented decline in groundwater level with each passing year. Most of the rain received during the monsoon season, instead of being stored and infiltrated, is subjected to flow as stormwater runoff. Hence, the objective of urban stormwater runoff modeling in this study is to see: (i) how present and future land use and rainfall will have an impact on surface runoff, and (ii) how decentralized or decentralized infiltration practices at the household level can reduce the impact.

2. Materials and Methods

2.1. Study Area

This study was conducted in Lucknow, the capital city of India's most populous state, Uttar Pradesh. It receives an average rainfall of 890 mm, mostly from the southwest monsoons between June and September. The city derives multiple benefits from the monsoon- and groundwater-fed Gomti River, which passes through center of city dividing it into Trans and Cis Gomti. As per a report by Sharma and Shukla [22], the drainage system in the city was established five decades ago and the coverage of stormwater drains is limited to one-third of the city. In the absence of proper stormwater drains, the rainwater forces itself through sewerage lines, resulting in the choking of sewer lines which are not maintained regularly, and leads to overflowing of drain water causing waterlogging and flooding in low lying areas. While Gomti is a major water supply source in the city, it receives its water from 26 drains at the study site, Kukrail Nala catchment (Figure 1).

Kukrail Nala, a fourth order tributary, is critical as it brings large volumes of water into Gomti River while passing through central Lucknow city. Kukrail Nala basin, with a river length of 26 km and an origin inside Kukrail reserve forest, falls in the Trans-Gomti area. The Kukrail Nala basin is expected to contribute large volumes of flood water in the downstream area due to basin characteristics and a slope of 6%. In recent years, Kukrail Nala basin, which falls in Trans-Gomati area, has seen rapid urbanization and concretization resulting in more flood water. In the past decade or so, Lucknow city, with its increased impervious surface due to unplanned urbanization and the alteration in rainfall pattern due to climate change, faces regular waterlogging in the downstream areas and groundwater depletion.

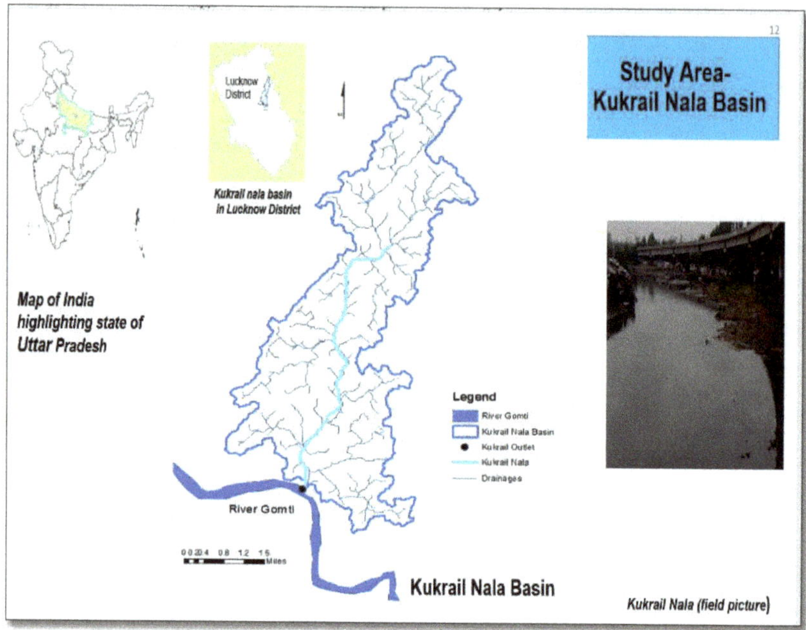

Figure 1. Location of Kukrail river basin inside Lucknow city, India.

2.2. Data Preparation

Land use land cover data is one of the key inputs for hydrological modeling. In this study, the watershed modeling system (WMS), which is a popular tool for automated watershed delineation, hydrologic and hydraulic modeling, floodplain mapping, and storm drain modeling, was used for testing impacts of various stormwater infiltration measures. The Water and Urban Initiative project of the United Nations University availed land use land cover data for the 2007 (past), 2016 (present), and 2030 (projected) periods. ASTER 30 m digital elevation model data was used for delineating the river basin and the estimation of several other hydrologic parameters.

Annual daily maximum rainfall of 50- and 100-year return periods for present and future climate conditions were applied for assessing the effects of increased extreme rainfall events. The estimation of extreme rainfall for the present climate was based on the Lucknow weather station. On the other hand, multiple GCMs (Global Climate Models) precipitation outputs were used for estimating extreme rainfalls for future conditions. Daily precipitation output of two GCMs: MIROC-5 and MRI-GCM3 with spatial resolution of 140 km and 110 km, respectively, were employed for deriving future rainfall data. The use of multiple emission scenarios and a general circulation model output are recommended due to uncertainty with climate projections and, hence, GCMs under different representative concentration pathways (RCPs) were used for the study [23]. For this study RCP 4.5, a representation of a stabilization scenario and mitigation policies, and RCP 8.5, a representation of an extreme, unstabilized, and destructive scenario, were used. The soil data was retrieved from Natural Resources Conservation Service (NRCS), United States Department of Agriculture, and the National Bureau of Soil Survey and Land Use Planning, India.

2.3. Stormwater Runoff Modeling

In this study, HEC-HMS, built into the WMS tool, was used for the stormwater runoff modeling (Figure 2). Based on the different land use and climate conditions for present and future periods, 10 different scenarios were established. The first six scenarios did not include any infiltration measures. The ModClark model inside the HEC-HMS, which is a distributed parameter model, was selected for simulating flood values. A distributed parameter model is one in which spatial variability of characteristics and processes are considered explicitly. The list of parameters, methods, and their name or values used for setting the ModClark stormwater runoff model in this study are shown in Table 1. The Kukrail Nala basin modeling was carried out at a 30 m grid cell scale. The grid size is user-specified and depends on the planning purpose. Travel time was calculated for each grid cell and scaled to the overall watershed time of concentration. The lagged runoff from each grid cell was routed through the linear reservoir. The outputs from each linear reservoir were combined to form an outflow hydrograph.

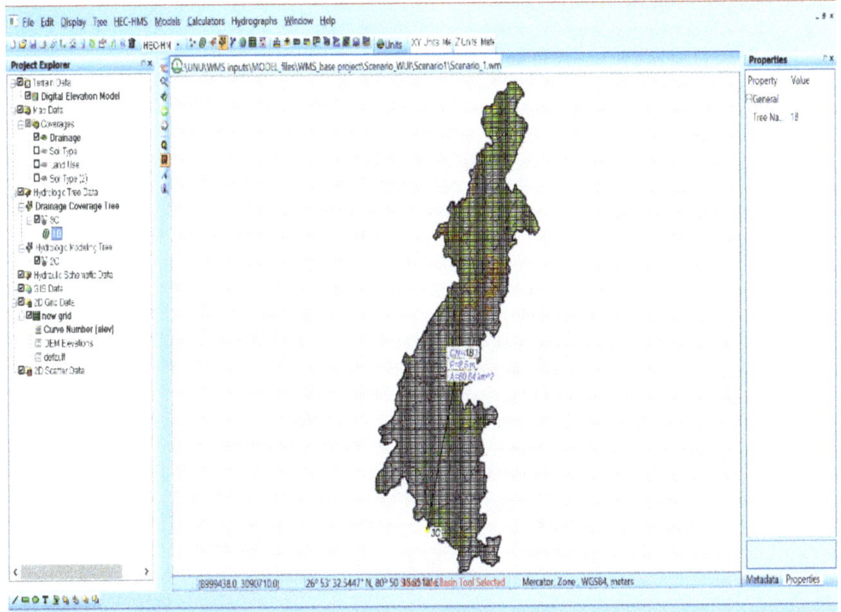

Figure 2. HEC-HMS gridded ModClark modeling in WMS platform.

Table 1. Hydro-meteorological modeling methods, parameters, and their values for stormwater runoff calculation.

Parameter/Method	Description/Value
Loss rate method	Gridded Soil Conservation Services (SCS) Curve Number
Initial abstraction	0.2
Potential retention scale factor	1
Transformation method	ModClark
Baseflow method	Linear reservoir
GW1 fraction (Faster responding interflow)	0.5
GW2 fraction (Slower responding baseflow)	0.2
Storm type	SCS type II 24 h distribution

2.3.1. Calibration of the Model

Calibration of the model was done to adjust certain parameter values of the model until the results matched acceptably with the observed data. In the precipitation-runoff model, it can be measured through the degree of variation between computed and observed hydrographs [24]. In this study, the model was calibrated by comparing the hydrograph of the simulated and observed data. The observed data was obtained from Lucknow Development Authority [25]. The maximum discharge that was observed in the Kukrail was 425 m^3/s. The model was calibrated by using parameters such as curve number, lag time, and Muskingum parameters on hydrograph shape.

2.3.2. Scenario Development

Based on the calibrated model, the simulation was done for 10 scenarios with different land use and rainfall values (Table 2). The land use was categorized as present (developing conditions) and denser urban core area (projected). The model was run for scenario 1 through 6 with no infiltration measures for the original curve number, as mentioned in [26]. For scenarios 7 to 10, certain grids with higher curve numbers in high density built-up areas were modified based on the soil type and infiltration facilities. Lower curve numbers were assigned to see the difference in runoff discharge.

Table 2. Description for various land use and climate change scenarios.

Scenario	Land Use Scenarios (Year)	Rainfall (mm)	Climate Scenarios	Return Period (Year)
1	2016	199	Present climate	50
2	2030	256.54	Average of RCP 4.5 and 8.5	50
3	2030	299.72	Extreme among RCPs 4.5 and 8.5	50
4	2016	226.06	Present climate	100
5	2030	290.06	Average of RCP 4.5 and 8.5	100
6	2030	337.8	Extreme among RCPs 4.5 and 8.5	100
7	2030 + infiltration measures	256.54	Average of RCP 4.5 and 8.5	50
8	2030 + infiltration measures	299.72	Extreme among RCPs 4.5 and 8.5	50
9	2030 + infiltration measures	290.06	Average of RCP 4.5 and 8.5	100
10	2030 + infiltration measures	337.8	Extreme among RCPs 4.5 and 8.5	100

3. Results

The following results were obtained for ten different scenarios (Figure 3, Table 2):

Scenario 1: the hydrograph with a daily maximum rainfall of 199 mm resulted in a peak discharge of 182 m^3/s. The peak discharge at 17:00 h depicted a lag period of 5 h.

Scenario 2: the daily maximum rainfall of 256.64 mm generated a peak discharge of 259 m^3/s. The excess rainfall of 104 mm at 12:00 h and a peak discharge at 17:00 h depicted the same lag time as in scenario 1.

Scenario 3: the daily maximum rainfall of 299.72 mm led to an excess rainfall of 123.4 mm and a peak discharge of 315 m^3/s. Analysis of the rainfall hyetograph and flood hydrograph pointed to a time lag of 4 h with extreme rainfall at 12:00 h and peak discharge at 16:00 h.

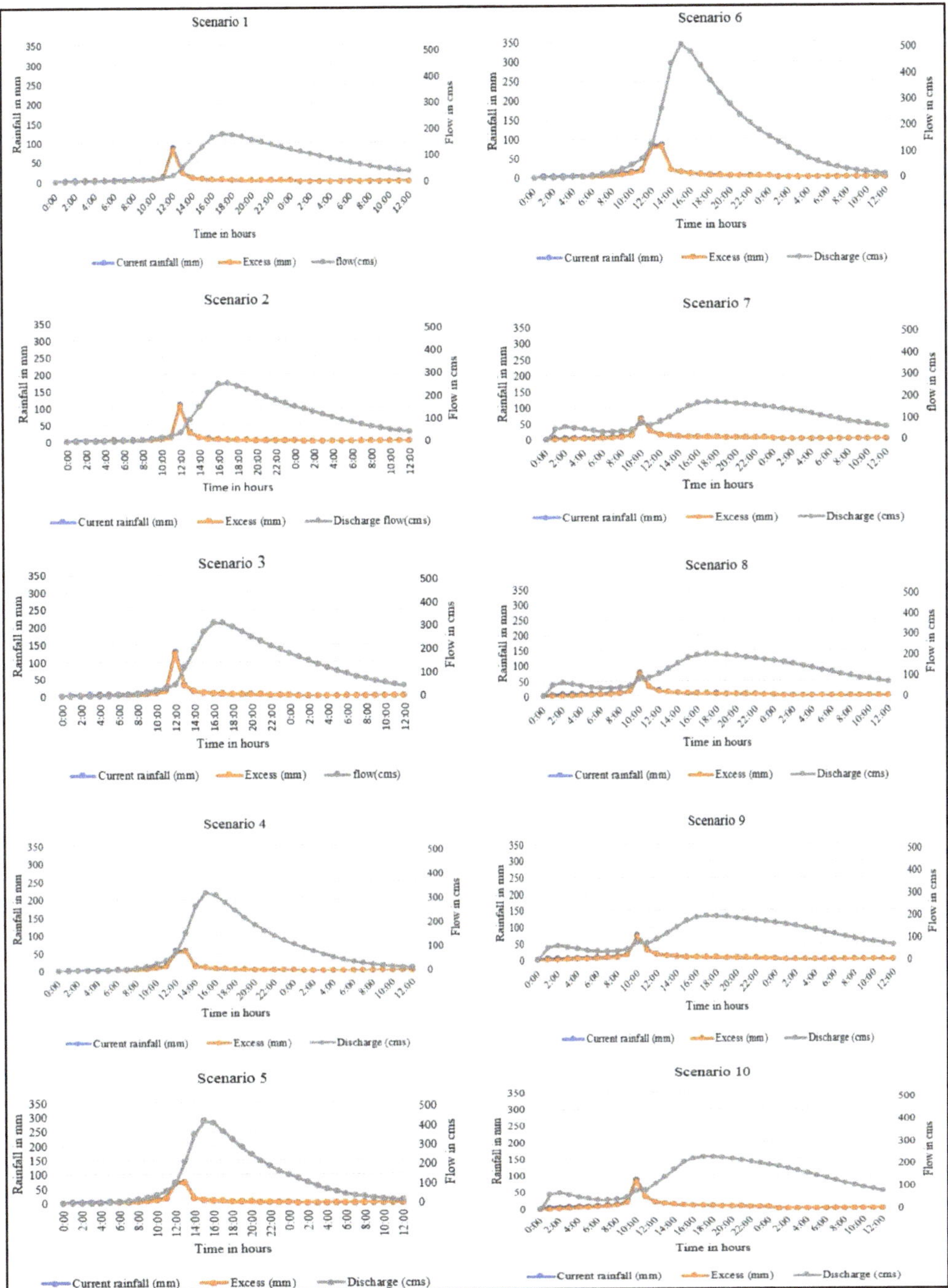

Figure 3. Hydrographs for ten different land use and climate change scenarios.

Scenario 4: the daily maximum rainfall of 226.06 mm generated a peak discharge of 324 m^3/s and a steep rising climb. The hydrograph suggested that the lag period was reduced to 2 h, with excess rainfall of 54 mm at 13:00 h and a peak discharge at 15:00 h.

Scenario 5: the daily maximum rainfall of 290.06 mm generated a peak discharge of 426 m^3/s. A steep rise in flow hydrograph, starting at 06:00 h, and a peak rise at 13:00 h depicted that the time of concentration reduced with a lag period of 3 h.

Scenario 6: the daily maximum rainfall of 337.8 mm, which is an excess of 81 mm, generated a high peak discharge of 509 m^3/s at 15:00 h. The flow hydrograph depicted that the lag period reduced to just 2 h with high runoff volumes.

Scenario 7: the daily maximum rainfall of 256.54 mm generated a discharge of 40 m^3/s at 02:00 h. The highest rainfall was recorded at 10:00 h. A peak discharge of 174 m^3/s was observed at 18:00 h with a lag time of 8 h.

Scenario 8: the daily maximum rainfall of 299.72 mm resulted in a discharge of 67 m^3/s at 02:00 h for alternative land use land cover condition. Extreme rainfall was recorded at 11:00 h. The peak discharge, 204 m^3/s, was observed at 17:00 h with a lag time of 7 h.

Scenario 9: the daily maximum rainfall of 290.06 mm resulted in a flow discharge of 56 m^3/s at 02:00 h. Extreme rainfall was recorded at 10:00 h and a peak discharge of 174 m^3/s was observed at 18:00 h with lag time of 8 h.

Scenario 10: the daily maximum rainfall of 337.8 mm resulted in a flow discharge of 72 m^3/s at 03:00 h. Extreme rainfall was recorded at 10:00 h. A peak discharge of 232 m^3/s was observed at 17:00 h with lag time of 7 h.

A Pearson product-moment correlation coefficient was computed to assess the relationship between the peak rainfall values of 50- and 100-year return periods for the years 2016 and 2030 with consideration of the infiltration facilities and runoff volume at the outlet of the Kukrail Nala basin (Figure 4a,b). For scenarios without infiltration facilities, a strong and positive correlation was found between the two variables' peak rainfall and runoff volumes. There was a positive correlation between the two variables with r = 0.98, n = 5, and p = 0.055. Despite of strong positive correlation between peak rainfall and runoff volumes, the relation was not significant as p > 0.05. On the other hand, for the scenarios with infiltration facilities, a scatterplot between peak rainfall and runoff volumes revealed a strong and positive correlation. The relation was quite significant as p = 0.003. Runoff volume for each scenario is shown is Table 3.

Table 3. Rainfall values, resulting runoff volumes, and lag time for different scenarios.

Scenario	Rainfall	Lag Time (Hours)	Runoff Volume
S1	199	5	154
S2	256.54	4.5	218.7
S3	299.72	4	249.1
S4	226.06	2	193.8
S5	290.06	3	255.7
S6	337.8	2	290.5
S7	256.54	8	207.5
S8	299.72	7	224.6
S9	290.06	8	244.8
S10	337.8	7	253.2

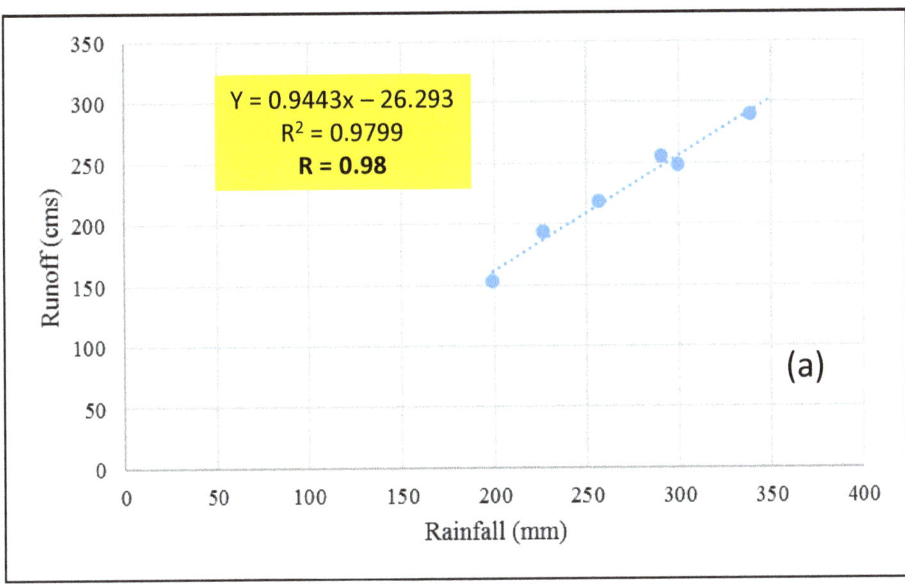

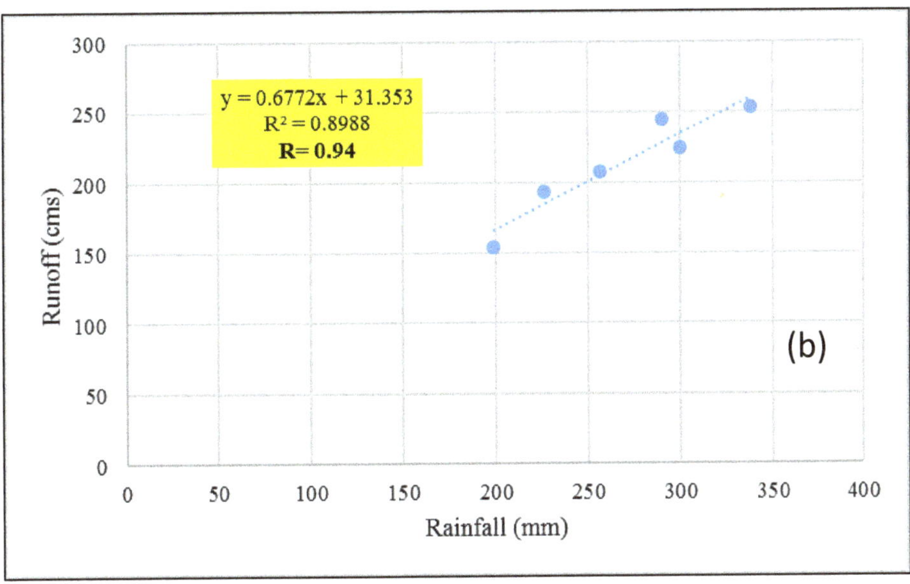

Figure 4. Rainfall runoff correlations (**a**) without infiltration facilities, and (**b**) with infiltration facilities.

4. Discussion

Each scenario has a different characteristic for its hydrograph. Six scenarios (1–6) did not consist of any infiltration facilities. Later, alternative stormwater capture measures (scenarios 7–10) were applied to test their impacts on the flow hydrograph for effectively reducing the runoff volume.

4.1. Without Infiltration Systems in Practice

With a 50-year return period and a peak discharge of 182 m^3/s for current conditions (scenario 1), the study revealed that peak discharge will increase by 42% and 73% for the increase in rainfall, corresponding to scenarios 2 and 3, respectively. This is in line with the already established correlation between increased imperviousness and surface runoff. An average increase of 2.5 times was reported in peak stormflow with the absence of stormwater drains as impervious coverage increases from 0% to 100% [27]. This would result in an increase in runoff volume by 34% in scenario 2 and 61% in scenario 3. However, the percent increase in runoff volumes from scenario 2 to 3 was 20% and depended largely on the calibrated climate models used. Meanwhile, for a 100-year return period, the peak discharge of 324.4 m^3/s under scenario 4 was projected to increase by 31.4% and 57% in scenario 5 and 6, respectively. This would result in an increase in runoff volumes by 31% and 49.9% in scenarios 5 and 6, respectively, from a business as usual scenario. However, the percent increase in runoff volumes from scenario 5 to scenario 6 was projected to be around only 13%. The slope, as observed in scenario 1 through 6, have similar observations as the steeper the slope, the less lag time and the higher the peak discharge and runoff volume. Sharma and Shukla [22] found similar observations in their studies. The percentage increase in the runoff volume will also depend on manmade stormwater drainage, which were not present in the land use land cover data. The increase in sudden peak and smaller lag times also represents the smaller time of concentration as scenario changes from stabilized to unstabilized ones.

4.2. With Infiltration Systems in Practice

A significant reduction in peak discharge and runoff volumes was observed by attenuating the excess precipitation through source control infiltration measures. It resulted in an overall decrease for the 50-year return period in both scenarios 7 and 8, and a similar observation was made under scenario 9 and scenario 10 for the 100-year return period. A decrease in peak discharge and runoff volumes by 48% and 51%, respectively, was observed for scenario 7 with infiltration facilitates, in comparison to scenario 2 with no infiltration measures. Similarly, under scenario 8 with infiltration measures, the peak discharge and runoff volumes were reduced by 54% and 10%, respectively, in comparison to scenario 3 with no infiltration measures. For the 100-year return period, a decrease in peak discharge and runoff volumes by 59.5% and 4%, respectively, was observed for scenario 9, in comparison to scenario 5 with no infiltration measures. Similarly, under scenario 10 with infiltration measures, the peak discharge and runoff volumes reduced by 54.4% and 12%, respectively, in comparison to scenario 6 with no infiltration measures. The decentralized stormwater facilities in the 50-year return period had extreme values that were more effective in reducing the runoff volumes as compared to the extreme values for the 100-year return period. The amount of infiltration will depend not just on type of decentralized measure, but also how well it is maintained, and the results of discharge will vary from a connected green roof system (more than 50% absorption on the roof) to pervious pavement (material of manufacturing). Additionally, maintenance and clearance of associated drainage will also impact the runoff discharge.

However, for managing flood water and stormwater reduction, we need to look for the best management practices, which are practically feasible, low cost, and easy to maintain, especially in resource-limited countries like India. The most important task is to estimate the carrying capacity and stability of the existing stormwater management system. Different possible management options are considered via demand supply gap analysis and cost-benefit analysis. Some of the most common solutions are gutter filters, infiltration trenches, permeable pavements, decentralized retention ponds, street sweeping on a regular basis to prevent the choking of sewerage pipes, and surface sand filters, etc. Another crucial component for the success of stormwater management practices is community awareness, as the community is largely unaware about the advantages of this management system and are not willing to participate in maintenance works.

4.3. Correlation Analysis

Although there is a positive relation between rainfall and runoff, this relationship was not significant for several scenarios without infiltration facilities. This result resonated with the findings of Miguez and Magalhaes [3], which mentioned that rainfall may not be the only criteria for high runoff volumes and other factors like geography, and geomorphology, etc., might also impact runoff volume. However, runoff volumes can be significantly reduced with infiltration facilities at times of peak rainfall events. It can be also said that infiltration measures are more suitable for an average climate with extreme rainfall of a 50-years return period. The infiltration rate may not be effective if rainfall under extreme climate with a 100-years return period occurs.

Thus, it can be said that the capacity of decentralized infiltration facilities to reduce the runoff volume and peak discharge depend not only on land use and rainfall, but also on the soil condition and topography of the watershed (slope and time of concentration). While there is a correlation between land use and rainfall with runoff volume, it might be strong or weak depending upon the condition of the area. The closer the distance between the facilities in a gradient, the more the infiltration. It should also be noted that the infiltration facilities are more effective for short-term planning.

4.4. BMP Implementation Challenges

Lucknow city has witnessed a major decline in water bodies which were once the natural:a sponge for water storage and infiltration. Low- to medium-cost housing with infiltration facilities could lead to incurring a higher per unit cost value, which might make it unaffordable for many low to middle-income families. The regular operational and maintenance cost of the facilities might influence the willingness of people to adopt it at a household level and hence was considered a threat. The current developmental activities in the peripheral area, resulting in higher abstraction of groundwater and lower infiltration, was also seen as a threat. Lastly, as a part of a developing nation, water management is not considered a priority when providing necessities for human survival, namely food, shelter, and water, are more pressing. In the absence of strong groundwater management policies, urbanizing areas are witnessing faster depletion of groundwater. While surface imperviousness plays a role in increased runoff, it is also believed that having a small built-up area of less than 200 square feet leaves one with no other option but to construct homes without any gardens. Most people are unaware of decentralized infiltration facilities as a result of loopholes in the regulatory body, due to a communication gap between various departments, and the inconsistency in the field data collection as well as a lack of data. The lack of communication in higher authorities trickles down to the public, where awareness regarding the source of stormwater control is low. The present conditions of the city, which have made it difficult to practice decentralized stormwater management, include a lack of technical expertise, soil conditions, and maintenance cost. Maintaining these facilities is a continuous challenge. Moreover, the government does not provide any subsidies for maintenance and there are no provisions (tax rebate, grants, etc.) for green infrastructure. If the available facilities are not properly maintained, causing runoff and urban flooding, it is less likely that the decentralized management system will work. There are no frameworks or committees to investigate policies specifically looking into subsidies for infiltration facilities. This could directly affect any implementation practices and, from an economic viewpoint, it is difficult to say if it will succeed in the future.

The monsoon is witnessing a shift in its pattern. While there is not much change in the annual rainfall received, the intensity and frequency has altered. At present, the drainage system and other traditional systems fail due to the heavy rainfall, which might cause the decentralized stormwater infiltration facilities to fail too.

The following 5 strategies were suggested as the best to promote the idea of stormwater infiltration practices:

WT1:a stakeholders workshop on integrated water resource management with the involvement of citizens;

WT2: preparation of strategic plans to promote decentralized stormwater infiltration practices and segregated and designated pipelines through the Master Plan, with a focus on climate change;

WT3: the involvement of real estate for developing solar plus water secured societies;

WT4: focusing on research and development by investing in it and expanding the base of technical experts among the growing population;

WT5: making the procedure of policymaking more transparent through coherent government policies and with a focus on climate change.

5. Conclusions

Replenishment of surface and ground waterbodies through rainfall is a natural phenomenon necessary to maintain the geo-hydrological balance of nature. Urbanization and climate change pose a severe threat to the quality and quantity of waterbodies. It is essential to recognize the associated risks of excess runoff from urban areas to enhance not just the quality of the environment, but an overall quality of life. Conventional centralized flood management systems are reported to have several limitations. There is much to be learned from non-conventional, decentralized stormwater management systems that municipalities have adopted to cope with urban floods and waterlogging in several parts of developed countries. However, these facilities need to be mainstreamed into the communities of vulnerable areas in the rapidly urbanizing cities of developing nations, where the cities are overburdened with more complex stormwater. To achieve that, research based on local conditions, practices, and management strategies should be understood in detail and incorporated with more advanced scientific knowledge to tackle the problems of conventional stormwater management systems.

This study delved into the hydrological assessment of decentralized stormwater management practices in an urban watershed of Lucknow city in Northern India. The study explored multiple scenarios based on current and future projections of land use and climate change to understand impacts of urbanization and precipitation on stormwater runoff. The results showed an increased runoff volume with the rapid urbanization and climate change. Implementation of infiltration facilities using the local landscape can have different impacts on the volume of runoff. The stormwater modeling study results showed that conversion of impervious surfaces to pervious ones by installing a decentralized facility can help infiltrate the water and immediately curb the peak runoff discharge. Decentralized stormwater runoff infiltration systems are a relatively new concept in most developing countries, and Lucknow city is no different. However, keeping in mind the growing scarcity of water resources and climate change, more studies around best strategies to conserve water quality and quantity should be encouraged at a policy level.

Author Contributions: Conceptualization, J.B., B.K.M. and P.K.; methodology, J.B., B.K.M. and P.K.; formal analysis, J.B.; writing—original draft preparation, J.B., B.K.M. and P.K.; writing—review and editing, J.B., B.K.M. and P.K. All authors have read and agreed to the published version of the manuscript.

Funding: This research received no external funding.

Institutional Review Board Statement: Not applicable.

Informed Consent Statement: Not applicable.

Data Availability Statement: Not applicable.

Acknowledgments: The authors express their gratitude to the Water Urban Initiative Project (WUI) of the United Nations University-Institute for the Advanced Study of Sustainability, Tokyo, Japan and Japan Educational Exchanges and Services (JEES) for providing financial assistance for the completion of this study.

Conflicts of Interest: The authors declare no conflict of interest.

References

1. Indiana Department of Environmental Management, 2007 Storm Water Runoff and Its Impact. Available online: http://www.in.gov/idem/stormwater/files/stormwater_manual_chap_01.pdf (accessed on 27 March 2016).
2. Chen, M.; Zhang, H.; Liu, W.; Zhang, W. The Global Pattern of Urbanization and Economic Growth: Evidence from the Last Three Decades. *PLoS ONE* **2014**, *9*, e103799. [CrossRef] [PubMed]
3. Miguez, G.M.; Magalhaes, P.C.L. Methods and techniques in urban engineering. In *Urban Flood Control, Simulation, and Management-an Integrated Approach*; De Pina, A.C., Ed.; 2010; pp. 131–160. Available online: http://www.scribd.com/doc/58441384/Methods-and-Techniques-in-Urban-Engineering (accessed on 20 October 2016).
4. Miller, J.D.; Kim, H.; Kjeldsen, T.R.; Packman, J.; Grebby, S.; Dearden, R. Assessing the impact of urbanization on storm runoff in a peri-urban catchment using historical change in impervious cover. *J. Hydrol.* **2014**, *515*, 59–70. [CrossRef]
5. Zhou, Q. A review of sustainable urban drainage systems considering the climate change and urbanization impacts. *Water* **2014**, *6*, 976–992. [CrossRef]
6. Saraswat, C.; Kumar, P.; Mishra, B.K. Assessment of stormwater runoff management practices and governance under climate change and urbanization: An analysis of Bangkok, Hanoi and Tokyo. *Environ. Sci. Policy* **2016**, *64*, 101–117. [CrossRef]
7. Windapo, A. Examination of Green Building Drivers in the South African Construction Industry: Economics versus Ecology. *Sustainability* **2014**, *6*, 6088–6106. [CrossRef]
8. Huang, H.; Cheng, S.; Wen, J.; Lee, J. Effect of growing watershed imperviousness on hydrograph parameters and peak discharge. *Hydrol. Process.* **2008**, *22*, 2075–2085. [CrossRef]
9. Gebre, S.L. Application of the HEC-HMS Model for Runoff Simulation of Upper Blue Nile River Basin. *J. Waste Water Treat. Anal.* **2015**, *6*, 1. [CrossRef]
10. Mohit, M.A.; Sellu, G.M. Mitigation of Climate Change Effects through Non-structural Flood Disaster Management in Pekan Town, Malaysia. *Procedia Soc. Behav. Sci.* **2013**, *85*, 564–573. [CrossRef]
11. Brown, R.R.; Farrelly, M.A. Delivering sustainable urban water management: A review of the hurdles we face. *Water Sci. Technol.* **2009**, *59*, 839–846. [CrossRef]
12. Barałkiewicz, D.; Chudzińska, M.; Szpakowska, B.; Świerk, D.; Gołdyn, R.; Dondajewska, R. Stormwater contamination and its effect on the quality of urban surface waters. *Environ. Monit. Assess.* **2014**, *186*, 6789–6803. [CrossRef]
13. Tang, S.; Luo, W.; Jia, Z.; Wu, Y. Evaluating Retention Capacity of Infiltration Rain Gardens and Their Potential Effect on Urban Stormwater Management in the Sub-Humid Loess Region of China. *Water Resour. Manag.* **2016**, *30*, 983–1000. [CrossRef]
14. Flower, K.; Knoben, W.; Peel, M.; Peterson, T.; Ryu, D.; Saft, M.; Seo, K.; Western, A. Many commonly used rainfall-runoff models lack long, slow dynamics: Implications for runoff projections. *Water Resour. Res.* **2020**, *56*, e2019WR025286.
15. Gogate, N.G.; Kalbar, P.P.; Raval, P.M. Assessment of stormwater management options in urban contexts using Multiple Attribute Decision-Making. *J. Clean. Prod.* **2017**, *142*, 2046–2059. [CrossRef]
16. Zahmatkesh, Z.; Burian, S.J.; Mohammad Karamouz, M.; Tavakol-Davani, H. Low-Impact Development Practices to Mitigate Climate Change Effects on Urban Stormwater Runoff: Case Study of New York City. *J. Irrig. Drain. Eng. ASCE* **2015**, *141*, 04014043. [CrossRef]
17. Hunt, W.; Jarrett, A.; Smith, J.; Sharkey, L. Evaluating bioretention hydrology and nutrient removal at three field sites in North Carolina. *J. Irrig. Drain. Eng.* **2006**, *132*, 600–608. [CrossRef]
18. Hatt, B.; Fletcher, T.; Deletic, A. Hydrologic and pollutant removal performance of stormwater biofiltration systems at the field scale. *J. Hydrol.* **2009**, *365*, 310–321. [CrossRef]
19. Zhou, T.; Song, F.; Lin, R.; Chen, X. The 2012 North China floods: Explaining an extreme rainfall event in the context of a longer-term drying tendency. *Bull. Am. Meteorol. Soc.* **2013**, *94*, S49–S51.
20. United Nations, Department of Economic and Social Affairs, Population Division (UN DESA). *World Urbanization Prospects: The 2014 Revision, 2015, (ST/ESA/SER.A/366)*; United Nations Publication: New York, NY, USA, 2015; p. 517.
21. Andimuthu, R.; Kandasamy, P.; Mudgal, B.V.; Jeganathan, A.; Balu, A.; Sankar, G. Performance of urban storm drainage network under changing climate scenarios: Flood mitigation in Indian coastal city. *Sci. Rep.* **2019**, *9*, 7783. [CrossRef]
22. Sharma, P.; Shukla., K. Integrated Drainage Management Plan for Urban Flooding: A Case Study of Lucknow City, Uttar Pradesh, India. In Proceedings of the 3rd World Conference on Applied Sciences, Engineering & Technology, Kathmandu, Nepal, 27–29 September 2014.
23. Mishra, B.K.; Rafiei Emam, A.; Masago, Y.; Kumar, P.; Regmi, R.K.; Fukushi, K. Assessment of future flood inundations under climate and land use change scenario in Ciliwung river basin, Jakarta. *J. Flood Risk Manag.* **2018**, *11*, S1105–S1115. [CrossRef]
24. Rafiq, F.; Ahmed, S.; Ahmed, S.; Ahmed, A.A. Urban Floods in India. *Int. J. Sci. Eng. Res.* **2016**, *7*, 721–734.
25. Lucknow Development Authority. Hydrological Study for Gomti River Front Development. 2013. Available online: https://ddmtn57ju2md7.cloudfront.net/media/bbifiles/1181924/38eeafb5-ca82-4c74-8c64-65170953eb36.pdf (accessed on 12 January 2016).
26. New Jersey Stormwater Best Management Practices Manual, 2004 Low Impact Development Techniques. Available online: http://www.njstormwater.org/bmp_manual/NJ_SWBMP_2print.pdf (accessed on 14 February 2017).
27. Barnes, M.L.; Welty, C.; Miller, A.J. Impacts of Development pattern on urban groundwater flow regime. *Water Resour. Res.* **2018**, *54*, 5198–5212. [CrossRef]

hydrology

Article

Estimation of Peak Discharges under Different Rainfall Depth–Duration–Frequency Formulations

Andrea Gioia [1], Beatrice Lioi [1], Vincenzo Totaro [1,*], Matteo Gianluca Molfetta [1], Ciro Apollonio [2], Tiziana Bisantino [3] and Vito Iacobellis [1]

1   Dipartimento di Ingegneria Civile, Ambientale, del Territorio, Edile e di Chimica, Politecnico di Bari, 70125 Bari, Italy; andrea.gioia@poliba.it (A.G.); beatrice.lioi@poliba.it (B.L.); matteogianluca.molfetta@poliba.it (M.G.M.); vito.iacobellis@poliba.it (V.I.)
2   DAFNE Department, Tuscia University, 01100 Viterbo, Italy; ciro.apollonio@unitus.it
3   Puglia Region Civil Protection Section, 70026 Modugno, BA, Italy; t.bisantino@regione.puglia.it
*   Correspondence: vincenzo.totaro@poliba.it

Abstract: One of the main signatures of short duration storms is given by Depth–Duration–Frequency (DDF) curves. In order to provide reliable estimates for small river basins or urban catchments, generally characterized by short concentration times, in this study the performances of different DDF curves proposed in literature are described and compared, in order to provide insights on the selection of the best approach in design practice, with particular reference to short durations. With this aim, 28 monitoring stations with time series of annual maximum rainfall depth characterized by sample size greater than 20 were selected in the Northern part of the Puglia region (South-Eastern Italy). In order to test the effect of the investigated DDF curves in reproducing the design peak discharge corresponding to an observed expected rainfall event, the Soil Conservation (SCS) curve number (CN) approach is exploited, generating peak discharges according to different selected combinations of the main parameters that control the critical rainfall duration. Results confirm the good reliability of the DDF curves with three parameters to adapt on short events both in terms of rainfall depth and in terms of peak discharge and, in particular, for durations up to 30 min, the three-parameter DDF curves always perform better than the two-parameter DDF.

Keywords: rainfall time series; DDF curves; SCS-CN method; peak discharge

Citation: Gioia, A.; Lioi, B.; Totaro, V.; Molfetta, M.G.; Apollonio, C.; Bisantino, T.; Iacobellis, V. Estimation of Peak Discharges under Different Rainfall Depth–Duration–Frequency Formulations. *Hydrology* **2021**, *8*, 150. https://doi.org/10.3390/hydrology8040150

Academic Editors: Davide Luciano De Luca and Andrea Petroselli

Received: 10 August 2021
Accepted: 28 September 2021
Published: 8 October 2021

Publisher's Note: MDPI stays neutral with regard to jurisdictional claims in published maps and institutional affiliations.

1. Introduction

The investigations about the increasing of extreme events at a global level carried out in last decades generated a debate on the need of revisiting the risk management approach, in particular with regard to rainstorms and floods. Concerns and possible consequences deriving from changes in an extreme rainfall regime led scientists and practitioners to investigate the influence of these events on the current design practice. One of the most used tools in water management are the Depth–Duration–Frequency (DDF) and Intensity–Duration–Frequency (IDF) curves, which have the valuable quality of being analytical relationships able to provide a design rainfall depth (or intensity) for an assigned duration and return period [1,2]. The underlying theoretical framework of these curves has been widely discussed and assessed in hydrological literature (e.g., [2–6]). However, concerns about past changes and possible evolutions of climate on the phenomenology of rainfall raised questioning about the opportunity of retaining the still valid adoption of stationary hypothesis during the IDF/DDF deriving procedure [7–10]. Implications arising from the adoption of nonstationary probability distributions for modelling changes in extremes and applications of related statistical tests for trend detection were discussed in several studies (e.g., [11–18]).

Starting from the middle of the 20th century, the first relationships between depth (or intensity) and duration of rainfall were studied above all in developed countries such as the USA, the UK and Ireland [19]. The results of these analyses can be applied in water resources engineering for the mitigation of hydraulic risk and are particularly needed for urban areas with poorly permeable surfaces and small catchment and, consequently, very short run-off times. Typically, two- or three-parameter DDF/IDF formulations are considered in design practice [5], in a balance between an enhanced model structure and parameter uncertainty [2]. Obtaining analytical results that are compatible for reliable design purposes, requires the availability of recorded time series for durations critical of the same order of magnitude of physical processes having place in the environment. As in the case of urban basins and small catchments, often mass transfer phenomena require durations less than 1 h, making essential the availability of rainfall series which span this time domain [3]. These data were very difficult to retrieve in the past, but refinements in recording technology together with new technologies (such as satellite and radar observations) are providing scientists and practitioners with new updated and consistent datasets of data. However, it should be noted that databases of rainfall time series in short durations (5, 15 and 30 min) are generally less extensive than hourly data. In this sense, short-duration rainfall data constitute a precious source of information for investigating if remodulation in the extreme rainfall regime is happening, and their use in the context of a test for trend detection (and the consequent adoption of a nonstationary probabilistic model) may deserve particular attention, because the low magnitude of the trend and the reduced sample size of recorded data may lead to low power statistical tests [14,15].

Disregarding the stationary/nonstationary problem (whose discussion is outside the aim of this paper), as remarked before, the increased availability of short-duration series constitutes an important support for building DDF/IDF curves that are more consistent with observed data and for improving design operations, often based on the classical two-parameter 'Montana curve'. With respect to Italian case studies, we mention the work of Di Baldassarre et al. [2] (Emilia Romagna and Marche regions) and of Rossi and Villani [20] (Campania region), where DDF curves with more than two parameters were applied.

The study of short storm durations that extend from a few minutes to an hour is interesting for several design purposes. They may trigger flash floods characterized by short durations and occurring on small catchments such as a part of cities such as residential areas or infrastructures. Design practice often leads to the need to define drainage and disposal systems for rainwater for small drainage surfaces intended for urban, commercial or industrial use. Therefore, the small size of the drainage area in addition to the low permeability of the soil, may produce hydrological responses characterized by short lag times, with a preponderant runoff with respect to the infiltration component. Furthermore, often rainfall and discharge observations are not available, addressing the practitioner to apply empirical approaches [21].

With the aim of providing a quantitative discussion on the influence of different DDF formulations on the evaluation of peak discharges, an extended analysis is conducted exploiting the widely used SCS-CN method (e.g., [22,23]), in order to test the ability of different DDF/IDF curves in reproducing the peak discharge corresponding to an observed rainfall event. In the SCS-CN method, the critical rainfall duration depends on three parameters: the catchment area, basin slope and CN. The application to urban basins is carried out through the choice of a parameters range, i.e., high values of the CN from 50 to 100, small areas from 0.1 to 10 km^2 and an average slope of the basin from 3% to 18%. For each combination of these parameter values, the corresponding concentration time is calculated using the SCS-CN method. After obtaining the concentration time values, the peak discharge is calculated with the observed rainfall depth compared with those obtained from different DDF formulations.

Furthermore, peak flow curves based on a sensitivity analysis were created to compare results obtained from the two-parameter and three-parameter DDF curves. The study also provides an evaluation of the relative error of DDF on peak discharge, highlighting the benefits of the use of three-parameter DDF curves in practical applications.

The paper is structured as follows: In Section 2 the case study of Northern Puglia (South-Eastern Italy) and dataset consistence are illustrated, while in Section 3, DDF and IDF curves used for the analysis are described, including steps for the evaluation of peak discharge obtained from the corresponding DDF curves and estimated by means of the SCS-CN method. In Section 4, results of illustrated applications are critically discussed and in Section 5 conclusions are reported.

2. Study Area and Dataset

Puglia is the region of peninsular Italy with the greatest coastal development and its territory is covered by 1.5% of mountains, 45.2% by hills and 53.3% by plains. The study area is located in the northern part of the Puglia region (Southern Italy) and, moving from South-West to North-East, is constituted by the Daunian Apennine, the Tavoliere plain and the Gargano promontory (Figure 1). The highest peak of the Daunian Mountains is Monte Cornacchia (1151 m a.s.l.) and is located on the border with the Campania region, followed by the 1055 m high Monte Calvo in the Gargano promontory.

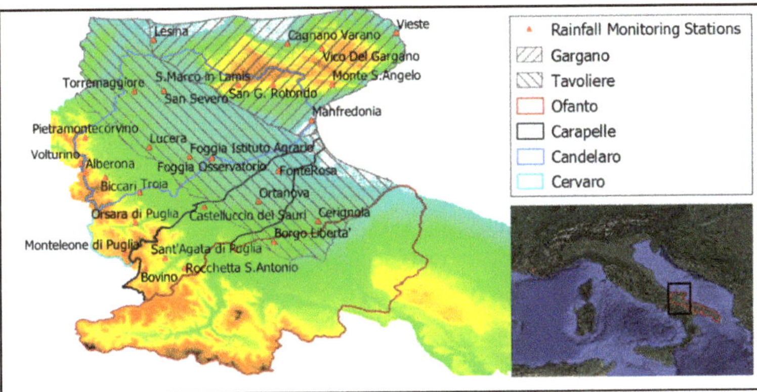

Figure 1. Study area.

The most important rivers inside and surrounding the study area are Fortore, Candelaro, Carapelle, Cervaro and Ofanto, whose main streams cross or delimit the Tavoliere plain. Due to the steep change in slope, from mountains to the plain, the regime of rivers in the area is torrential, and urban areas are prone to flash floods that may be triggered by frontal events, convective storms or Mediterranean cyclones [24,25]. Moreover, low-lying coastal areas are often affected by inundation phenomena due to the combined effect of heavy rainfalls and severe storm surges [26].

The analyses were carried out exploiting rainfall data recorded by a gauges network managed by the Civil Protection of Puglia Region. For this study, annual maximum series with a duration of 5, 15 and 30 min and 1, 3, 6, 12 and 24 h were employed, with a recording period comprised between 1921 and 2019. Data are freely available and can be downloaded by accessing the website of the Civil Protection of Puglia Region [27]. In order to provide more reliable evaluations, only stations with time series greater than 20 data for all durations were selected. In Figure 1, positions of selected stations in the study area are illustrated, while in Table 1 their main characteristics are reported.

Table 1. Rainfall monitoring stations.

Rainfall Monitoring Stations	Altitude (m a.s.l.)	River Basin/Area	Installation Year
Alberona	663	Candelaro	1917
Biccari	484	Candelaro	1922
Borgo Liberta'	235	Ofanto	1924
Bovino	623	Carapelle	1917
Cagnano Varano	165	Gargano	1921
Castelluccio dei Sauri	190	Carapelle	1922
Cerignola	118	Tavoliere	1921
Foggia Osservatorio	99	Candelaro	1873
Foggia Istituto Agrario	85	Candelaro	1949
Fonte Rosa	25	Tavoliere	1925
Lesina	6	Gargano	1928
Lucera	246	Candelaro	1917
Manfredonia	68	Tavoliere	1900
Monte Sant'Angelo	799	Gargano	1920
Monteleone di Puglia	828	Cervaro	1920
Orsara di Puglia	689	Cervaro	1921
Ortanova	53	Carapelle-Tavoliere	1921
Pietramontecorvino	441	Candelaro	1928
Rocchetta Sant'Antonio	724	Carapelle	1922
Sant'Agata di Puglia	703	Carapelle	1917
San Giovanni Rotondo	619	Candelaro-Gargano	1923
San Marco in Lamis	566	Candelaro-Gargano	1917
San Severo	114	Candelaro-Tavoliere	1928
Torremaggiore	195	Candelaro-Tavoliere	1917
Troia	469	Candelaro	1907
Vico Del Gargano	459	Gargano	1921
Vieste	96	Gargano	1921
Volturino	615	Candelaro	1964

3. Methodology

3.1. Formulations of DDF/IDF Curves

As recalled in the Introduction, several formulations have been used for characterizing DDF/IDF relationships. Following Di Baldassarre et al. [2], we limited our analysis to the DDF with two and three parameters, considering the consistency of the available dataset inadequate for other formulations.

Among two-parameter DDF curves, we focused our attention on the following:

$$h = ad^b \tag{1}$$

Known also as 'Montana curve', Equation (1) expresses the rainfall depth h (mm) by means of a power law of storm duration d (h), using parameters a and b. Despite its simplicity, it can be considered the standard relationship implemented when approaching the problem of DDF estimation. However, several criticisms have been moved to this formulation. Turning to the intensity i, defined as the ratio between the rainfall depth h and its duration, this value goes to infinity when the duration d of the event approaches to zero [2]. In the context of the two-parameter formulation, starting from Equation (1), two specific case studies were identified for the purposes of this work: in the first, we indicated with D2P-1 Equation (1) when parameters a and b were estimated considering both the hourly and sub-hourly data, while in the second case, we indicated with D2P-2 Equation (1) when parameters a and b were estimated considering only the hourly data. This choice was motivated by the need of providing a comparison with the practice of using hourly data, widely diffused in the real world due to the limited amount of short-duration rainfall data diffused in the past decades.

In the realm of three-parameter formulations, the following DDF equations were analyzed starting from the available datasets, respectively, called D3P-1, D3P-2 and D3P-3:

$$h = \frac{ad}{(d+c)^b} \tag{2}$$

$$h = \frac{ad}{(d^b + c)} \tag{3}$$

$$h = \left(c + \frac{a}{d+b}\right)d \tag{4}$$

with a, b and c parameters to be estimated. All parameters of DDFs were estimated by using the least squares technique. Equations (2)–(4) are the same formulations of DDF with three parameters investigated in Di Baldassarre et al. [2]; our choice was also motivated by the opportunity of having a similar investigation of a geographically similar area, which ranges from the Apennine chain to the Adriatic Sea.

3.2. Error Measures

Stands that $h = h(d)$, a quantitative evaluation of the goodness of estimate of rainfall depth h_j obtained using each of the selected DDF curves in comparison with the observed average rainfall depth $\overline{h}_j$ was carried out for all the $j = 1, 2, \ldots, N$ monitoring stations and for all $i = 1, 2, \ldots, M$ durations of recorded event. Analysis is performed by applying: (i) the relative error $\varepsilon_j(d_i)$ for each monitoring rainfall station and duration; (ii) the mean absolute percentage error $E\%$ calculated for each duration and averaged over all the monitoring rainfall stations and (iii) the Root Mean Square Error ($RMSE$), integrated over all the durations and monitoring station. In particular, the observed rainfall depth $\overline{h}_j$ was evaluated considering the average value on the time series for each duration and for each monitoring station. The evaluation of these error indicators represents a reliability test as also analyzed in Di Baldassarre et al. [2], but, in this case, was considered independent from the return period. The relationships was hereafter reported:

$$\varepsilon_j(d_i) = \frac{h_j(d_i) - \overline{h}_j(d_i)}{\overline{h}_j(d_i)} \tag{5}$$

$$E\,(d_i)\% = \frac{100}{N} \cdot \sum_{j=1}^{N} \frac{\left|h_j(d_i) - \overline{h}_j(d_i)\right|}{\overline{h}(d_i)} \tag{6}$$

$$RMSE = \frac{1}{M} \cdot \sum_{i=1}^{M} \sqrt{\sum_{j=1}^{N} \frac{\left[h_j(d_i) - \overline{h}_j(d_i)\right]^2}{N}} \tag{7}$$

3.3. SCS-CN Method

In order to test the effect of the investigated DDF/IDF curves in reproducing the design peak discharge corresponding to an observed rainfall event, the Soil Conservation (SCS) curve number (CN) method was applied. The SCS-CN is a conceptual method, widely used in many hydrologic applications for predicting runoff from watersheds. The first publication of SCS-CN method dates back to 1956 and was proposed by the U.S. Department of Agriculture in the National Engineering Handbook of Soil Conservation Service [23]. This method was originally elaborated to predict runoff volumes for a given rainfall event and for the evaluation of storm runoff in small agricultural watersheds [28]. Rallison [22] describes the origin and development of SCS method, based on infiltrometer tests and measures of rainfall and runoff. The SCS-CN method was developed well beyond its original scope and was adopted for different land uses and climate conditions [29–36]. This approach has been taken as procedure by many users (professionals or public administrations) in numerous hydrological applications for design flood estimation and/or for runoff evaluation for a particular storm event (e.g., [37]). More details about the theoretical background are given in Pilgrim et al. [38].

The peak discharge was computed using a triangular hydrograph method developed by Mockus [39] and reported in Pilgrim et al. [38], with rainfall characterized by duration equal to D. Considering the definition of lag time, t_l (distance between the centroid of excess precipitation and the peak of the hydrograph) provided by the Natural Resource Conservation Service (NRCS, [40]), the peak discharge, q_p, was reached in a duration T_P (i.e., time of rise).

Application of SCS-CN Method

In order to test the performances of the three-parameter DDF curves, in this work, the SCS-CN method was used for calculating the runoff volume and the triangular hydrograph method for evaluating the expected value of peak discharge, using both the three three-parameter and the two classic two-parameter expected DDF curves described in Section 3.1. In this way, according to Equation (8), the evaluation of peak discharge depended on the rainfall volume, V, drainage basin area, A, and lag-time, t_l; according to Equation (10), lag time t_l depends on the mainstream length, L, the river basin's slope, s, and the Curve Number, CN.

The peak discharge q_p (m^3/s) is given by:

$$q_p = \frac{0.208\,AV}{0.5D + t_l} \tag{8}$$

A (km^2) is the area of the drainage basin and V (mm) is the depth of runoff.

Assuming D equal to a concentration time and considering the empirical relationship between the concentration time and lag-time proposed by Ward and Elliot [41]:

$$D = t_c = t_l/0.6 \tag{9}$$

$$t_l = 0.342 \frac{L^{0.8}}{s^{0.5}} \left(\frac{1000}{CN} - 9\right)^{0.7} \tag{10}$$

L is connected to basin area through an empirical equation introduced by Hack [42] and hereafter reported:

$$L = 1.4\,A^{0.6} \tag{11}$$

This relation was also discussed in recent scientific literature such as Maritan [43] and Rinaldo [44].

Different combinations of parameter values A, s and CN were selected (considering usually observed ranges in small river basin) leading to the definition of different lag time values, according to Equation (10) and, consequently, rainfall durations (D), assumed equal to the concentration time t_c. Thus, for each selected duration, in according to Equation (9), the expected rainfall depth, evaluated through the D2P-1, D2P-2, D3P-1, D3P-2 and D3P-3 curves, was used for calculating the runoff and, consequently, the expected peak discharge, q_p, according to Equation (8).

The different combinations of these parameters simulated the conditions of urban basin characterized by small area and high impermeability. For this reason, a basin area with range from 0.1 to 10 km^2, CN values from 50 to 100 and slope from 3% to 18% were chosen.

4. Results and Discussion

For each of the 28 investigated rainfall monitoring stations, all DDF/IDF formulations described in Section 3.1 were fitted. Results are shown in graphical form in Figure 2. only for six rainfall monitoring stations selected, each falling within a basin or zone of the study area. The curves D2P-1, D2P-2, D3P-1, D3P-2 and D3P-3 were plotted on the left side of each subplot, in addition to the observed (mean) values of rainfall depth for all durations; in the right part of the subplots, the relative IDF curves (I2P-1, I2P-2, I3P-1, I3P-2 and I3P-3) are shown on the basis of the estimated parameters of the corresponding DDF curves.

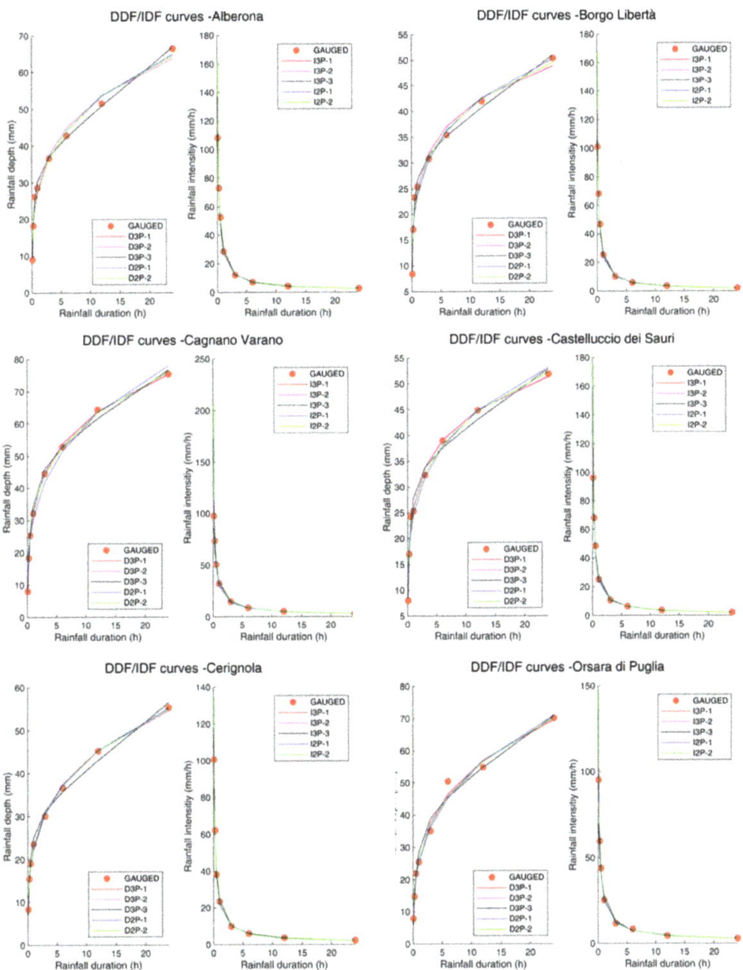

Figure 2. Fitted DDF and IDF curves.

Focusing on the short durations from 5 min to 1 h, the intensity evaluated using three-parameter DDF curves was lower and more reasonable than that obtained with two-parameter DDF curves, as shown in Figure 3. Additionally, in this case, shown only are the results of the same rainfall monitoring stations selected for the Figure 2.

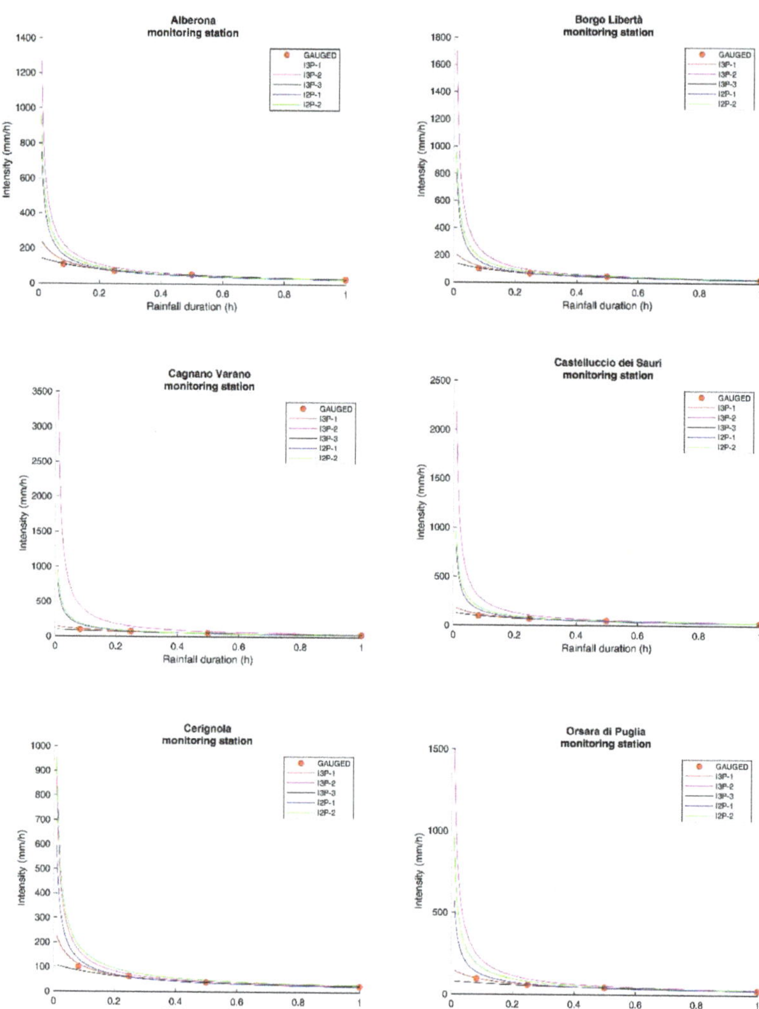

Figure 3. IDF curves for short durations.

4.1. Results of DDFs Error Measures

In Figure 4, the box plots of the relative errors (with reference to all the investigated monitoring stations) for the considered DDF equations are reported. Focusing on the boxplot of D2P-1 and D2P-2, it was interesting to observe the poorer performance of the two-parameter curves for short durations; the worst, between them, was the D2P-2 equation, which presented a greater variability of the relative error, for short durations (5, 15 and 30 min). For durations equal to or greater than one hour, the D2P-2 showed less uncertainty than D2P-1. The comparison between D2P-1 and D2-P-2 was further explored through Figure 5. The three-parameter DDF curves seemed to have a far better performance than the two-parameter DDF curves especially for durations of 5 and 15 min.

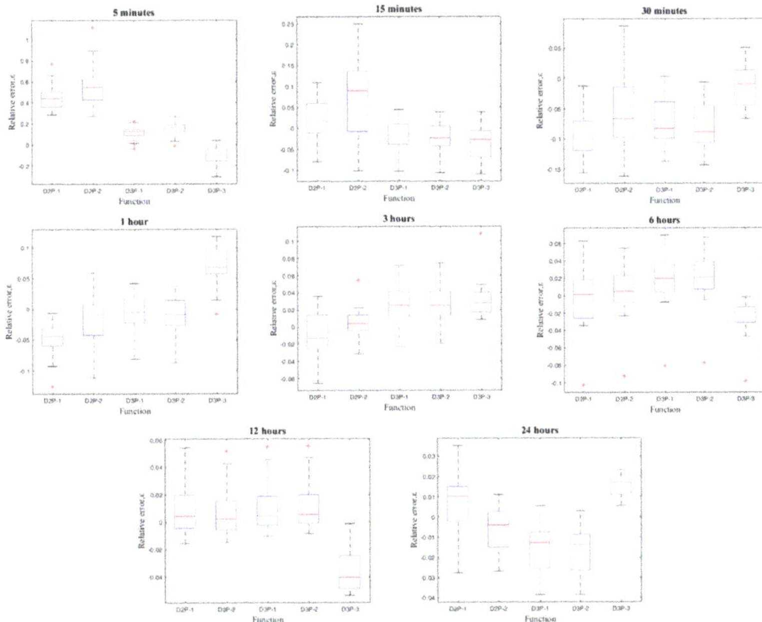

Figure 4. Boxplots of the relative errors evaluated for sub hourly and hourly rainfall durations, with reference to all considered DDFs.

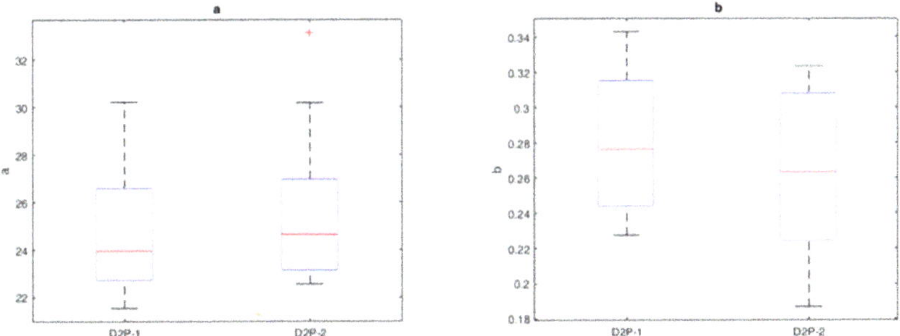

Figure 5. Boxplot with comparison of D2P-1 and D2P-2 estimated parameters.

Figure 5 shows the boxplot for the D2P-1 and D2P-2 parameters (a and b). The subplot on the left compares the parameter aestimated considering in the first case sub-hourly and hourly data and in the second case only hourly data; on the right the same comparison is shown for the parameterb. It is interesting to note that the estimated parameter, a, which represented the expected rainfall depth of duration equal to one hour, presented a higher uncertainty when estimated from a greater range of durations (i.e., including short durations). The parameter *b* had, in both cases, an extremely low variability.

The Mean Absolute Percentage Error (E%) and the Root Mean Square error (RMSE) averaged over all the durations and the monitoring stations are, respectively, reported in Tables 2 and 3.

Table 2. Mean Absolute Percentage Error.

	D2P-1	D2P-2	D3P-1	D3P-2	D3P-3
E $(d_5)\%$	44.58	55.26	11.82	15.36	11.63
E $(d_{15})\%$	4.40	9.54	3.11	3.21	3.99
E $(d_{30})\%$	9.26	7.47	7.34	8.00	2.82
E $(d_{60})\%$	4.76	3.03	2.26	2.43	7.17
E $(d_{180})\%$	2.40	1.19	2.73	2.80	3.12
E $(d_{360})\%$	2.15	1.87	2.57	2.67	2.38
E $(d_{720})\%$	1.54	1.45	1.48	1.48	3.48
E $(d_{1440})\%$	1.41	0.96	1.55	1.64	1.54

Table 3. Root Mean Square Error.

	D2P-1	D2P-2	D3P-1	D3P-2	D3P-3
RMSE	1.64	1.68	1.14	1.21	1.33

It is possible to note from Table 2, that the worst performance for durations of 5 and 15 min in terms of the mean absolute percentage error were given by D2P-2, followed by D2P-1, both characterized by two parameters. On the other hand, for rainfall events with durations greater than or equal to one hour, the performance of both D2Ps improved, while the D2P-2 remained the best performing DDF, among all. The better performance of DDF with three parameters which is visible in Table 3 in terms of RMSE, was mainly attributed to their better fit of short rainfall events.

4.2. Results of Applications of SCS-CN Method

As previously explained, the application of the SCS-CN method was conducted for understanding the effect of DDF curves with three parameters on the expected peak discharge for short-duration rainfall events. All the monitoring stations were examined, but for the sake of brevity, only one example case was reported. In particular the results for the Alberona monitoring station are shown and the graphs were obtained as explained in Section 3.3. Figures 6–10 report the peak discharge values, obtained using rainfall input from the DDF equations and different combinations of quantities A, s and CN on which the rainfall duration was set. As remarked before, the rainfall duration D was assumed equal to the concentration time t_c. Each figure, corresponding to a selected DDF curve was composed of six sub-plots which referred to different selected values of slope s. Each curve in the graph was obtained for a fixed drainage area, A, varying the CN value; in particular, according to Equations (9) and (10) the concentration time decreased with an increasing CN. The different curves reported in each subplot, were generated changing the catchment area; therefore, a curve beam was obtained where each color corresponded to a value visible in the legend of the plot. Red points, called QoR "Q-Peak evaluated with observed Rainfall", represented the peak discharge values obtained, for each observed duration, by applying the same SCS-CN method for each combination of the three quantities (CN, A and s) and considering the gauged input rainfall volume. It was also necessary to observe that, for a chosen value of basin area and basin slope, a minimum concentration time was obtained corresponding to CN = 100.

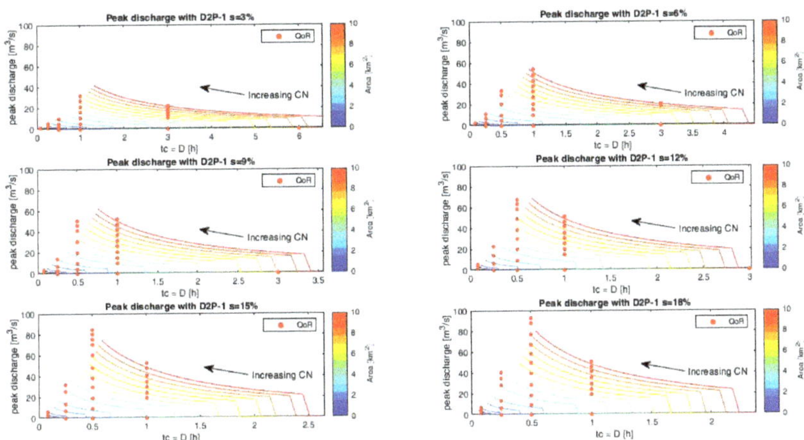

Figure 6. Peak curves obtained with D2P-1 vs. Q-Peak (red dots) evaluated with observed Rainfall.

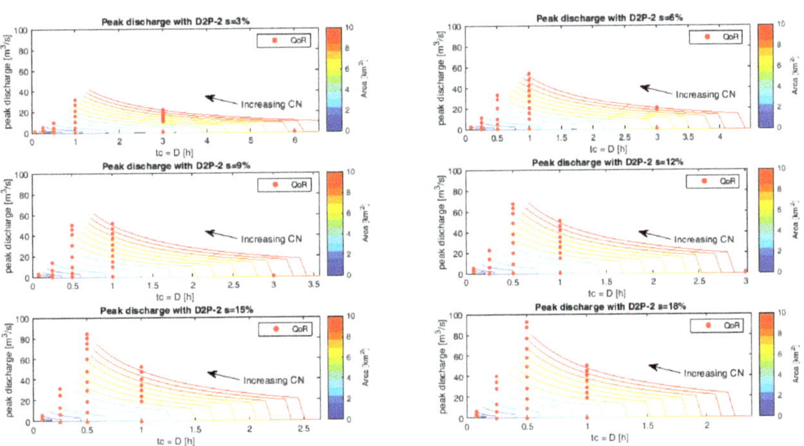

Figure 7. Peak curves obtained with D2P-2 vs. Q-Peak (red dots) evaluated with observed Rainfall.

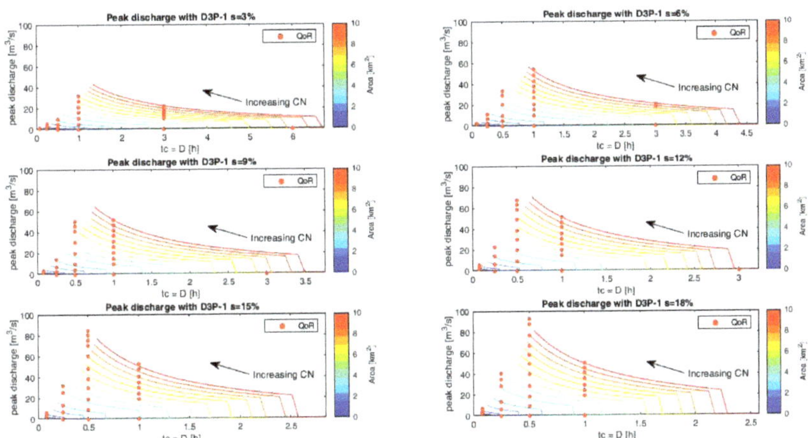

Figure 8. Peak curves obtained with D3P-1 vs. Q-Peak (red dots) evaluated with observed Rainfall.

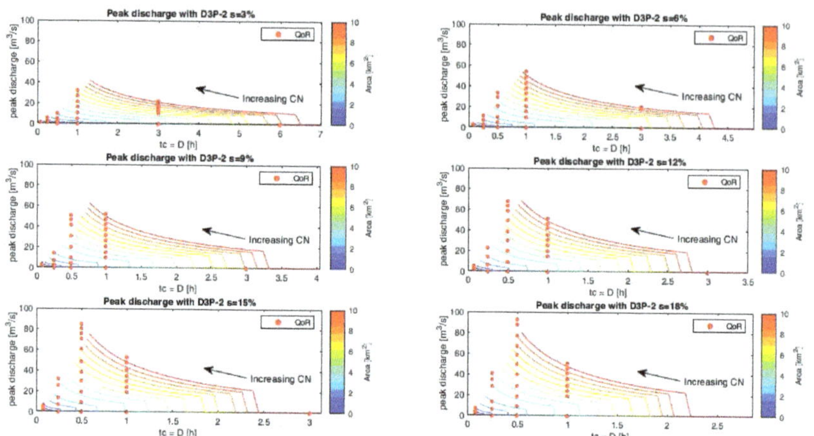

Figure 9. Peak curves obtained with D3P-2 vs. Q-Peak (red dots) evaluated with observed Rainfall.

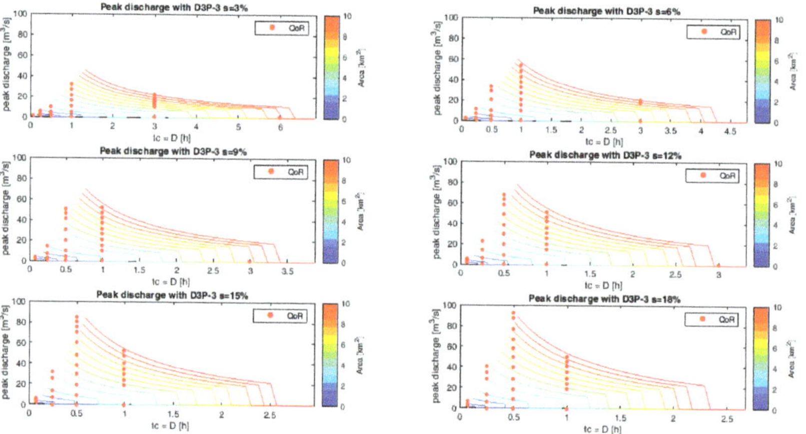

Figure 10. Peak curves obtained with D3P-3 vs. Q-Peak (red dots) evaluated with observed Rainfall.

Figures 6–10 illustrate the results on peak discharge, for the case study of Alberona monitoring station, applying the SCS-CN method with different DDF curves highlighting the capability to adapt to the observed peak discharge, in particular for durations lower than one hour. The Figures highlight that, for short durations, a comparison was possible only for the basins with a high value of CN, and performances may depend significantly from the basin size and average slope. In general, these results confirmed the good reliability of the DDF curves with three parameters to adapt on short events, both in terms of rainfall depth and in terms of peak discharge.

To evaluate, in terms of error, the fit of the known curves to the QoR, the mean absolute percentage error was calculated on all the stations for the different slopes examined, focusing on the short durations (5, 15, 30 and 60 min). These results are shown in graphical form in Figure 11. The best fit corresponded to the least error and showed that, for durations up to 30 min, the three-parameter DDF curves always performed better than the two-parameter DDF curves. Among the three-parameter DDFs, the worst performance was given by D3P-2 with E% between 20% and 30%, while the best one was provided by the D3P-3 with errors always below 10%. Figure 11 also shows that there was a significant dependence on the slope of the basin in the calculation of the peak flow related to the lag time and, hence, on the propagation of the error from the rainfall input to the peak flow output. For durations equal to or greater than an hour, the best DDF curve may have changed with duration, looking both to the rainfall depth (as in Figure 4) and to the peak discharge. Nevertheless, in design practice, for concentration times equal to or greater than one hour, the traditional two-parameter curve 2DP-2 remained a robust choice.

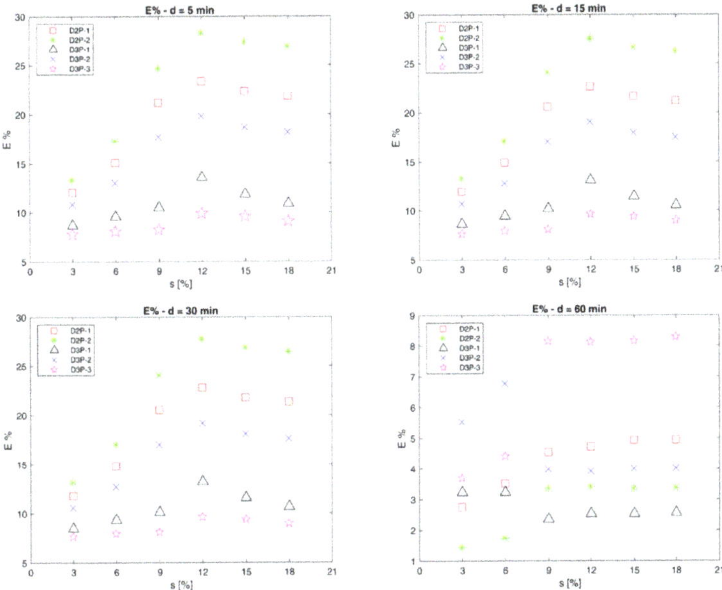

Figure 11. E% on Peak discharge for short durations.

Figure 12 shows the boxplot on relative errors on peak discharge calculated for different DDFs and all durations in the range 5 min to 24 h, related to the Alberona monitoring station. All of them presented a positive bias with respect to values obtained with the observed expected rainfall depth. It can be noted that for the gradually increasing slopes, the D3P-3 presented a reduced variability of the relative error, in the range 0–0.1, with respect to the other mentioned DDF curves. Table 4 also shows the mean absolute percentage error for each slope considered, confirming what was stated for Figure 12, but, quantitatively, we found the close dependence between the mean absolute percentage error and the basin slope which generally tended to increase as the slope increased.

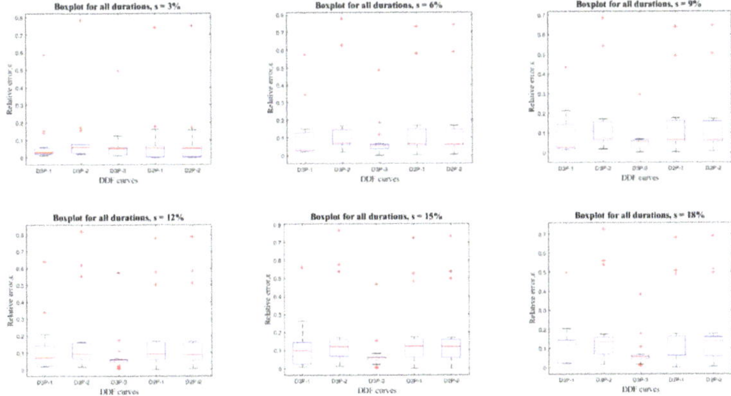

Figure 12. Boxplot on peak discharge.

Table 4. Mean Absolute Percentage Error on peak discharge.

	D2P-1	D2P-2	D3P-1	D3P-2	D3P-3
$E_{(s=3\%)}\%$	8.90	8.93	7.53	10.24	6.75
$E_{(s=6\%)}\%$	12.91	12.94	9.36	13.95	7.48
$E_{(s=9\%)}\%$	13.31	13.31	8.90	14.11	5.99
$E_{(s=12\%)}\%$	16.46	16.49	11.45	17.37	7.89
$E_{(s=15\%)}\%$	15.94	15.94	11.07	16.76	7.07
$E_{(s=18\%)}\%$	15.47	15.52	10.71	16.35	6.78

5. Conclusions

In small river basins (or urban catchments), characterized by short concentration times, the traditional approach based on the use of two-parameter Depth–Duration–Frequency curves for the evaluation of the rainfall design fails for durations close to zero, as the rainfall intensity diverges and tends to have unreliably high values. In this context, recent literature [2,5] proposes the use of three-parameter DDFs, being more reliable in the design practice for durations closing to zero.

In this study, the aforementioned approaches for the rainfall design evaluation were described and compared, in order to provide users with general information useful for the selection of the best approach in design practice. With this aim, a number of rainfall monitoring stations, with time series characterized by sample size greater than 20, were selected in the study area located in the Northern part of the Puglia region (South-Eastern Italy).

Among the analyzed three-parameter DDF curves, the one that gave the best performance by evaluating the cumulative relative errors over short durations for all monitoring station was D3P-3, as seen in the previous Section. By the evaluation of the peak discharge using the SCS-CN method, it was possible to obtain a series of peak discharges according to the different selected combinations of the three parameters CN, A and s, using the investigated DDF curves (D2P-1, D2P-2, D3P-1, D3P-2 and D3P-3). The range of values adopted for each of these three parameters was obtained considering wide ranges covering those usually observed in urban catchments, characterized by small dimensions. The best results evaluated in terms of relative errors were obtained with D3P-3 curves for durations below one hour. On the other hand, the use of an enlarged dataset and of equations with three parameters, in most cases, did not produce a reduction in the output uncertainty, when the critical rainfall duration was equal to or greater than one hour. Then, in such a case, the classical two-parameter power law estimated on rainfall data in the range 1–24 h remained a reliable choice. This investigation may be useful for practitioners and designers for selecting the best approach for the definition of the rainfall design depending on the duration of the critical precipitation for the investigated river basin.

Several insights can arise from this study. Among these, there is the opportunity of revisiting DDF/IDF relations in the light of changes in the rainfall extremes phenomenology, with the consequence of the possibility of implementing a nonstationary parametric procedure. A wide number of procedures is currently available for this type of modelling. At the same time, a growing number of papers criticized the traditional tools for assessing nonstationarity, both with parametric and non-parametric tests [14–17], in particular with respect to their statistical power. This is not a trivial issue because of, as recognized by Vogel et al. [18], when moving in the field of infrastructure decision committing a type II error means incurring in under-preparedness.

Possible developments in this field of research can also be carried out by using, in addition to the monitoring stations that provide ground data, radar systems which allow to obtain data on a large spatial scale, calibrating the maps with the precise data of the rain gauges on the ground. This procedure is well defined in [45], where the radar maps were obtained by combining empirical and physical adjustments and comparing the IDF curves obtained by fitting the GEV on the series of annual maximums lasting twenty minutes, one hour and four hours. In [46], from the estimation of the radar quantities of precipitation on case studies in Germany, the IDF curve was studied with a time scale of five minutes and a spatial scale of 1 km. Instead, in [47], the comparison between the traditional two-parameter IDF curve and the satellite maps was studied for the first time, the power law that links the rainfall depth to the duration of the event, therefore, showing similarities. This may represent an interesting direction addressing the comparison of investigated models based on ground data with grid models that exploit radar or satellite data.

Author Contributions: Conceptualization, A.G., B.L., V.T., M.G.M., C.A., T.B. and V.I.; methodology, A.G., B.L., V.T., M.G.M. and V.I.; software, A.G., B.L. and V.T.; validation, A.G., B.L. and C.A.; formal analysis, A.G., B.L. and M.G.M.; investigation, A.G. and B.L.; data curation, A.G., B.L. and T.B.; writing—original draft preparation, A.G., B.L., V.T. and M.G.M.; writing—review and editing, C.A., T.B. and V.I.; visualization, B.L. and V.T.; supervision, A.G., V.T., C.A., T.B. and V.I.; funding acquisition, T.B. and V.I. All authors have read and agreed to the published version of the manuscript.

Funding: This research received no external funding.

Institutional Review Board Statement: Not applicable.

Informed Consent Statement: Not applicable.

Data Availability Statement: Rainfall data used in this study are freely available and can be downloaded by accessing the website of the Civil Protection of Puglia Region (https://protezionecivile.puglia.it/centro-funzionale-decentrato/ (accessed on 1 July 2021)).

Conflicts of Interest: The authors declare no conflict of interest.

References

1. Sun, Y.; Wendi, D.; Kim, D.E.; Liong, S.Y. Deriving intensity–duration–frequency (IDF) curves using downscaled in situ rainfall assimilated with remote sensing data. *Geosci. Lett.* **2019**, *6*, 17. [CrossRef]
2. Di Baldassarre, G.; Brath, A.; Montanari, A. Reliability of different depth-duration-frequency equations for estimating short-duration design storms. *Water Resour. Res.* **2006**, *42*, 1–6. [CrossRef]
3. García-Bartual, R.; Schneider, M. Estimating maximum expected short-duration rainfall intensities from extreme convective storms. *Phys. Chem. Earth Part B Hydrol. Ocean. Atmos.* **2001**, *26*, 675–681. [CrossRef]
4. Burlando, P.; Rosso, R. Scaling and multiscaling models of depth-duration-frequency curves for storm precipitation. *J. Hydrol.* **1996**, *187*, 45–64. [CrossRef]
5. Koutsoyiannis, D.; Kozonis, D.; Manetas, A. A mathematical framework for studying rainfall intensity-duration-frequency relationships. *J. Hydrol.* **1998**, *206*, 118–135. [CrossRef]
6. Veneziano, D.; Furcolo, P. Multifractality of rainfall and scaling of intensity-duration-frequency curves. *Water Resour. Res.* **2002**, *38*, 42-1–42-12. [CrossRef]
7. Agilan, V.; Umamahesh, N.V. Is the covariate based non-stationary rainfall IDF curve capable of encompassing future rainfall changes? *J. Hydrol.* **2016**, *541*, 1441–1455. [CrossRef]
8. Silva, D.F.; Simonovic, S.P.; Schardong, A.; Goldenfum, J.A. Assessment of non-stationary IDF curves under a changing climate: Case study of different climatic zones in Canada. *J. Hydrol. Reg. Stud.* **2021**, *36*, 100870. [CrossRef]
9. Vu, T.M.; Mishra, A.K. Nonstationary frequency analysis of the recent extreme precipitation events in the United States. *J. Hydrol.* **2019**, *575*, 999–1010. [CrossRef]
10. Ganguli, P.; Coulibaly, P. Does Nonstationarity in Rainfall Requires Nonstationary Intensity-Duration-Frequency Curves? *Hydrol. Earth Syst. Sci. Discuss* **2017**, *21*, 6461–6483. [CrossRef]
11. Salas, J.D.; Obeysekera, J. Revisiting the Concepts of Return Period and Risk for Nonstationary Hydrologic Extreme Events. *J. Hydrol. Eng.* **2014**, *19*, 554–568. [CrossRef]
12. Coles, S. *An Introduction to Statistical Modeling of Extreme Values*; Springer: London, UK, 2001.
13. Cooley, D. Return Periods and Return Levels under Climate Change. In *Extremes in a Changing Climate*; AghaKouchak, A., Easterling, D., Hsu, K., Schubert, S., Sorooshian, S., Eds.; Springer: Dordrecht, The Netherlands, 2013; pp. 97–113. [CrossRef]
14. Gioia, A.; Bruno, M.F.; Totaro, V.; Iacobellis, V. Parametric assessment of trend test power in a changing environment. *Sustainability* **2020**, *12*, 3889. [CrossRef]
15. Totaro, V.; Gioia, A.; Iacobellis, V. Numerical investigation on the power of parametric and nonparametric tests for trend detection in annual maximum series. *Hydrol. Earth Syst. Sci.* **2020**, *24*, 473–488. [CrossRef]
16. Yue, S.; Pilon, P.; Cavadias, G. Power of the Mann-Kendall and Spearman's rho tests for detecting monotonic trends in hydrological series. *J. Hydrol.* **2002**, *259*, 254–271. [CrossRef]
17. Wang, F.; Shao, W.; Yu, H.; Kan, G.; He, X.; Zhang, D.; Ren, M.; Wang, G. Re-evaluation of the Power of the Mann-Kendall Test for Detecting Monotonic Trends in Hydrometeorological Time Series. *Front. Earth Sci.* **2020**, *8*, 14. [CrossRef]
18. Vogel, R.M.; Rosner, A.; Kirshen, P.H. Brief Communication: Likelihood of societal preparedness for global change: Trend detection. *Nat. Hazards Earth Syst. Sci.* **2013**, *13*, 1773–1778. [CrossRef]

19. Yarnell, D.L. *Rainfall Intesity-Frequency Data*; U.S. Department of Agricolture: Washington, DC, USA, 1935; p. 35.
20. Rossi, F.; Villani, P. *Valutazione Delle Piene in Campania*; CNR-GNDCI: Salerno, Italy, 1995.
21. Grimaldi, S.; Nardi, F.; Piscopia, R.; Petroselli, A.; Apollonio, C. Continuous hydrologic modelling for design simulation in small and ungauged basins: A step forward and some tests for its practical use. *J. Hydrol.* **2021**, *595*, 125664. [CrossRef]
22. Rallison, R.E. Origin and evolution of the SCS runoff equation. In *Symposium on Watershed Management 1980*; American Society of Civil Engineers: New York, NY, USA, 1980; Volume II, pp. 912–924.
23. SCS. Section 4: Hydrology. In *National Engineering Handbook*; Soil Conservation Service, USDA: Washington, DC, USA, 1956.
24. De Luca, C.; Furcolo, P.; Rossi, F.; Villani PVitolo, C. Extreme rainfall in the Mediterranean. In Proceedings of the International Workshop Advances in Statistical Hydrology, Taormina, Italy, 23–25 May 2010.
25. Alpert, P.; Neeman, B.U.; Shay-El, Y. Climatological analysis of Mediterranean cyclones using ECMWF data. *Tellus A Dyn. Meteorol. Oceanogr.* **1990**, *42*, 65–77. [CrossRef]
26. Bruno, M.F.; Saponieri, A.; Molfetta, M.G.; Damiani, L. The DPSIR Approach for Coastal Risk Assessment under Climate Change at Regional Scale: The Case of Apulian Coast (Italy). *J. Mar. Sci. Eng.* **2020**, *8*, 531. [CrossRef]
27. Protezione Civile Puglia—Centro Funzionale Decentrato. Available online: https://protezionecivile.puglia.it/centro-funzionale-decentrato/ (accessed on 1 July 2021).
28. Soulis, K.X. Soil Conservation Service Curve Number (SCS-CN) Method: Current Applications, Remaining Challenges, and Future Perspectives. *Water* **2021**, *13*, 192. [CrossRef]
29. Baltas, E.A.; Dervos, N.A.; Mimikou, M.A. Technical note: Determination of the SCS initial abstraction ratio in an experimental watershed in Greece, Hydrol. *Earth Syst. Sci.* **2007**, *11*, 1825–1829. [CrossRef]
30. Mishra, S.K.; Singh, V.P. Another look at SCS-CN method. *J. Hydrol. Eng. ASCE* **1999**, *4*, 257–264. [CrossRef]
31. Holman, P.; Hollis, J.M.; Bramley, M.E.; Thompson, T.R.E. The contribution of soil structural degradation to catchmentflooding: A preliminary investigation of the 2000 floodsin England and Wales. *Hydrol. Earth Syst. Sci.* **2003**, *7*, 755–766. [CrossRef]
32. Hua, J. Application of SCS model in Lanhe watershed. *J. Taiyuan Univ. Technol.* **2003**, *34*, 735–762. (In Chinese)
33. Romero, P.; Castro, G.; Gómez, J.A.; Fereres, E. Curve number values for olive orchards under different soil management. *Soil Sci. Soc. Am. J.* **2007**, *71*, 1758–1769. [CrossRef]
34. Lewis, M.J.; Singer, M.J.; Tate, K.W. Applicability of SCS curve number method for a California Oak Woodlands Watershed. *J. Soil Water Conserv.* **2000**, *55*, 226–230.
35. Soulis, K.X.; Valiantzas, J.D. SCS-CN parameter determination using rainfall-runoff data in heterogeneous watersheds—The two-CN system approach. *Hydrol. Earth Syst. Sci.* **2012**, *16*, 1001–1015. [CrossRef]
36. Xianzhao, L.; Jiazhu, L. Application of SCS Model in Estimation of Runoff from Small Watershed in Loess Plateau of China. *Chin. Geogr. Sci.* **2008**, *18*, 235–241. [CrossRef]
37. Hoesein, A.A.; Pilgrim, D.H.; Titmarsh, G.W.; Cordery, I. Assessment of the US Conservation Service method for estimating design floods. In *New Directions for Surface Water Modeling (Proceedings of the Baltimore Symposium, May 1989)*; International Association of Hydrological Sciences: Wallingford, CT, USA; Oxfordshire, UK, 1989.
38. Pilgrim, D.H.; Cordery, I. Flood Runoff. In *Handbook of Hydrology*; Maidment, D.R., Ed.; McGraw-Hill: New York, NY, USA, 1992.
39. Mockus, V. *Use of Storm and Watershed Characteristics in Synthetic unit Hydrograph Analysis And Application*; USSCS: Washington, DC, USA, 1957.
40. Folmar, N.D.; Miller, A.C.; Woodward, D.E. History and Development of the NRCS Lag Time Equation. *J. Am. Water Resour. Assoc.* **2007**, *43*, 829–838. [CrossRef]
41. Ward, A.D.; Elliot, W.J. *Environmental Hydrology*; CRC Press: New York, NY, USA, 1995.
42. Hack, J.T. *Studies of Longitudinal Stream Profiles in Virginia and Maryland*; United States Geological Survey Professional: Washington, DC, USA, 1957; pp. 259-B, 45–97.
43. Maritan, A.; Rinaldo, A.; Rigon, R.; Giacometti, A.; Rodriguez-Iturbe, I. Scaling Laws for River Networks. *Phys. Rev. E* **1996**, *53*, 1510. [CrossRef]
44. Rinaldo, A.; Banavar, J.R.; Maritan, A. Trees, Networks, and Hydrology. *Water Resour. Res.* **2006**, *42*. [CrossRef]
45. Marra, F.; Morin, E. Use of radar QPE for the derivation of Intensity–Duration–Frequency curves in a range of climatic regime. *J. Hydrol.* **2015**, *531*, 427–440. [CrossRef]
46. Pöschmann, J.M.; Kim, D.; Kronenberg, R.; Bernhofer, C. An analysis of temporal scaling behaviour of extreme rainfall in Germany based on radar precipitation QPE data. *Nat. Hazards Earth Syst. Sci.* **2021**, *21*, 1195–1207. [CrossRef]
47. Breña-naranjo, J.A.; Pedrozo-acuña, A.; Rico-ramirez, M.A. World's greatest rainfall intensities observed by satellites. *Atmos. Sci. Lett.* **2015**, *16*, 420–424. [CrossRef]

Article

Evaluation of Global Precipitation Products over Wabi Shebelle River Basin, Ethiopia

Kindie Engdaw Tadesse [1,2,*], **Assefa M. Melesse** [3], **Adane Abebe** [4], **Haileyesus Belay Lakew** [5] **and Paolo Paron** [6]

1   Africa Center of Excellence for Water Management, Addis Ababa University, Addis Abeba P.O. Box 1176, Ethiopia
2   Department of Hydraulic and Water Resource Engineering, Gondar University, Gondar P.O. Box 196, Ethiopia
3   Department of Earth and Environment Sciences, Florida International University, Miami, FL 33199, USA; melessea@fiu.edu
4   Facality of Water Resource and Irrigation Engineering, Arba Minch University, Addis Ababa P.O. Box 21, Ethiopia; adaneabe@gmail.com
5   Ethiopian Space Science and Technology Institute, Addis Ababa P.O. Box 385, Ethiopia; haileyesusbelay@gmail.com
6   Department of Water Resource and Ecosystems, IHE, 2611 AX Delfit, The Netherlands; p.paron@un-ihe.org
*   Correspondence: kindie.engdaw@aau.edu.et; Tel.: +251-910-269-49

Citation: Tadesse, K.E.; Melesse, A.M.; Abebe, A.; Lakew, H.B.; Paron, P. Evaluation of Global Precipitation Products over Wabi Shebelle River Basin, Ethiopia. *Hydrology* **2022**, *9*, 66. https://doi.org/10.3390/hydrology9050066

Academic Editors: Davide Luciano De Luca, Andrea Petroselli and Mohammad Valipour

Received: 16 March 2022
Accepted: 15 April 2022
Published: 19 April 2022

Publisher's Note: MDPI stays neutral with regard to jurisdictional claims in published maps and institutional affiliations.

Abstract: This study presents three global precipitation products and their downscaled versions (CHIRPSv2, TAMSATv3, PERSIANN_CDR, CHIRPS_D, PERSIANNN_CDR_D, and TAMSAT_D) estimated with observed values from 1983 to 2014. Performance evaluation of global precipitation products and their downscaled versions is important for accurate use of those measured values in water resource management, climate, and hydrological applications, particularly in the data-sparse Wabi Shebelle River Basin, Ethiopia. Categorical and quantitative evaluation index techniques were applied. The spatial downscaled global precipitation products outperformed raw spatial resolution estimates in all statistical indicators. TAMSAT-D had acceptable performance ratings in terms of RMSE, CC, and scatter plots (R^2). CHIRPSv2 showed the least performance at a daily timestep. Performance of global precipitation products and their downscaled versions increased when daily data were aggregated to the monthly data. CHIRPS-D performed better than other products with a minimum error value (RMSE) and higher CC at a monthly timestep. On the other hand, PERSIANN_CDR_D showed a relatively good performance with a lower, positive Pbias and higher POD values compared to other products for daily and monthly timescales. For spatial mismatch analysis, the bias and RMSE from reference data (individual rain gauge station vs. the average of all available eight stations) against satellite rainfall estimates (PERSIANN_CDR) had a significantly different weight, which could be related to the position of the gauge station to provide the "true" spatial rainfall amount. Overall, TAMSATv3 and CHIRPSv2 and their downscaled version satellite estimates showed good performance at daily and monthly timesteps, respectively. PERSIANN_CDR performed best with low Pbias and the highest POD values. Thus, this study decided that the downscaled version of CHIRPSv2 and PERSIANN_CDR-D satellite estimates could be applicable as an alternative to gauge data on a monthly timestep for hydrological and drought-monitoring applications, respectively.

Keywords: satellite; product; precipitation; rain gauge; evaluation; downscaling

1. Introduction

Ranfall is an essential and fundamental primary input for the hydrologic cycle, as well as for hydro-meteorological modeling [1–4]. On the other hand, rainfall data are constrained by poor networks and uneven distribution because of the insufficient budget for operation and installation of rain gauge networks for most parts of the developing world [5]. The current meteorological network are inadequate and have poor maintenance for water resource assessment and climate studies in most tropical regions including Ethiopia [6]. The Wabi Shebelle River Basin (hereinafter WSRB) suffers from the scarce and uneven distribution of the gauge network, inadequate and low-accuracy precipitation, and incoherence of rainfall records [7]. The study area also experienced sociopolitical instability during the civil war in the region, resulting in precipitation measurements not being taken continuously. The gauge network spatial distribution is below the World Meteorological Organization (WMO) recommendations as a guideline for checking the adequacy of a meteorological network for the different physiographic units (one station per 575 km^2 interior plains and 250 km^2 for the mountainous regions) [8]. Therefore, satellite precipitation products are considered a significant alternative source for obtaining precipitation datasets for the nonexistence of observed data in filling spatiotemporal gaps [9,10]. Several higher global precipitation products exist at the regional and global level, including Tropical Applications of Meteorology using SATellite version three (hereinafter TAMSATv3) [11], Climate Hazards Group Infrared Precipitation with station data version two (hereinafter CHIRPSv2) [12], and Multisource and Precipitation Estimation from Remotely Sensed Information using Artificial Neutral Network Climate Data Record (PERSIANN-CDR) [13]. However, the products from these algorithms and assimilation models need to be evaluated as their precision is impacted by gauge density, orography, rainfall regime, temporal and spatial resolution, and algorithms used [14]. Global precipitation products (hereinafter GPPs) are impacted by exposure to important errors [15]. Such errors can be due to upscaling/downscaling the raw spatial resolution of global precipitation products with complicated terrain [16–21] and temporal sampling constraints [22]. Before being used as input for the hydro-meteorological modeling, global precipitation products and their downscaled versions need to be evaluated against ground measurements. For investigations of climate extremes, climate change, and water potential assessment for local-scale applications, global precipitation data with fine spatial and temporal resolution going back in time (30+ years) are required for data-scarce local basins. The Wabi Shebelle river basin is Ethiopian's largest river basin in terms of its catchment area, but the surface water potential resource was reported as the minimum of all river basins in the master plan study [23]. The issue with this basin is that no compressive research has been conducted to determine water potential using accurate spatial and temporal rainfall datasets.

Several studies have been undertaken to evaluate the performance of global precipitation estimates, which were concentrated in the Blue Nile Basin (for instance, [6,10,24–33]) and central parts of the country [22,34–38], with a coarser resolution and limited time period. The performance of the global precipitation product is highly affected by spatial resolution, which is largely uncertain because of the scale discrepancy with point measurements. The authors of [39] validated 10 satellite precipitation estimates across 120 relatively dense gauge network highlands of Ethiopia. TAMSAT, CMORPH, and TRMM-3B 42 from the first (high-resolution) group had a strong performance. The authors of [40] evaluated CHIRPSv2, TRMM 3B43v7, CMORPH, ARC2, and TAMSAT across various rainfall regimes (eastern, central, western, and southern) of Ethiopia. The CHIRPSv2 precipitation products at a monthly timescale performed comparatively better across all rainfall regimes. In addition, eight satellite-based rainfall estimates were assessed over the Tekeze-Atbara Transboundary River Basin. TRMM, RFEv2, and CHIRPS precipitation products had superior performance to other products at all spatial resolutions (basin, sub-basin, and point) [14]. Similar research focused mainly on a grid-to-point technique in data-sparse regions [22,25,34,41,42]. Furthermore, this method can induce uncertainty in the performance of global precipitation products due to the comparison of two datasets on different spatial scales regardless of the location of gauges in the pixel. Uncertainty due to the mismatch of satellite-based spatial resolution scale with point measurements may affect the application of GPPs for climate study and hydrological modeling [43]. Although global precipitation products exist at a coarser resolution (larger pixel size) than required by climate studies and hydrological applications, they have to be downscaled to fine resolution for matching the sampling of GPPs with gauge data [44].

A comprehensive evaluation of global precipitation products and their downscaled versions, particularly with a spatial mismatch at different timescales, is needed for a better understanding of watershed hydrology; however, this has not been performed for the Wabi Shebelle River Basin to the best of our knowledge. Uncertainty related to the grid-to-point method can be addressed by avoiding the spatial mismatch between the global precipitation product and corresponding station measurement by downscaling the coarse resolution to a fine resolution.

Therefore, this study attempted (i) a comprehensive evaluation of native and downscaled global precipitation products against ground reference rainfall data, and (ii) a quantification of the uncertainty associated with a grid-to-point approach for the spatial scale of global precipitation products at a selected pixel scale.

2. Study Area Description and Dataset

2.1. Study Area Description

The Wabi Shebelle River Basin (WSRB) is one of the largest basins in Ethiopia, located in the southeastern part of the country. It originates from the Arsi and Bale Mountain ranges 4000 m above sea level and drains to the Indian Ocean after crossing Somalia. The basin's absolute location is within the latitudes 4°45′–9°45′ N and longitudes 38°45′–45°30′ E. The WSRB is characterized by bimodal rainfall seasons due to the southern and northern movement of the intertropical convergence zone (ITCZ) from March to May and from July to September. According to the master plan hydrology report, the highest mean annual rainfall recorded is 1467 mm in Seru Wereda of the Arsi Zone. The lowest mean annual rainfall recorded is 220 mm in the Kelafo Area of the Somali Region [23]. In general, the spatial and temporal distribution of rainfall is not evenly distributed; it is clustered in the upper and urban areas of the basin, and tends to decrease with decreasing altitude as shown in Figure 1.

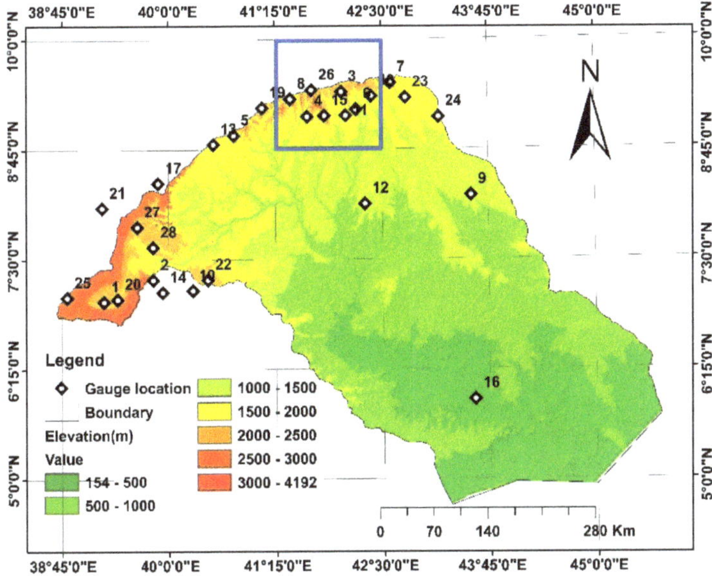

Figure 1. Study area description with rain gauge station distribution and topography elevation of Wabi Shebelle River Basin, Ethiopia.

2.2. Datasets

2.2.1. Rain Gauge Data

There are about 74 meteorological stations within and around the basin which are not evenly distributed spatially, clustered in upper and urban areas. The rainfall dataset for WSRB was taken from the National Mereological Agency (NMA), covering the period 1983 to 2014. Long-term meteorological data for the WSRB are more complete in upstream parts of the basin, and these stations were taken to analyze precipitation in the area as shown (Figure 2) below.

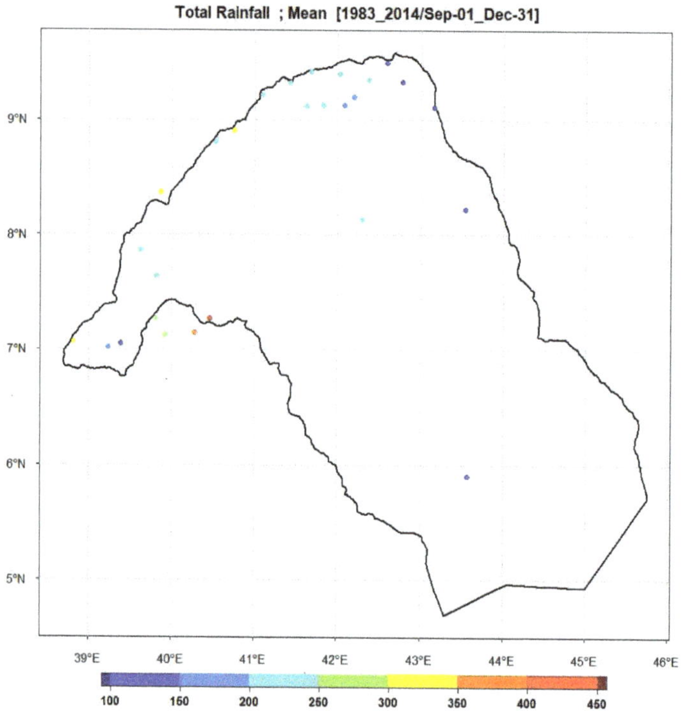

Figure 2. Annual rainfall in Wabi Shebelle River Basin.

Rain gauge stations for this study were carefully chosen on the basis of their quality control process for climate data (verification of in situ station's geographical coordinates, checking for false zeros, checking for the presence of outliers, and homogeneity testing) using the Climate Data Tool (CDT) https://github.com/rijaf-iri/CDT (accessed on 27 June 2020). Twenty-seven out of 74 gauging stations with a percentage of available (non-missing) and continuous data greater than 80% were selected for the comparison of different GPPs in the study area.

2.2.2. Global Precipitation Products

Global precipitation data with fine spatial and temporal resolution provide optional homogeneous timeseries information for data-scarce areas, going back in time (30+ years) as far as possible for hydrological applications and climate studies [45]. Global precipitation data are a combined product of reanalysis, rain gauge data, and remote sensing estimates.

For this desired specific objective, three global precipitation products and their down-scaled versions, with different temporal and spatial scales, were taken as inputs (Table 1). The selection of the GPPs was based on public availability, ease of estimation, global coverage, multiyear period, and previous record of estimate performance.

Table 1. Global precipitation products used for Wabi Shebelle River Bain.

Category	Input Data	Spatial Res. (Degree)	Temporal Res.	Start Date	Designation
Raw GPPs	CHIRPSv3	0.05	1 day	1981	CHIRPSv3
	PERSIANN_CDR	0.25	3 h	1983	PERSIANN_CDR
	TAMSATv3	0.04	1 day	1983	TAMSATv3
Downscaled GPPs	CHIRPSv3-Downscale	0.01 *	1 day	1981	CHIRPS_D
	PERSIANN_CDR_Downscale	0.01 *	3 h	1983	PERSIANN_CDR_D
	TAMSATv3-Downscale	0.01 *	1 day	1983	TAMSAT_D

* The raw spatial resolution of the selected GPPs was downscaled to 1 km.

The Climate Hazards Group Infrared Precipitation with station data version two (hereinafter CHIRPSv2) was developed by the United States Geological Survey (USGS) and University of California, Santa Barbara (USCB); it merges estimates using blending satellite, global climatology, and gauge observation data from the Global Telecommunication System (GTS). The CHIRPSv2 dataset incorporates 0.05° spatial resolution with ground reference measurements to generate a daily sequence of data points for an area coverage of 50° S–50° N since 1981 [12].

The Tropical Applications of Meteorology using SATellite version three (hereinafter TAMSATv3) estimate, developed by Reading University in the UK, features Meteosat thermal infrared (TIR) fine-resolution observations on a daily timescale employing attuned cold cloud duration (CCD) data measurements for Africa by downscaling pentadal total measurements. The TAMSATv3 estimate incorporates 0.0375° spatial resolution through ground rainfall measurements to generate timeseries for all of Africa from January 1983 to date [11].

The Precipitation Estimation from Remotely Sensed Information using Artificial Neutral Network Climate Data Record (PERSIANN-CDR) system was developed by the Center for Hydrometeorology and Remote Sensing (CHRS) at the University of California, Irvine (UCI); it uses a neutral function classification procedure to determine the product of precipitation amount for each 0.25° × 0.25° grid in an IR temperature spectrum offered by a geostationary satellite. The rainfall product features an area coverage of 60° S–60° N globally from 1983 to 2015 [13].

3. Methodology

This study evaluated the performance of three global precipitation product and their downscaled versions (CHIRPSv2, TAMSATv3, PERSIANN_CDR, CHIRPS_D, TAMSAT_D, and PERSIANN_CDR_D) at different spatial and temporal scales against 27 ground gauge stations from 1983 to 2014. Categorical and quantitative evaluation index techniques were applied to WSRB, Ethiopia.

3.1. Grid-to-Point Approach

There are two typical approaches for evaluating global precipitation products, i.e., the grid-to-grid and point-to-grid methods. The first method requires the interpolation of gauge data to grid data, whereby gauge-gridded data are compared with grid data from global precipitation estimates; however, converting points to gridded interpolated data induces an error resulting from the interpolation of an uneven geospatial distribution [46–50]. The second approach involves an immediate comparison of station rainfall data to the respective pixel in which the gauges are located [23,34,41,42]. In an area such as the Wabi Shebelle River Basin, with a scarcely and unevenly distributed gauge network, a pixel-to-point approach is the first choice to assess the GPPs independently, considering the gauge network as representative measurements irrespective of grids from nominated GPP, without considering the location of the station in the grid. Although global precipitation products exist at coarser resolution (larger pixel size) than required by climate studies and hydrological applications, they have to be downscaled to 1 km fine spatial resolution for evaluation with point gauge rainfall in the desired application. The spatial downscaling method and satellite rainfall estimate are the two most critical aspects in determining the accuracy of downscaled findings. In the Upper Tekezie River Basin, bilinear downscaling performed marginally better than the nearest-neighbor method to integrate satellite products with observed rainfall [51]. Other studies also preferred the bilinear downscaling method for smooth interpolated satellite-derived rainfall [52,53]. Therefore, bilinear downscaling was the approach chosen to downscale the spatial resolution of pixels for this study area.

The downscaled global precipitation product is more accurate than the original coarser resolution [54,55]. Therefore, the pixel value of raw spatial resolution GPPs in their downscaled version ($0.01° \times 0.01°$) was compared to gauge measurements.

The grid-to-point method can induce uncertainty in the performance of global satellite precipitation products due to the comparison of two datasets on different spatial scales regardless of the location of gauges in the pixel. The PERSIANN_CDR ($0.25° \times 0.25°$) pixel contains multiple rain gauge stations (greater than 3), which allows investigating the spatial mismatch global precipitation products against station observations for the eastern upper course (blue-colored grid box in study area map.

3.2. Evaluation Performance Indices

The quantitative and categorical evaluation indicator methods were carefully selected according to robustness, common usage, and recommendation in previous studies [39]. These performance indicators are described at https://www.cawcr.gov.au/projects/verification/ (accessed on 12 May 2017), implemented within the Climate Data Tool (CDT). Performance was assessed through quantitative evaluation indicators such as the coefficient of determination (R^2) (Equation (1)), percentage bias (Pbias) (Equation (3)), bias (Equation (4)), Pearson's correlation coefficient (CC) (Equation (2)), and root-mean-square error (RMSE) (Equation (5)). CC justifies the relationship between the exact values of two variables (independent and dependent). Values range between zero (no correlation) and one (perfect correlation). R^2 measures how well the independent variables explain the dependent variable in a regression. Values range between zero (no correlation) and one (perfect correlation). Bias describes the extent to which the observed value is underestimated or overestimated. The RMSE represents how closely the satellite observation predicts the measured value.

$$R^2 = 1 - \frac{\sum_{i=1}^{n}(S_i - G_i)^2}{\sum_{i=1}^{n}(S_i - \overline{S})^2}, \tag{1}$$

$$CC = \frac{\sum_{i=1}^{n}(S_i - \overline{S})(G_i - \overline{G})}{\sqrt{\sum_{i=1}^{n}(S_i - \overline{S})^2 \cdot (G_i - \overline{S})^2}}, \tag{2}$$

$$Pbias = \frac{\sum_{i=1}^{n}(S_i - G_i)}{\sum_{i=1}^{n} G_i} \times 100\%, \tag{3}$$

$$Bias = \frac{\sum_{i=1}^{n}(S_i - G_i)}{\sum_{i=1}^{n} G_i}, \tag{4}$$

$$RMSE = \sqrt{\frac{\sum_{i=1}^{n}(S_i - G_i)^2}{n}}, \tag{5}$$

where G_i and S_i represent the gauge and global precipitation data on the i-th day, i is the index, and $\overline{S}$ & $\overline{G}$ are the average values of S_i and G_i, respectively.

The ability of global precipitation estimates to determine the existence of precipitation rates was tested using the probability of detection (POD) (Equation (6)). POD was employed to evaluate the likelihood of the observed precipitation event being correctly detected by the satellite estimate. A dichotomous estimate that says "yes, an event will happen" or "no, an event will not happen" was used to quantify the metrics, as shown in Table 2. For this application, a rainfall threshold value of 1 mm was applied to decide the occurrence of a rainy or non-rainy day [25,33].

$$POD = \frac{\text{hits}}{\text{hits} + \text{misses}}, \tag{6}$$

where the absolute score of *POD* varies from 0–1.

Table 2. Contingency table of ainy and non-rainy event prediction by global precipitation products.

		GPPs		
		Yes	No	Total
Station	No	Misses	Correct negative	Observed no
measurement	Yes	Hit	False alarm	Observed yes
	Total	Estimate yes	Estimate no	Total

4. Result and Discussion

4.1. Comparison of Global Precipitation Products at Temporal Scale

This section presents a comparison of three global precipitation products and their downscaled versions vs. station data measurements according to the essential subject of gauge representativeness to identify the most reliable products for water resource assessment, climate studies, and hydrological applications across the data-scarce WSRB at different temporal scales for the period from 1983 to 2014.

4.1.1. Daily Comparison

The raw and downscaled global precipitation data were evaluated with observed rainfall at a daily timescale. Global precipitation products and their downscaled versions presented weak performance according to the majority of statistical indicator indices. The downscaled GPPs outperformed the original coarser resolutions as can be seen in Table 3. This result is similar to previous findings [54,55]. This might be due to the accuracy of the original precipitation product and the spatial downscaling method [39]. The RMSE in global precipitation products and their downscaled versions was highest in the southern and northeastern parts of the basin, with values ranging from 4 to 13 mm, as can be seen in Figure 3b. TAMSAT-D performed better than other products with a minimum RMSE for a value of 6.926 mm. The value of Pearson's correlation coefficient (CC) showed a poor relationship for all global precipitation products, but the CC value was relatively higher in the southern and northern parts of the basin, with values between 0.05 and 0.5, as can be seen from Figure 3a. TAMSAT-D showed the best agreement with a higher CC (0.332). The highest coefficient ($R^2 = 0.039$) was obtained by TAMSTAv3 and TAMSAT_D, as can be seen Figure 5a. The high performance of daily rainfall estimates from TAMSTAv3 and its downscaled version could be due to the loss of localized convective precipitation with the specified threshold value of the study area. This discovery is in line with the findings of previous investigations. CHIRPSv2 and its downscaled version showed the worst performance, as can be seen in Table 3. This could be attributed to the areal discrepancy of gauge observations and satellite estimates, as well as of the retrieval algorithms in disaggregating pentadal data to daily values [56]. On the other hand, PERSIANN_CDR_D showed a relatively good performance with a lower, positive Pbias compared to other products (underestimate), with a value of 3.09%, as presented in Table 3. The spatial distribution of Pbias for PERSIANN_CDR_D (Figure 3c) showed better performance than most stations.

Table 3. Daily statistical indicators of validation.

Product	CHIRPSv2	PERSIANN-CDR	TAMSATv3	CHIRPS_D	TAMSAT_D	PERSIANN-CDR_CDR
Correlation	0.245	0.298	0.33	0.25	0.331	0.307
Pbias	−6.739	3.3	−4.974	−7.064	−4.92	3.09
RMSE	8.297	7.183	6.948	8.141	6.926	7.027
POD	0.334	0.684	0.569	0.363	0.575	0.691

The ability of GPPs to detect the occurrence of precipitation events was also evaluated. In general, the downscaled products had better rainfall capability detection than the raw spatial resolution products in terms of the POD categorical statistical indicator. In this context, PERSIANN_CDR-D revealed a higher POD (0.691) than the PERSIANN_CDR precipitation product, as presented in Table 3. Both the raw and the downscaled precipitation products provided reasonably good PODs, varying between 0.25 and 0.893, as shown in Figure 4a. The highest POD and low Pbias indicate that PERSIANN_CDR-D is suitable for capturing the behavior of extreme precipitation events in the Wabi Shebelle River Basin, Ethiopia. The same result was also confirmed by [57]. This could be due to the adjustment of PERSIANN_CDR using GPCP monthly 2.50 precipitation products [13]. CHIRPSv3 showed extremely poor performance according to the categorical statistical indicator values.

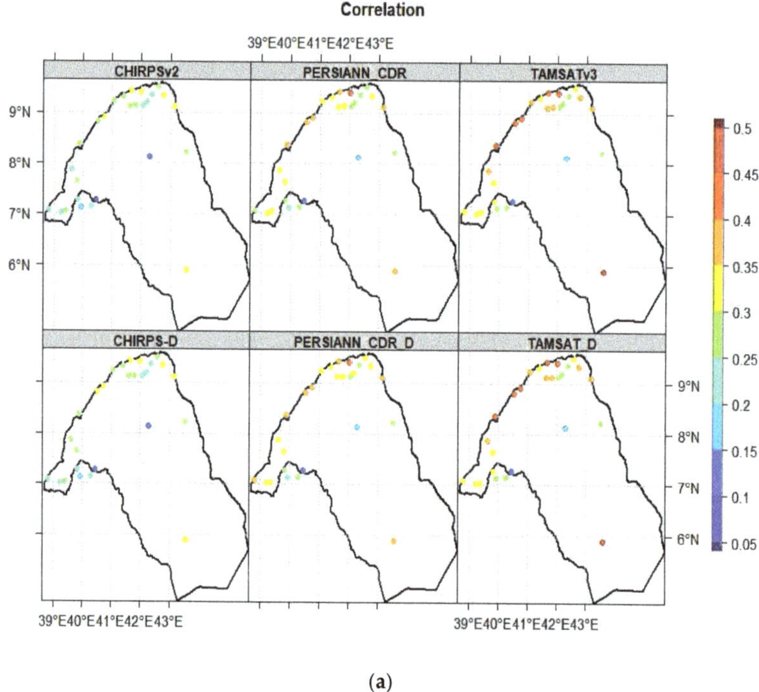

(a)

Figure 3. *Cont.*

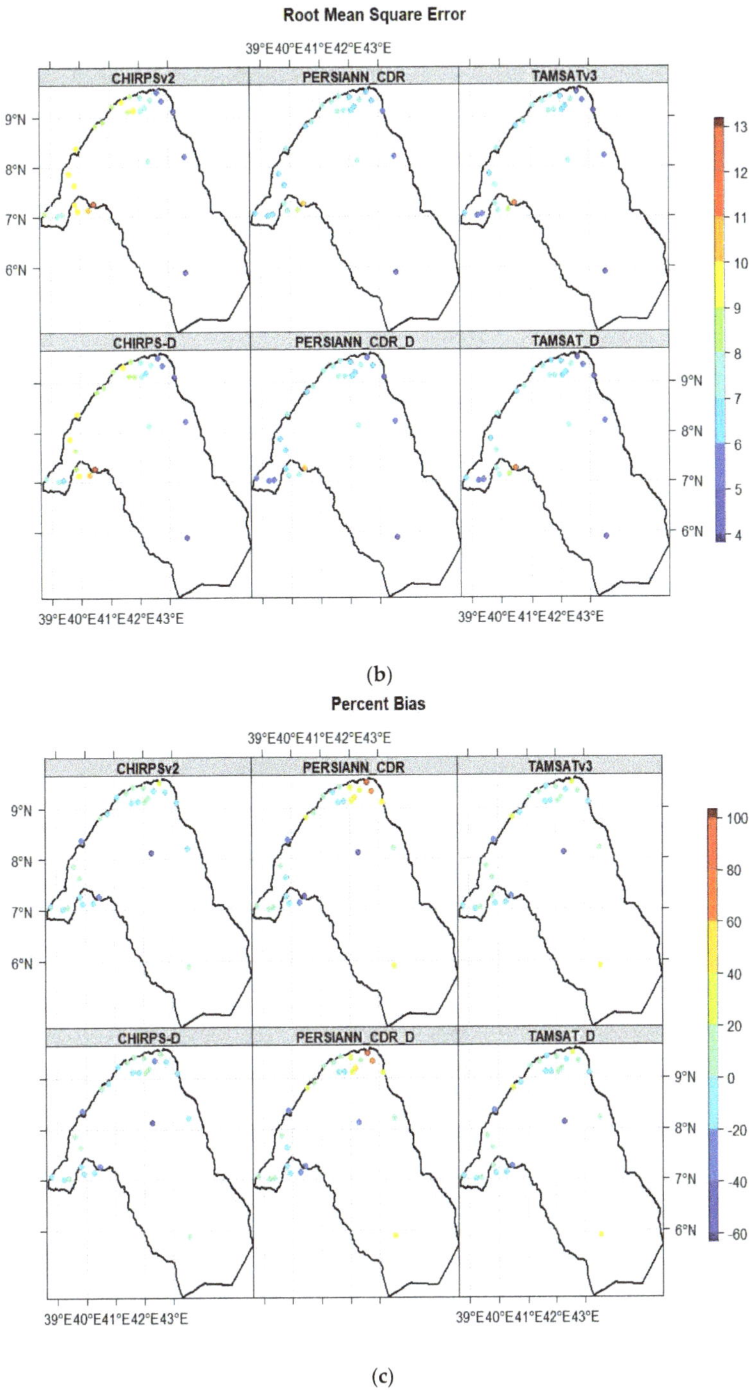

Figure 3. The statistical parameter correlation coefficient (**a**), root-mean-square error (**b**), and percentage bias (**c**) for each station at a daily timescale.

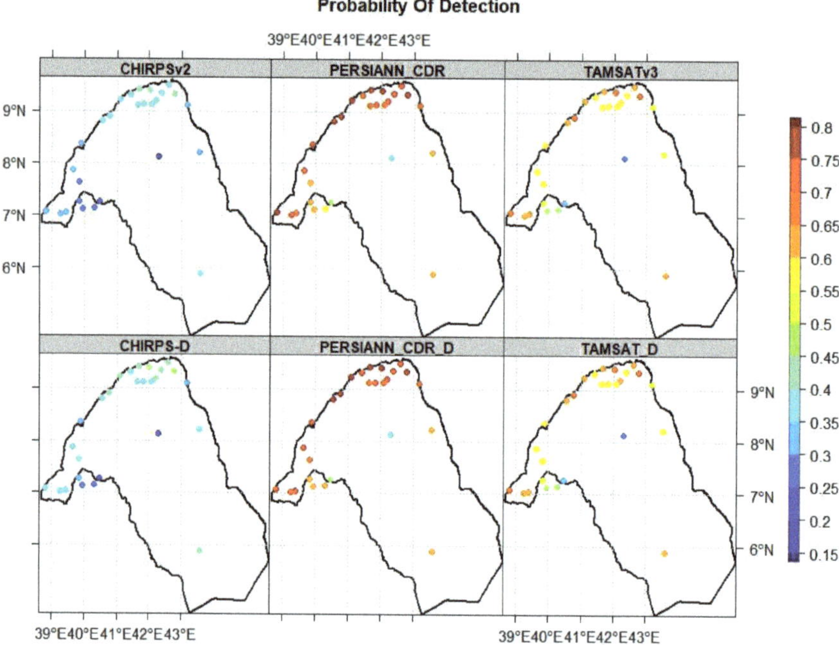

Figure 4. The probability of detection for each station at a daily timescale.

It can be observed that, in general, the downscaled and raw products presented poor agreement with the ground reference data ($r < 0.5$). The scatter plots and cumulative distribution functions using the average daily timeseries gauge rainfall data against the GPPs were examined (Figure 5a,b). A relatively high coefficient ($R^2 = 0.039$) was obtained by TAMSTAv3 and TAMSAT_D, and whereas CHIRPSv2 scored the lowest value. PERSIANN_CDR (downscaled and raw) showed reasonable agreement with the ground reference ($R^2 = 0.03$). Furthermore, all products were comparatively symmetric to a 45° inclination. According to the CDFs (Figure 5b), all products were not comparatively denser for a 45° inclination. Furthermore, TAMSAT_D and PERSIANN_CDR_D revealed the worst correspondence with the station CDFs. This shows that these products underestimated the distribution for rainfall ≤ 10 mm/day, whereas CHIRPSv2 and CHIRPS_D overestimated the distribution for rainfall ≤ 10 mm/day.

4.1.2. Monthly Comparison

The accuracy of the global precipitation products in replicating precipitation was further investigated at a monthly timescale, as shown in Figure 6 and Table 4. The results indicate that the performance of GPPs and their downscaled versions increased when daily data were aggregated to monthly data. These findings were also confirmed by [10,58], which evaluated the performance accuracy of aggregated global precipitation products toward a coarser temporal resolution. For example, one study [36] investigated several global precipitation products over Burkina Faso with different temporal resolutions. The results indicated that the categorical and volumetric indicators significantly increased upon aggregating the timescale. Similarly, the authors of [59] evaluated the CHIRPS satellite precipitation estimates over eastern parts of the continent. In the comparison of CHIRPS estimates with ARC2 and TAMSTA, the findings exhibited reasonably better reference estimates at decadal and monthly timescales, with a better skill of detection and lower bias, while TAMSAT performed better at a daily timescale.

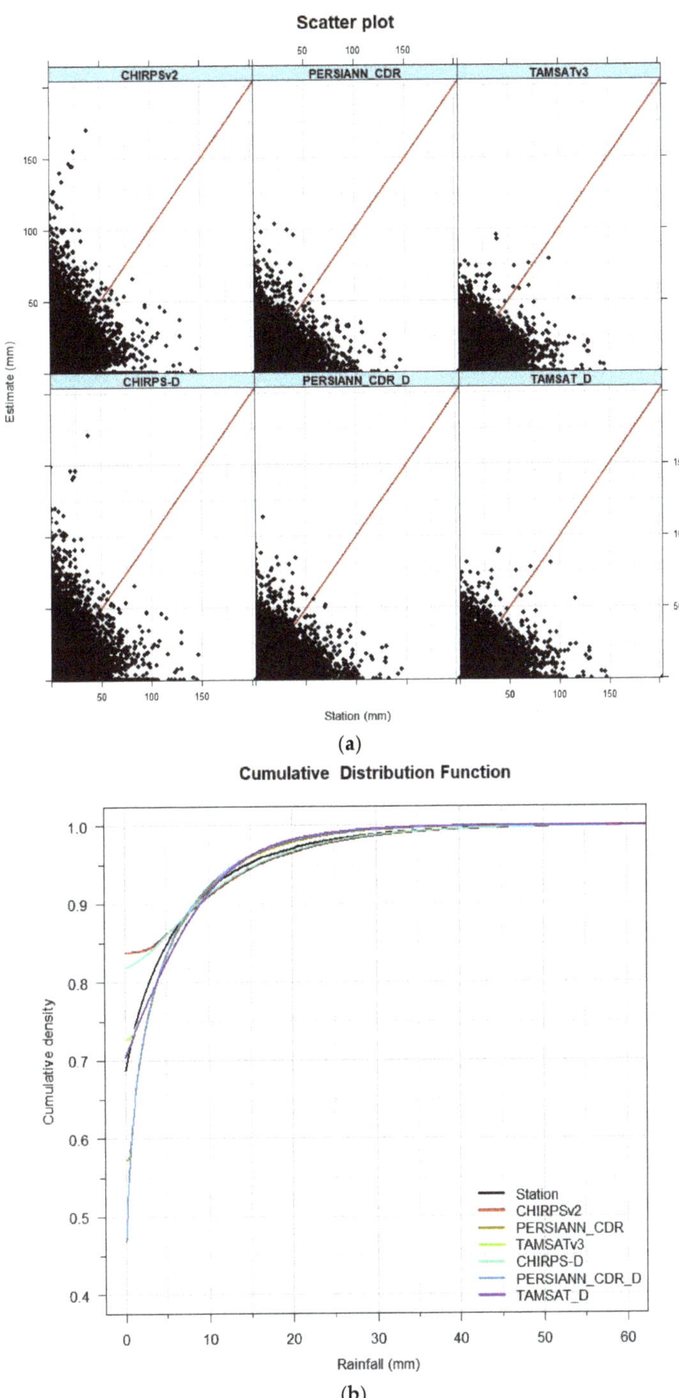

Figure 5. (**a**) Scatter plot of cumulative distribution function; (**b**) plots of observed rainfall against the global precipitation products.

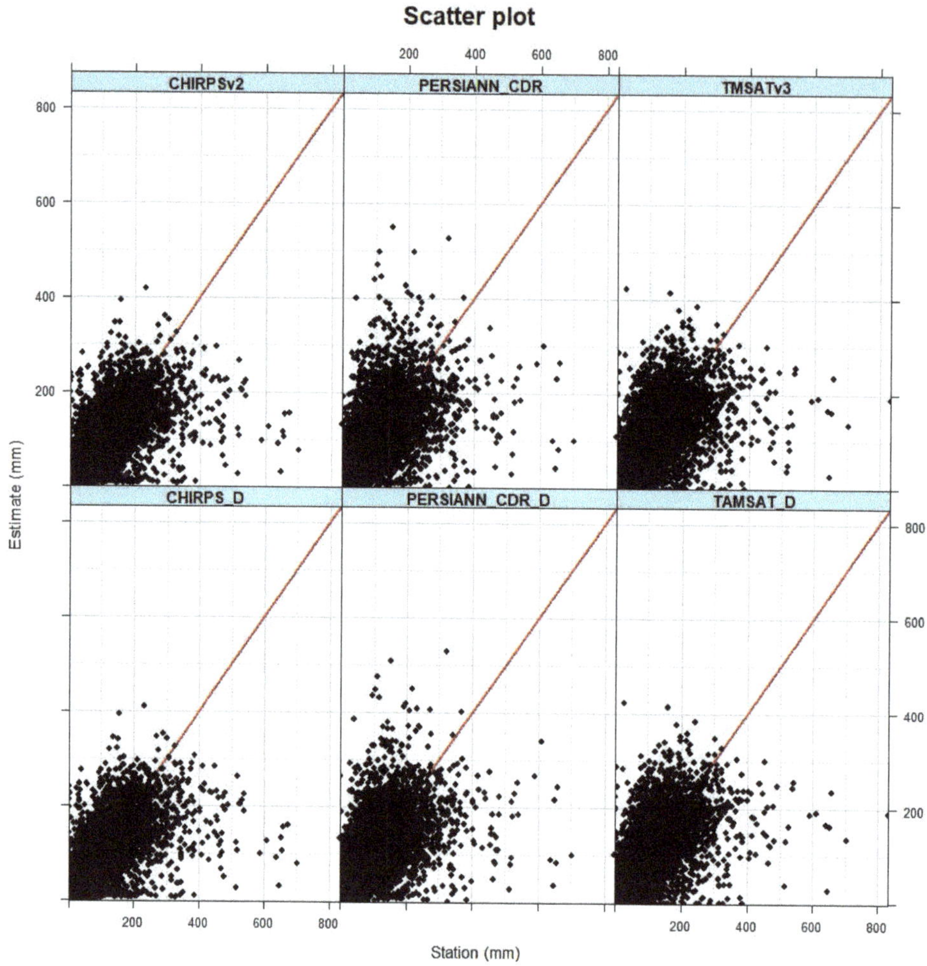

Figure 6. Scatter charts of observed precipitation versus global precipitation products at monthly timescale.

Table 4. Monthly statistical indicators of GPPs evaluated.

Product	CHIRPSv2	PERSIANN-CDR	TAMSATv3	CHIRPS_D	TAMSAT_D	PERSIANN-CDR_D
Correlation	0.745	0.659	0.699	0.748	0.709	0.664
PBIAS	−6.729	2.388	−6.066	−7.055	−3179	1.999
RMSE	53.958	63.739	58.894	53.734	57.884	62.0345
POD	0.984	0.986	0.897	0.986	0.897	0.993

The downscaled GPPs outperformed their original coarser-resolution counterparts according to all statistical indicators of accuracy. CHIRPS-D performed better than other products with a minimum error value (RMSE = 53.734 mm) and higher correlation (CC = 0.748). The value of Pearson's correlation coefficient (CC) showed a good relationship for raw and downscaled global precipitation products. Scatter plots using average monthly timeseries gauge rainfall data against the three GPPs and their downscaled versions were generated. The highest coefficient (R^2 = 0.418) was obtained by CHIRPS_D. As the time resolution increased from days to months, the rainfall amount estimated by CHIRPSv2 became increasingly accurate. The best performance of CHIRPSv2 and its downscaled version could be due to the elimination of error as the data were aggregated to a coarser timescale. These findings are consistent with earlier investigations of CHIRPSv2 rainfall data at a monthly timescale [59,60].

PERSIANN-CDR showed the lowest values for RMSE, CC, and R^2, as can be seen from Table 4 and Figure 6, on the daily timescale. PERSIANN_CDR_D showed relatively good performance with a lower, positive Pbias compared to other products (overestimate), with a value of 1.999%. PERSIANN_CDR_D resulted in the highest POD value of 0.993. CHIRPS_D and PERSIANN_CDR had the second better probability of detection (POD), whereas the TAMSAT group had the lowest value (Table 4). This implies that the performance of satellite estimates was influenced by the algorism and data source used.

4.2. Uncertainty Associated with a Pixel-To-To Point Method

In addition to the spatiotemporal investigation, the significant effect of the position of the stations in a pixel on the evaluation of the global precipitation product was analyzed, as shown in Figure 1 (blue-colored grid box). Furthermore, attempts were made to compare a pixel of selected GPPs (PERSIANN_CDR) against reference data, using the spatial average of all existing station data versus individual gauge stations within a pixel. Findings show that the minimum RMSE was obtained for PERSIANN_CDR when comparing the spatial average over each gauge station in the blue-colored box, with an average value of 4.667, as presented in Table 5.

Table 5. Statistical indicators of global precipitation products and station measurements within the grid.

STATION	PERSIANN_CDR			
	LON	STATS/LAT	BIAS	RMSE
Average over the pixel			1.126	4.667
Girawa	41.83	9.13	0.938	7.86
Gursum	42.38	9.35	1.223	7.962
Fedis	42.08	9.13	1.317	6.897
Kulubi	41.68	9.42	1.026	7.519
Alemaya	42.03	9.4	1.314	6.557
Bisidimo	42.2	9.2	1.404	6.924
Deder	41.43	9.32	0.991	7.676
Bedeno	41.63	9.12	0.967	7.223

PERSIANN_CDR achieved a reasonable maximum bias (overestimated by 12.6%) for the spatial average in the comparison of two datasets at the pixel level. On the other hand, the maximum bias ranged from, 40% and 31% using individual gauge stations Bisidimo and Fedis, respectively. In the comparison between the spatial average and the individual stations, Deder exhibited the smallest bias, while other stations changed the direction of the bias, with the exception of the Grawa and Bedeno gauge stations.

Generally, in terms of bias and RMSE, spatial averages estimated using rainfall data (eight stations) exhibited considerably different values to the referenced individual rain gauges in terms of magnitude. This magnitude difference may be related to the positions of the gauge stations and the uncertainty due to the representativeness of an individual rain gauge in providing the "true" spatial rainfall amount. Furthermore, the authors of [17,20,22,61] examined the variability and gauge representativeness of rainfall retrieved from the global precipitation product and showed the effect of network density on performance assessment. Therefore, it is essential to apply appropriate representative gauge data for the evaluation of products. Uncertainty related to the grid-to-point method can be addressed by avoiding the spatial mismatch between global precipitation products and the corresponding station measurements by downscaling the coarse resolution to a fine resolution [44]. In addition, installing additional rain gauges is strongly recommended within the grid [25].

5. Conclusions

In the current study, a total of six GPPs, three from raw global precipitation products (CHIRPSv2, TAMSATv3, and PERSIANN_CDR) and three from downscaled global precipitation products (CHIRPS_D, TAMSAT_D, and PERSIANN_CDR_D), were used. A bilinear method was applied to downscale the coarse spatial resolution of GPPs to 1 km resolution pixels. Categorical and quantitative evaluation index techniques were applied to WSRB, Ethiopia. The primary objective of the study was to assess the performance of the global precipitation products and their downscaled versions at different temporal scales compared to ground gauge stations.

The results indicated that the performance of global precipitation products is affected by factors such as the gauge density, spatiotemporal scale, and type of satellite algorithm. The daily evaluations were executed poorly in the majority of gauge stations. According to the evaluation parameters at the daily timescale, the downscaled GPPs performed best in terms of all statistical indicators. The evaluation assessment clearly indicated that TAMSAT_D was the best performer in terms of RMSE, CC, and scatter plots (R^2). On the other hand, PERSIANN_CDR_D showed a relatively good performance with a lower, positive Pbias and higher POD values compared to other products. CHIRPSv2 showed the worst performance at a daily timescale. The results indicated that the performance of the GPPs and their downscaled versions increased when daily data were aggregated to monthly data. Therefore, CHIRPS-D performed better than other products with a minimum error value (RMSE) and higher CC and R^2. However, PERSIANN_CDR_D presented a low Pbias and the highest POD values on daily and monthly timescales. In spatial mismatch analysis, the bias and RMSE estimated using rainfall data from individual rain gauges exhibited different magnitudes over the spatial average for PERSIANN_CDR, indicating that individual gauge data could not accurately estimate the product.

Overall, the performance of downscaled global precipitation products was better than that of the coarser-resolution products according to all statistical parameters. TAMSAT-D and CHIRPS-D products were the best-performing GPPs in reproducing the daily and monthly rainfall data, respectively. PERSIANN_CDR also accurately captured the extreme rainfall over the study area. This study provides a relatively long consistent and homogeneous timeseries rainfall dataset for climatology analysis and hydrological applications with a 1 km resolution for the study area. Although satellite precipitation products provide information at a high spatial resolution, they are lower in precision. On the other hand, gauges provide accurate point measurements but have limited spatial representativity. Therefore, for future studies, we recommend merging the downscaled product to improve the data availability in terms of accuracy, spatial distribution, and accumulated rainfall volume over the data-scarce Wabi Shebelle River Basin, Ethiopia, with a complex terrain, as well as other regions with a similar climate and topographical location.

Author Contributions: K.E.T. and A.M.M. conceptualized the study, framed the design, conducted data analysis, and wrote the manuscript. H.B.L., P.P. and A.A. conceptualized the study and provided satellite data. All authors have read and agreed to the published version of the manuscript.

Funding: This research received no external funding.

Institutional Review Board Statement: Not appilicable.

Informed Consent Statement: Not appilicable.

Data Availability Statement: All data models and code generated or used during the paper in the summited article.

Acknowledgments: This research was financially supported by the Africa Center of Excellence for Water Management (ACEWM) hosted by Addis Ababa University. The authors acknowledge the National Meteorological Agency (NMA) for providing a relevant data.

References

1. Gioia, A.; Lioi, B.; Totaro, V.; Molfetta, M.G.; Apollonio, C.; Bisantino, T.; Iacobellis, V. Estimation of Peak Discharges under Different Rainfall Depth–Duration–Frequency Formulations. *Hydrology* **2021**, *8*, 150. [CrossRef]
2. Abitew, T.A. *On the Use of Remotely Sensed Products in Spatially Distributed Hydrological Models of Data Scarce Tropical Basins;* VUB—Crazy Copy Center Productions: Ixelles, Belgium, 2017; ISBN 9789492312556.
3. Soo, E.Z.X.; Jaafar, W.Z.W.; Lai, S.H.; Islam, T.; Srivastava, P. Evaluation of Satellite Precipitation Products for Extreme Flood Events: Case Study in Peninsular Malaysia. *J. Water Clim. Chang.* **2018**, *10*, 871–892. [CrossRef]
4. Hobouchian, M.P.; Salio, P.; García Skabar, Y.; Vila, D.; Garreaud, R. Assessment of Satellite Precipitation Estimates over the Slopes of the Subtropical Andes. *Atmos. Res.* **2017**, *190*, 43–54. [CrossRef]
5. Behrangi, A.; Khakbaz, B.; Chun, T.; Aghakouchak, A.; Hsu, K. Hydrologic Evaluation of Satellite Precipitation Products over a Mid-Size Basin. *J. Hydrol.* **2011**, *397*, 225–237. [CrossRef]
6. Worqlul, A.W.; Maathuis, B.; Adem, A.A.; Demissie, S.S.; Langan, S.; Steenhuis, T.S. Comparison of Rainfall Estimations by TRMM 3B42, MPEG and CFSR with Ground-Observed Data for the Lake Tana Basin in Ethiopia. *Hydrol. Earth Syst. Sci.* **2014**, *18*, 4871–4881. [CrossRef]
7. Gebere, S.B.; Alamirew, T.; Merkel, B.J.; Melesse, A.M. Performance of High Resolution Satellite Rainfall Products over Data Scarce Parts of Eastern Ethiopia. *Remote Sens.* **2015**, *7*, 11639–11663. [CrossRef]
8. WMO. *Guide to Hydrological Practice; Data Acqision and Processing, Analysis, Forcing and Other Appilication;* World Meteorological Organization: Geneva, Switzerland, 1994.
9. Nigatu, Z.M.; Rientjes, T.; Haile, A.T. Hydrological Impact Assessment of Climate Change on Lake Tana's Water Balance, Ethiopia. *Am. J. Clim. Chang.* **2016**, *05*, 27–37. [CrossRef]
10. Bayissa, Y.; Tadesse, T.; Demisse, G.; Shiferaw, A. Evaluation of Satellite-Based Rainfall Estimates and Application to Monitor Meteorological Drought for the Upper Blue Nile Basin, Ethiopia. *Remote Sens.* **2017**, *9*, 669. [CrossRef]
11. Maidment, R.I.; Grimes, D.; Allan, R.P.; Tarnavsky, E.; Marcstringer, M.; Hewison, T.; Roebeling, R.; Black, E. The 30 Year TAMSAT African Rainfall Climatology and Time Series (TARCAT) Data Set. *J. Geophys. Res.* **2014**, *119*, 10–619. [CrossRef]
12. Funk, C.; Peterson, P.; Landsfeld, M.; Pedreros, D.; Verdin, J.; Shukla, S.; Husak, G.; Rowland, J.; Harrison, L.; Hoell, A.; et al. The Climate Hazards Infrared Precipitation with Stations—A New Environmental Record for Monitoring Extremes. *Sci. Data* **2015**, *2*, 150066. [CrossRef]
13. Ashouri, H.; Hsu, K.-L.; Sorooshian, S.; Braithwaite, D.K.; Knapp, K.R.; Cecil, L.D.; Nelson, B.R.; Prat, O.P. PERSIANN-CDR: Daily Precipitation Climate Data Record from Multisatellite Observations for Hydrological and Climate Studies. *Bull. Am. Meteorol. Soc.* **2015**, *96*, 69–83. [CrossRef]
14. Gebremicael, T.G.; Mohamed, Y.A.; van der Zaag, P.; Berhe, A.G.; Haile, G.G.; Hagos, E.Y.; Hagos, M.K. Comparison and Validation of Eight Satellite Rainfall Products over the Rugged Topography of Tekeze-Atbara Basin at Different Spatial and Temporal Scales. *Hydrol. Earth Syst. Sci. Discuss.* **2017**, 1–31. [CrossRef]
15. Gumindoga, W.; Haile, A.T.; Makurira, H.; Reggiani, P. Bias Correction Schemes for CMORPH Satellite Rainfall Estimates in the Zambezi River Basin. *Hydrol. Earth Syst. Sci. Discuss.* **2016**, 1–36. [CrossRef]
16. Dinku, T.; Hailemariam, K.; Maidment, R.; Tarnavsky, E.; Connor, S. Combined Use of Satellite Estimates and Rain Gauge Observations to Generate High-Quality Historical Rainfall Time Series over Ethiopia. *Int. J. Climatol.* **2014**, *34*, 2489–2504. [CrossRef]
17. Tang, G.; Behrangi, A.; Long, D.; Li, C.; Hong, Y. Accounting for Spatiotemporal Errors of Gauges: A Critical Step to Evaluate Gridded Precipitation Products. *J. Hydrol.* **2018**, *559*, 294–306. [CrossRef]
18. Dai, Q.; Yang, Q.; Zhang, J.; Zhang, S. Impact of Gauge Representative Error on a Radar Rainfall Uncertainty Model. *J. Appl. Meteorol. Climatol.* **2018**, *57*, 2769–2787. [CrossRef]
19. Peleg, N.; Ben-Asher, M.; Morin, E. Radar Subpixel-Scale Rainfall Variability and Uncertainty: Lessons Learned from Observations of a Dense Rain-Gauge Network. *Hydrol. Earth Syst. Sci.* **2013**, *17*, 2195–2208. [CrossRef]
20. Hu, L.; Nikolopoulos, E.I.; Marra, F.; Morin, E.; Marani, M.; Anagnostou, E.N. Evaluation of MEVD-Based Precipitation Frequency Analyses from Quasi-Global Precipitation Datasets against Dense Rain Gauge Networks. *J. Hydrol.* **2020**, *590*, 125564. [CrossRef]
21. Atiah, W.A.; Amekudzi, L.K.; Aryee, J.N.A.; Preko, K.; Danuor, S.K. Validation of Satellite and Merged Rainfall Data over Ghana, West Africa. *Atmosphere* **2020**, *11*, 859. [CrossRef]
22. Mekonnen, K.; Melesse, A.M.; Woldesenbet, T.A. Effect of Temporal Sampling Mismatches between Satellite Rainfall Estimates and Rain Gauge Observations on Modelling Extreme Rainfall in the Upper Awash Basin, Ethiopia. *J. Hydrol.* **2021**, *598*, 126467. [CrossRef]
23. MoWR (Ministry of Water Resources). *Integrated Development of Wabi Shebelle River Basin Master Plan Study in Climatology;* Addis Ababa University: Addis Ababa, Ethiopia, 2004.
24. Sahlu, D.; Nikolopoulos, E.I.; Moges, S.A.; Anagnostou, E.N.; Hailu, D. First Evaluation of the Day-1 IMERG over the Upper Blue Nile Basin. *J. Hydrometeorol.* **2016**, *17*, 2875–2882. [CrossRef]
25. Gebremichael, M.; Bitew, M.M.; Hirpa, F.A.; Tesfay, G.N. Accuracy of Satellite Rainfall Estimates in the Blue Nile Basin: Lowland Plain versus Highland Mountain. *Water Resour. Res.* **2014**, *50*, 8775–8790. [CrossRef]

26. Abera, W.; Brocca, L.; Rigon, R. *Evaluation of Different SREs and Bias Correction in the Upper Blue Nile*; UBN: Lagos, Nigeria, 2015.
27. Belete, M.; Deng, J.; Wang, K.; Zhou, M.; Zhu, E.; Shifaw, E.; Bayissa, Y. Evaluation of Satellite Rainfall Products for Modeling Water Yield over the Source Region of Blue Nile Basin. *Sci. Total Environ.* **2019**, *708*, 134834. [CrossRef]
28. Abera, W.; Brocca, L.; Rigon, R. Comparative Evaluation of Different Satellite Rainfall Estimation Products and Bias Correction in the Upper Blue Nile (UBN) Basin. *Atmos. Res.* **2016**, *178–179*, 471–483. [CrossRef]
29. Habib, E.; Haile, A.T.; Sazib, N.; Zhang, Y.; Rientjes, T. Effect of Bias Correction of Satellite-Rainfall Estimates on Runoff Simulations at the Source of the Upper Blue Nile. *Remote Sens.* **2014**, *6*, 6688–6708. [CrossRef]
30. Haile, A.T. *Rainfall Variability and Estimation for Hydrologic Modeling: A Remote Sensing Based Study at the Source Basin of the Upper Blue Nile River*; University of Twente: Enschede, The Netherlands, 2010; p. 209.
31. Haile, A.T.; Habib, E.; Rientjes, T. Evaluation of the Climate Prediction Center (CPC) Morphing Technique (CMORPH) Rainfall Product on Hourly Time Scales over the Source of the Blue Nile River. *Hydrol. Processes* **2013**, *27*, 1829–1839. [CrossRef]
32. Worqlul, A.W.; Collick, A.S.; Tilahun, S.A.; Langan, S.; Rientjes, T.H.M.; Steenhuis, T.S. Comparing TRMM 3B42, CFSR and Ground-Based Rainfall Estimates as Input for Hydrological Models, in Data Scarce Regions: The Upper Blue Nile Basin, Ethiopia. *Hydrol. Earth Syst. Sci. Discuss.* **2015**, *12*, 2081–2112. [CrossRef]
33. Haile, A.T.; Yan, F.; Habib, E. Accuracy of the CMORPH Satellite-Rainfall Product over Lake Tana Basin in Eastern Africa. *Atmos. Res.* **2015**, *163*, 177–187. [CrossRef]
34. Mekonnen, K.; Melesse, A.M.; Woldesenbet, T.A. Spatial Evaluation of Satellite-Retrieved Extreme Rainfall Rates in the Upper Awash River Basin, Ethiopia. *Atmos. Res.* **2021**, *249*, 105297. [CrossRef]
35. Koriche, S.A. Remote Sensing Based Hydrological Modelling for Flood Early Warning in the Upper and Middle Awash River Basin. In Proceedings of the Ethiopian Conference Center, Addis Ababa, Ethiopia, 19–20 October 2012; p. 67.
36. Dinku, T.; Funk, C.; Peterson, P.; Barbara, S.; Maidment, R.; Tadesse, T. Validation of the CHIRPS Satellite Rainfall Estimates over Eastern of Africa: Validation of the CHIRPS Satellite Rainfall Estimates over Eastern Africa. *Q. J. R. Meteorol. Soc.* **2018**, *144*, 292–312. [CrossRef]
37. Young, M.P.; Williams, C.J.R.; Christine Chiu, J.; Maidment, R.I.; Chen, S.H. Investigation of Discrepancies in Satellite Rainfall Estimates over Ethiopia. *J. Hydrometeorol.* **2014**, *15*, 2347–2369. [CrossRef]
38. Hirpa, F.A.; Gebremichael, M.; Hopson, T. Evaluation of High-Resolution Satellite Precipitation Products over Very Complex Terrain in Ethiopia. *J. Appl. Meteorol. Climatol.* **2010**, *49*, 1044–1051. [CrossRef]
39. Dinku, T.; Ceccato, P.; Grover-Kopec, E.; Lemma, M.; Connor, S.J.; Ropelewski, C.F. Validation of Satellite Rainfall Products over East Africa's Complex Topography. *Int. J. Remote Sens.* **2007**, *28*, 1503–1526. [CrossRef]
40. Lemma, E.; Upadhyaya, S.; Ramsankaran, R.A.A.J. Investigating the Performance of Satellite and Reanalysis Rainfall Products at Monthly Timescales across Different Rainfall Regimes of Ethiopia. *Int. J. Remote Sens.* **2019**, *40*, 4019–4042. [CrossRef]
41. Gella, G.W. Statistical Evaluation of High Resolution Satellite Precipitation Products in Arid and Semi-Arid Parts of Ethiopia: A Note for Hydro-Meteorological Applications. *Water Environ. J.* **2019**, *33*, 86–97. [CrossRef]
42. Kimani, M.W.; Hoedjes, J.C.B.; Su, Z. An Assessment of Satellite-Derived Rainfall Products Relative to Ground Observations over East Africa. *Remote Sens.* **2017**, *9*, 430. [CrossRef]
43. Peleg, N.; Marra, F.; Fatichi, S.; Paschalis, A.; Molnar, P.; Burlando, P. Spatial Variability of Extreme Rainfall at Radar Subpixel Scale. *J. Hydrol.* **2018**, *556*, 922–933. [CrossRef]
44. Zorzetto, E.; Marani, M.; Water Resources Research. Undefined Downscaling of Rainfall Extremes from Satellite Observations. *Wiley Online Libr.* **2019**, *55*, 156–174. [CrossRef]
45. Bui, H.T.; Ishidaira, H.; Shaowei, N. Evaluation of the Use of Global Satellite—Gauge and Satellite—Only Precipitation Products in Stream Flow Simulations. *Appl. Water Sci.* **2019**, *9*, 53. [CrossRef]
46. Mahbod, M.; Shirvani, A.; Veronesi, F. A Comparative Analysis of the Precipitation Extremes Obtained from Tropical Rainfall-Measuring Mission Satellite and Rain Gauges Datasets over a Semiarid Region. *Int. J. Climatol.* **2019**, *39*, 495–515. [CrossRef]
47. Nastos, P.T.; Kapsomenakis, J.; Douvis, K.C. Analysis of Precipitation Extremes Based on Satellite and High-Resolution Gridded Data Set over Mediterranean Basin. *Atmos. Res.* **2013**, *131*, 46–59. [CrossRef]
48. Aghakouchak, A.; Behrangi, A.; Sorooshian, S.; Hsu, K.; Amitai, E. Evaluation of Satellite-Retrieved Extreme Precipitation Rates across the Central United States. *J. Geophys. Res. Atmos.* **2011**, *116*, 3. [CrossRef]
49. Mantas, V.M.; Liu, Z.; Caro, C.; Pereira, A.J.S.C. Validation of TRMM Multi-Satellite Precipitation Analysis (TMPA) Products in the Peruvian Andes. *Atmos. Res.* **2015**, *163*, 132–145. [CrossRef]
50. Gebrechorkos, S.H.; Hülsmann, S.; Bernhofer, C. Evaluation of Multiple Climate Data Sources for Managing Environmental Resources in East Africa. *Hydrol. Earth Syst. Sci.* **2018**, *22*, 4547–4564. [CrossRef]
51. Gebremedhin, M.A.; Lubczynski, M.W.; Maathuis, B.H.P.; Teka, D. Novel Approach to Integrate Daily Satellite Rainfall with In-Situ Rainfall, Upper Tekeze Basin, Ethiopia. *Atmos. Res.* **2021**, *248*, 105135. [CrossRef]
52. Ulloa, J.; Ballari, D.; Campozano, L.; Samaniego, E. Two-Step Downscaling of TRMM 3b43 V7 Precipitation in Contrasting Climatic Regions with Sparse Monitoring: The Case of Ecuador in Tropical South America. *Remote Sens.* **2017**, *9*, 758. [CrossRef]
53. Chen, S.; Zhang, L.; She, D.; Chen, J. Spatial Downscaling of Tropical Rainfall Measuring Mission (TRMM) Annual and Monthly Precipitation Data over the Middle and Lower Reaches of the Yangtze River Basin, China. *Water* **2019**, *11*, 568. [CrossRef]
54. Chu, W.; Gao, X.; Phillips, T.J.; Sorooshian, S. Consistency of Spatial Patterns of the Daily Precipitation Field in the Western United States and Its Application to Precipitation Disaggregation. *Geophys. Res. Lett.* **2011**, *38*, 1–6. [CrossRef]

55. Chen, F.; Gao, Y.; Wang, Y.; Qin, F.; Li, X. Downscaling Satellite-Derived Daily Precipitation Products with an Integrated Framework. *Int. J. Climatol.* **2019**, *39*, 1287–1304. [CrossRef]
56. Belay, A.S.; Fenta, A.A.; Yenehun, A.; Nigate, F.; Tilahun, S.A.; Moges, M.M.; Dessie, M.; Adgo, E.; Nyssen, J.; Chen, M.; et al. Evaluation and Application of Multi-Source Satellite Rainfall Product CHIRPS to Assess Spatio-Temporal Rainfall Variability on Data-Sparse Western Margins of Ethiopian Highlands. *Remote Sens.* **2019**, *11*, 2688. [CrossRef]
57. Miao, C.; Ashouri, H.; Hsu, K.L.; Sorooshian, S.; Duan, Q. Evaluation of the PERSIANN-CDR Daily Rainfall Estimates in Capturing the Behavior of Extreme Precipitation Events over China. *J. Hydrometeorol.* **2015**, *16*, 1387–1396. [CrossRef]
58. Dembélé, M.; Zwart, S.J. Evaluation and Comparison of Satellite-Based Rainfall Products in Burkina Faso, West Africa. *Int. J. Remote Sens.* **2016**, *37*, 3995–4014. [CrossRef]
59. Malede, D.A.; Agumassie, T.A.; Kosgei, J.R.; Pham, Q.B.; Andualem, T.G. Evaluation of Satellite Rainfall Estimates in a Rugged Topographical Basin Over South Gojjam Basin, Ethiopia. *J. Indian Soc. Remote Sens.* **2022**, *50*, 1–14. [CrossRef]
60. Ayehu, G.T.; Tadesse, T.; Gessesse, B.; Dinku, T. Validation of New Satellite Rainfall Products over the Upper Blue Nile Basin, Ethiopia. *Atmos. Meas. Tech.* **2018**, *11*, 1921–1936. [CrossRef]
61. Tessema, K.B.; Haile, A.T.; Amencho, N.W.; Habib, E. Effect of Rainfall Variability and Gauge Representativeness on Satellite Rainfall Accuracy in a Small Upland Watershed in Southern Ethiopia. *Hydrol. Sci. J.* **2020**, 1–15. [CrossRef]

 *hydrology*

Article

Application of Machine Learning and Process-Based Models for Rainfall-Runoff Simulation in DuPage River Basin, Illinois

Amrit Bhusal, Utsav Parajuli, Sushmita Regmi and Ajay Kalra *

School of Civil, Environmental, and Infrastructure Engineering, Southern Illinois University, 1230 Lincoln Drive, Carbondale, IL 62901-6603, USA; amrit.bhusal@siu.edu (A.B.); utsav.parajuli@siu.edu (U.P.); sushmita.regmi@siu.edu (S.R.)
* Correspondence: kalraa@siu.edu; Tel.: +1-(618)-453-7008

Abstract: Rainfall-runoff simulation is vital for planning and controlling flood control events. Hydrology modeling using Hydrological Engineering Center—Hydrologic Modeling System (HEC-HMS) is accepted globally for event-based or continuous simulation of the rainfall-runoff operation. Similarly, machine learning is a fast-growing discipline that offers numerous alternatives suitable for hydrology research's high demands and limitations. Conventional and process-based models such as HEC-HMS are typically created at specific spatiotemporal scales and do not easily fit the diversified and complex input parameters. Therefore, in this research, the effectiveness of Random Forest, a machine learning model, was compared with HEC-HMS for the rainfall-runoff process. Furthermore, we also performed a hydraulic simulation in Hydrological Engineering Center—Geospatial River Analysis System (HEC-RAS) using the input discharge obtained from the Random Forest model. The reliability of the Random Forest model and the HEC-HMS model was evaluated using different statistical indexes. The coefficient of determination (R^2), standard deviation ratio (RSR), and normalized root mean square error (NRMSE) were 0.94, 0.23, and 0.17 for the training data and 0.72, 0.56, and 0.26 for the testing data, respectively, for the Random Forest model. Similarly, the R^2, RSR, and NRMSE were 0.99, 0.16, and 0.06 for the calibration period and 0.96, 0.35, and 0.10 for the validation period, respectively, for the HEC-HMS model. The Random Forest model slightly underestimated peak discharge values, whereas the HEC-HMS model slightly overestimated the peak discharge value. Statistical index values illustrated the good performance of the Random Forest and HEC-HMS models, which revealed the suitability of both models for hydrology analysis. In addition, the flood depth generated by HEC-RAS using the Random Forest predicted discharge underestimated the flood depth during the peak flooding event. This result proves that HEC-HMS could compensate Random Forest for the peak discharge and flood depth during extreme events. In conclusion, the integrated machine learning and physical-based model can provide more confidence in rainfall-runoff and flood depth prediction.

Keywords: rainfall-runoff; HEC-HMS; HEC-RAS; random forest; flood; forecast

Citation: Bhusal, A.; Parajuli, U.; Regmi, S.; Kalra, A. Application of Machine Learning and Process-Based Models for Rainfall-Runoff Simulation in DuPage River Basin, Illinois. *Hydrology* **2022**, *9*, 117. https://doi.org/10.3390/hydrology9070117

Academic Editors: Davide Luciano De Luca and Andrea Petroselli

Received: 30 May 2022
Accepted: 24 June 2022
Published: 27 June 2022

1. Introduction

Floods are some of the most common and costly natural catastrophes in the world [1–3]. The magnitude and frequency of extreme flooding events have increased considerably worldwide over the previous few decades [4]. Climate change, urbanization, and other anthropogenic activities are causing a flood risk globally [5–7]. Water-related natural hazards, such as floods, droughts, and landslides, have become the new normal due to the uncertainty in rainfall patterns and magnitudes caused by climate change and urbanization [8]. Flooding is projected to become more common in the coming years as the frequency of extreme precipitation events increases [9–11].

Flood severity has increased, resulting in a large number of flood fatalities, massive economic losses, and social consequences [12]. Given the negative consequences of flooding, developing floodplain management plans to avoid and mitigate flood damage is critical [13]. The estimation of Intensity–Duration–Frequency (IDF) curves and the monitoring of rainfall intensity are also critical factors in precisely calculating the flood hydrograph and the peak discharges [14,15]. The flood risk assessment depends on a precise estimation of peak runoff, calculated by rainfall-runoff simulation [16]. Accurate rainfall-runoff simulation is a prominent topic in hydrology research [17]. Precise rainfall-runoff modeling is essential for planning and applying flood control strategies in vulnerable areas to reduce the dangers to human life and infrastructure during high-precipitation events. Different hydrology models have been used in the past to perform a rainfall-runoff simulation in a watershed. The Hydrologic Modeling System (HMS), designed by the Hydrologic Engineering Center (HEC) of the United States Army Corps of Engineers, is a popular rainfall-runoff analysis tool worldwide [18].

Process-based physical models are typically employed to calculate runoff in a particular catchment area. By integrating regional variability in the watershed, a physical-based model such as HEC-HMS can compute an actual hydrology system [19]. Hydrology modeling using the HEC-HMS model can be used to investigate urban floods, flood frequency, flood warning systems, and the effectiveness of spillways and detention ponds over a watershed [20]. The HEC-HMS model is made up of four essential components. An analytical method is first applied to compute direct discharge and reach routing. Secondly, a basin model with interactive components is employed for depicting hydrology aspects within a catchment. Third, data are entered, edited, managed, and stored via a system. Fourth, the simulation results are reported and illustrated using a functional system [21]. Finally, the calibration procedure, which compares simulated results to observed data, can help to enhance the model's precision and predictability. With the regional and temporal variety of catchment features, rainfall patterns, and the number of variables applied in modeling physical processes, the connection between precipitation and discharge using HEC-HMS is challenging [22]. A physical-based model such as HEC-HMS necessitates a large amount of data, such as land use and land cover data, soil group data, and infiltration data, and a significant amount of time to calibrate to ensure the correctness of the model [23]. Furthermore, there are drawbacks to using a physical-based hydrology model, owing to the difficulties in completely understanding the complicated, nonlinear, and inter-related hydrology [24,25]. A hydrology model that uses HEC-HMS may be unsuitable for a larger watershed with scarce data. Therefore, as a complement to the physical model, recently, the application of machine learning and data-driven models has been used across hydrology domains [26,27].

Machine learning (ML) is a kind of artificial intelligence that can make an accurate prediction by training and testing datasets. ML provides a solution to a real-world problem by studying previously observed data and has been effective in generating accurate results [28]. ML provides adequate computation power [29,30] and is used in a wide variety of research and applications in hydrology. Some examples of ML applications in the hydrology domain are rainfall-runoff prediction [31–33], flood forecasting [34–36], sedimentation studies [37–39], water quality prediction [40–43], groundwater prediction [44,45], river temperature prediction [46–49], and rainfall estimation [50,51]. In recent years, ML algorithms have significantly improved and are also widely used for rainfall-runoff simulation [52,53] thanks to the rapid advancement of computer technology. Recently, many researchers have performed rainfall-runoff predictions using different machine learning and data-driven models. Some examples of these models are long short-term memory [54,55], artificial neural networks [56,57], support vector machines [58,59], and the Random Forest model [16,60]. Random Forest is a popular machine learning tool, and Breiman first developed it in 2001 [61]. Random Forest has recently acquired popularity as a powerful predictive modeling tool, and many researchers are using it in their fields as a potential method [62]. It is a classification and regression tree-based ensemble learning algorithm [61]. A bootstrap sample is used to train each tree, and optimal variables at each split are chosen from a random subset of all variables. Random Forest offers the highest accuracy of any contemporary method and works quickly on large datasets [63].

Previous studies showed that Random Forest's performance surpassed other machine learning and data-driven tools such as artificial neural networks, regression models, and support vector machines in multiple comparative studies in hydrology [63–67]. However, Random Forest is the least used for hydrology analysis among the data-driven and machine learning models [68]. Among the few applications of Random Forest, most of these studies focused on flood risk hazards [16,69] and mapping [70]. Therefore, this study evaluated the effectiveness of the Random Forest model for rainfall-runoff simulation. In addition, the main objective of this research is to determine the suitability of the Random Forest model for rainfall-runoff simulation in a scarce-data region. Therefore, this research also used a satellite precipitation product as an input variable for rainfall-runoff simulation and determined its appropriateness in hydrology research. Furthermore, this study assessed the appropriateness of using Random Forest generated discharge for hydraulic modeling using the Hydrologic Analysis Center's River Analysis Model (HEC-RAS).

HEC-RAS is the most widely accepted model [71] for analyzing channel flow and floodplain characterization [72]. Users can compute one-dimensional steady and unsteady flow, two-dimensional unsteady flow, sediment transport, and water quality models by using HEC-RAS [72]. Regularizing geometric data and identifying and analyzing hydraulic structures, such as weirs, culverts, reservoirs, pump stations, bridges, levees, and gates, blockage and ineffective regions, land use, the Manning roughness coefficient, streambed slopes, and ice cover are achievable with HEC-RAS [73]. The model employs geometric data and geometric and hydraulic computer algorithms to model natural and artificial streams. HEC-RAS requires fundamental inputs such as river discharge, channel geometry, bank lines, flow paths, and channel resistance. The discharge generated by Random Forest was employed as an input parameter in this study. While the HEC-RAS model has a wide variety of capabilities, the current research considered its capability to execute 1D river flow and calculate the flood depth at the most downstream section of the study reach.

The integration of different models in the sectors of hydrology and hydraulic domains is gaining global attention and is crucial for flood risk management techniques [74]. The novelty of this research is to assess the effectiveness of the Random Forest model for rainfall-runoff simulation using satellite precipitation products in a data-scarce region. This research work also evaluated the integration of machine learning and a HEC-RAS model for calculating water depth at the proposed study location during the study period. The following is an outline of this paper. Section 2 describes the study area, data preparation, and a physical-based and Random Forest model. Section 3 presents the results of this research, Section 4 provides a discussion of the results, and Section 5 provides the major conclusions from the current analysis.

2. Data and Methods

This section describes the methodology used for hydrology and hydraulic analysis in this research. Random Forest, HEC-HMS, and HEC-RAS are the three models used in this study. HEC-HMS and the Random Forest model were applied for hydrology analysis, and HEC-RAS was used for the hydraulic analysis. The complete workflow of the methodology used in this research work is shown in Figure 1. First, this study started with extracting and preprocessing the data on basin characteristics, such as digital elevation model (DEM), land use and land cover (LULC), and soil group data, and meteorological data, such as daily precipitation and discharge data. The integrated use of Arc-Hydro, HEC-GeoHMS, and HEC-HMS was employed for hydrology analysis in the upstream catchment area. Similarly, Random Forest, a machine learning algorithm, was used to predict the runoff for the training and testing period. After the preparation of the hydrology model, a comparison was performed between the machine learning model (Random Forest Regression) and the physical model (HEC-HMS) using the different statistical indexes. Finally, the runoff obtained from the machine learning model was used as an input variable in the HEC-RAS model to calculate the water depth at the downstream location. In conclusion, the modeling approach determined the effectiveness of Random Forest Regression for hydrology and the integrated Random Forest and HEC-RAS model for hydraulic analysis.

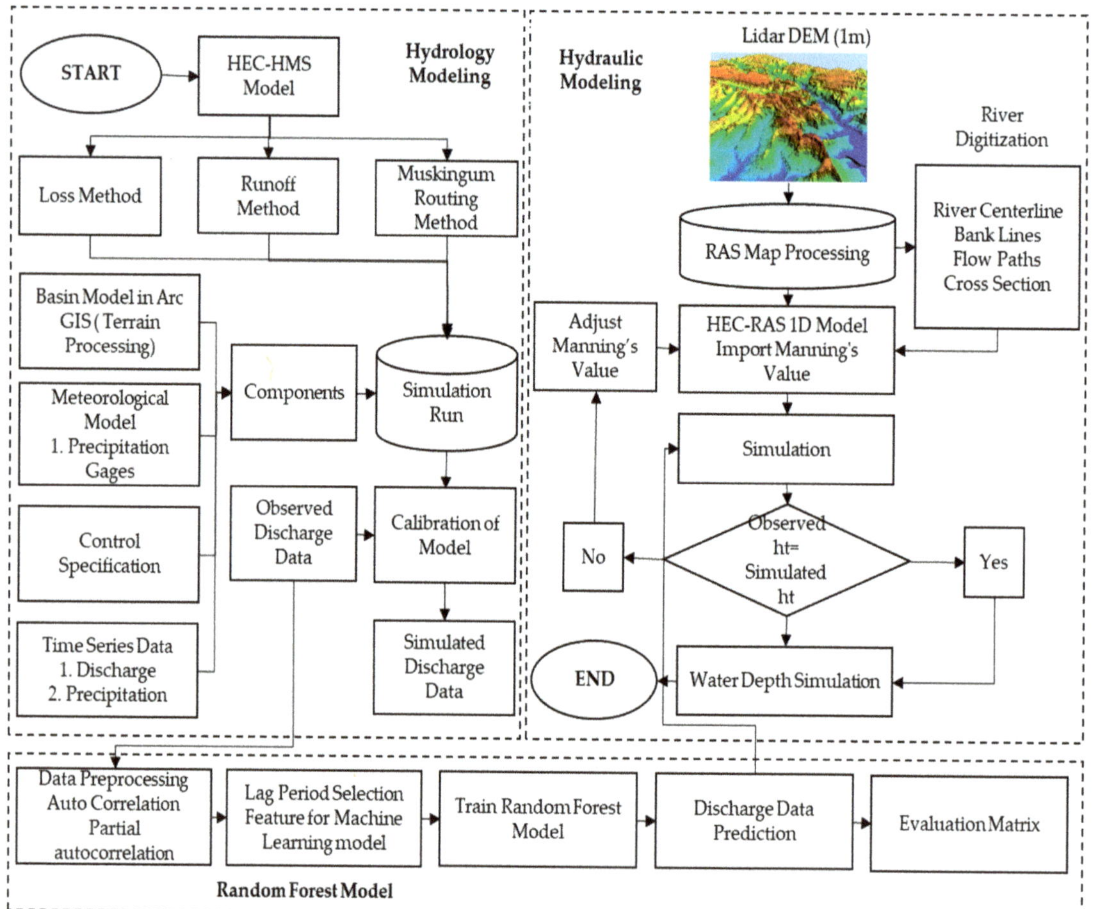

Figure 1. Figure portraying the flowchart of hydrology analysis using Random Forest and HEC-HMS and hydraulic analysis using HEC-RAS.

2.1. Study Area

This research used the East Branch DuPage watershed as a study area. Over the last twenty years, the study area has observed significant urbanization [75]. The study area has a history of high-flooding events (1996, 2008, 2013, and, most recently, 2020). In the year 2020, there was significant flooding due to the 178 mm of total precipitation over a period of five days. The study watershed has an area of 62.2 km^2 at the USGS gauging station, which is around Downers Grove, Illinois. The study area has an elevation ranging from 204 m to 250 m above mean sea level. Geographically, northern latitudes from 41°50′ to 41°57′ and western longitudes from 87°59′ to 88°6′ bound the study catchment area, as shown in Figure 2. The study area is highly residential, with an average imperviousness percentage of about 40%. The range of imperviousness percentages in the watershed is shown in Figure 3. The average soil permeability over the watershed is 62 mm/h [76]. The catchment consists of USGS gauge station 05540160 at the watershed outlet. The river reach for the hydraulic station lies between the gauging stations 05540160 and 05540228. The study reach is around 5221 m between two gauging stations. The proposed study area does not have any existing precipitation gauging station. The history of flooding events and the unavailability of observed precipitation data in this watershed are the two main reasons for proposing this watershed as a study area.

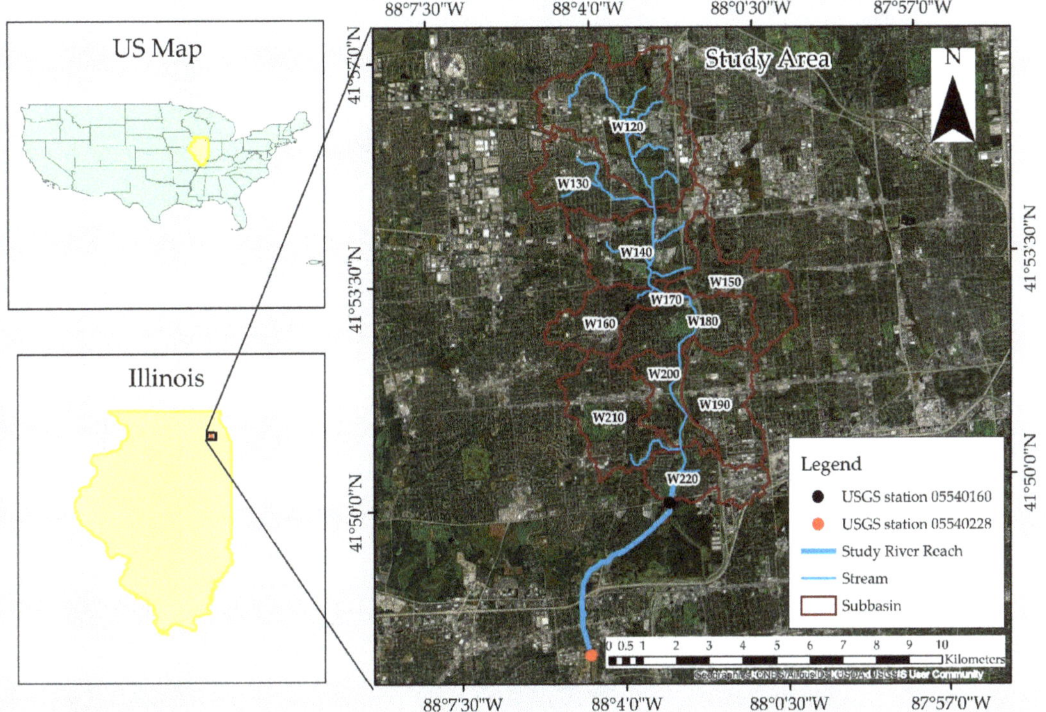

Figure 2. The East Branch DuPage Catchment around Downers Grove, Illinois, with the river system.

2.2. Data

Watershed characteristics datasets, such as land use and land cover, soil group, and DEM datasets, and meteorological model data, such as rainfall and discharge data, are all important data required for hydrology and hydraulic simulation. These datasets were used to estimate hydrology parameters and sub-basin characteristics and to prepare geometric data for hydrology and hydraulic analysis. The data types used in this research and their sources are detailed in Table 1.

2.3. Preprocessing Data

This section describes the extraction of basin characteristics and the meteorological data that were used for the hydrology analysis.

2.3.1. Digital Elevation Model (DEM)

DEM data are spatial data that provide the characteristics of the watershed. A 10 m DEM was retrieved from a United States Department of Agriculture (USDA) website and was clipped for the study catchment using Arc-Map in Arc-GIS.

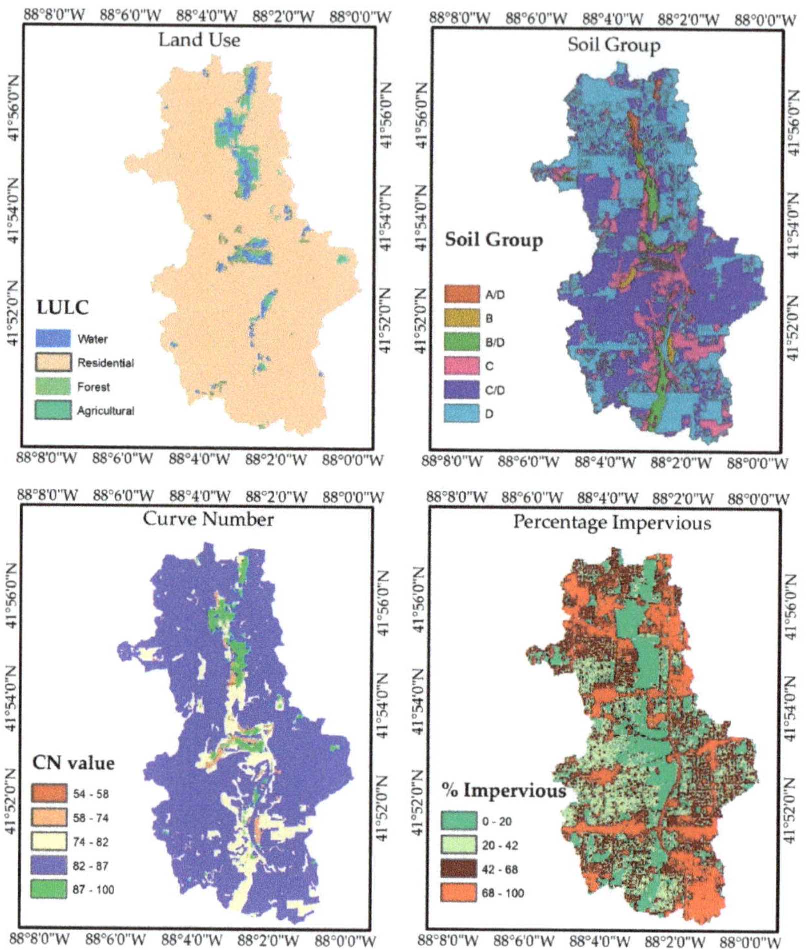

Figure 3. Map depicting characteristics of the study area.

Table 1. Data used for this research with their sources.

Data	Source
Precipitation	Precipitation Estimation from Remotely Sensed Information Using Artificial Neural Networks–Cloud Classification System (PERSIANN-CCS).
Soil	United States Department of Agriculture (USDA)
Land Use Land Cover	United States Geological Survey (USGS)
Runoff Data	United States Geological Survey (USGS) water data

2.3.2. Basin Characteristics

LULC data and soil map data were extracted from a USGS and USDA website, respectively. Both datasets were imported into ArcMap to clip for a study boundary and converted to the Shapefile from the raster. Composite curve number values were generated considering pervious and impervious areas. The average curve number of the watershed was 83.4, and the curve number values ranged from 54 to 100, corresponding to high infiltration to water bodies, respectively. The basin characteristics of the study area are shown in Figure 3.

2.3.3. Precipitation Data

Rainfall data are essential meteorological data for hydrology simulations. The study area does not consist of any observed precipitation station; therefore, in this study, precipitation data were obtained from a grid from the Precipitation Estimation from Remotely Sensed Information Using Artificial Neural Networks–Cloud Classification System (PERSIANN-CCS). The Center for Hydrometeorology and Remote Sensing (CHRS) develops it at the University of California, Irvine, and it is a real-time global high-resolution ($0.04° \times 0.04°$ pixel) satellite precipitation product [77]. The daily time series precipitation data were extracted from a grid using a python environment from 2006 to 2021.

2.4. Hydrologic Modeling Using Arc-GIS and HEC-HMS

HEC-GeoHMS is an extension of Arc-GIS that helps users to extract the essential data to develop the HEC-HMS project. The user must pick an outlet position on the river to begin the extraction procedure. HEC-GeoHMS utilizes terrain preprocessing tools for flow analysis. HEC-GeoHMS can enhance the sub-basin and stream delineations, collect physical attributes of sub-basins and rivers, predict model attributes, and create input files for HEC-HMS. Terrain preprocessing and model development were carried out as shown in Figure 4.

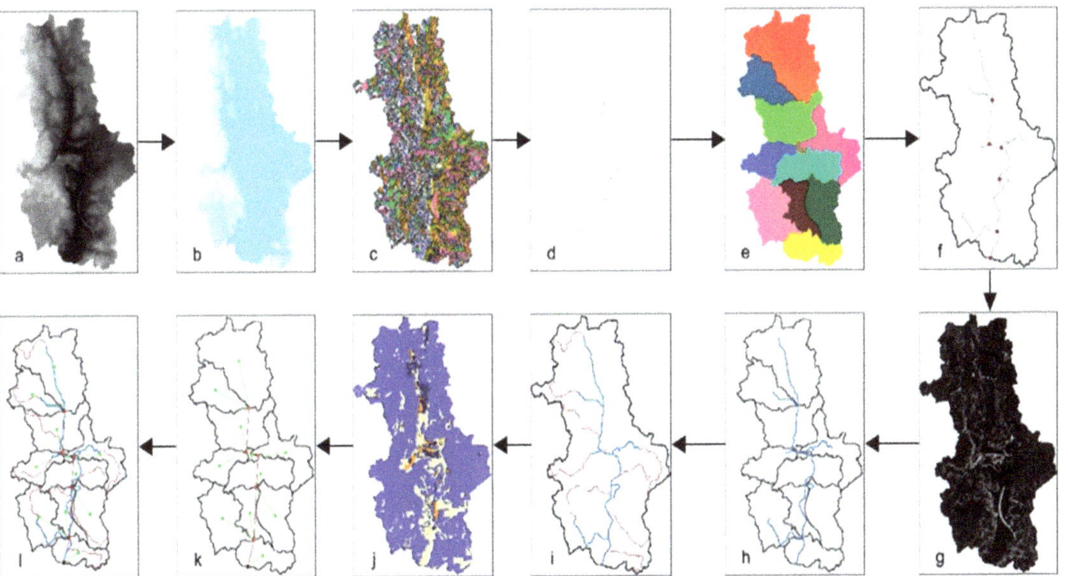

Figure 4. Preprocessing and model development: (**a**) DEM file; (**b**) Fill Sinks; (**c**) Flow Accumulation; (**d**) Flow Direction; (**e**) Stream Definition and Catchment Polygon; (**f**) Drainage Point and Line Processing; (**g**) Slope; (**h**) Basin and River Merge; (**i**) Lonest Flow Path; (**j**) CN Lag; (**k**) Sub-basin Nodes and River Links; (**l**) HEC-HMS input file.

2.4.1. Loss Method: SCS-CN for Rainfall-Runoff

The Soil Conservation Service curve number (SCS-CN) is a loss model that can compute the volume of the river flows [78]. Surface runoff excess depends on the precipitation, soil, and LULC of a particular watershed. Equation (1) is a mathematical expression used to determine the surface runoff.

$$Q = \frac{(P - I_a)^2}{(P - I_a) + S)} \tag{1}$$

where

Q = Runoff (inches);
P = Rainfall depth (inches);
I_a = Initial abstraction, and $I_a = 0.2\ S$;
S = Potential maximum retention.

The potential maximum retention in inches, S, is calculated using Equation (2):

$$S = \frac{1000}{CN} - 10 \tag{2}$$

2.4.2. Transform Method: SCS Unit Hydrograph

The SCS Unit Hydrograph transforms excess precipitation into a runoff. The SCS proposed the Unit Hydrograph, which is used in the HEC-HMS model. It is a parametric model based on the average Unit Hydrograph, which is created from gauged precipitation and discharge data of various agricultural watersheds collected across the United States. It assumes that a Unit Hydrograph depicts the constant properties of a watershed. The lag time is the sole input variable for this method. It is the time distance between the center of excess rainfall and the hydrograph peak, and HEC-HMS computes it for each sub-basin using Equation (3).

$$T_{lag} = \frac{(S+1)^{0.7} L^{0.8}}{1900 * Y^{0.5}} \tag{3}$$

where

T_{lag} = lag time (h);
L = hydraulic length of the watershed (ft);
Y = slope of the watershed (%);
S = maximum retention in the watershed (inches).

2.4.3. Routing Method: Muskingum Routing

Discharges from sub-basins were routed through the reaches to the outlet of the watershed using the Muskingum routing method. X and K are the two main parameters used in this method. Theoretically, the K parameter is the wave's travel time through the reach. These parameters can be approximated using observed inflow and outflow hydrographs. The X parameter is a weight coefficient of discharge, whose value fluctuates between 0 and 0.5. The interval between the inflow and outflow hydrographs of an identical station can be used to determine the parameter K. In this model, routing methods parameters were used to calibrate the model.

2.5. Hydrologic Modeling Using Random Forest

This study investigated the capacity of a Random Forest algorithm for predicting the daily discharge using the meteorological and hydrology features. Nonlinear interactions between a dependent variable and several independent variables can be represented using regression tree ensembles such as the Random Forest technique. Despite the popularity of the Random Forest algorithm in a myriad of environmental science fields, its application in the water sector needs to be further explored [79]. Random Forest is the type of supervised machine learning algorithm that can be used for classification and prediction. Random Forest uses the different tree predictors, and the random vector determines their values [61]. Random Forest is a collection of decision trees, where each tree is slightly different from the others. Ensemble learning combines all the decision trees and the average values predicted by each decision tree, solving the regression problem. This algorithm addresses the problem of training data overfitting in decision trees [80]. Random Forest has good performance in large datasets, and its features do not need to be scaled [81]. It is advantageous for features with different scales. Random Forests are appealing for both classification and regression tasks, are computationally fast, are efficient for unstable prediction, and perform well with high-dimensional features [82,83]. This algorithm's key idea is that each tree might make a fair prediction on its part; however, overfitting seems to occur on some of the data. If numerous trees are built, they will work and overfit in various ways. The average of these results will assist in the reduction of overfitting while holding onto the predictive power of decision trees.

Model Development

Many decision trees with bootstrap aggregation are used to minimize the overfitting issue [84]. A Random Forest Regressor consisting of 100 decision trees, as n-estimators, were applied to this dataset. The max depth parameter defines the maximum depth of the tree. The max depth of the model was fixed to be 100. The max depth by default was 'None', which signifies that the nodes were enlarged until all the leaves had fewer than min_samples_split samples. Min_samples_split means the total number of samples needed to break the internal node. Since we were trying to maintain the number of decision trees at only 100, max features was set to 'auto', which means that max features was equal to n features (the number of features seen during the model fitting). The parameter max-leaf nodes = None refers to an unlimited number of leaf nodes, leaving the decision trees to grow to best fit the model. All of the daily hydrology and meteorological feature samples from 2006 to 2021 were used for training and testing the algorithm. A total of 80% of the dataset was used for the training, and 20% of the dataset was used for the testing of the Random Forest model.

A box plot of daily discharge was created to visualize the patterns of daily discharge as shown in Figure 5c. Daily runoff was checked by plotting the autocorrelation and partial autocorrelation factors. Figure 5a,b show the autocorrelation plot and the partial autocorrelation plot of historical daily runoff observations, respectively. These plots helped us identify a suitable lag period for flow prediction in a watershed [84]. Five sets of discharge values at a lag time of 1 to 5 days were selected to predict the discharge. Similarly, six sets of precipitation at 1 to 5 days of lag time were selected. Table 2 represents the combination of input features used to train the Random Forest Regression. In addition, the cumulative precipitation for 5 days and the day on which the rainfall was greater than 12.7 mm were considered as additional features for predicting the runoff at the outlet of the watershed. NumPy, Pandas, Matplotlib, stats model, Sklearn, and seaborn are the python libraries that were used during data processing, training, and visualization.

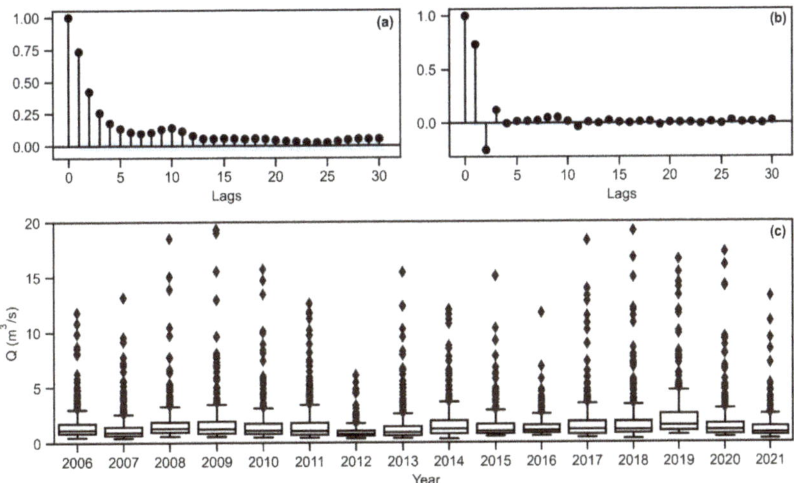

Figure 5. (**a**) Autocorrelation plot of the historical runoff observations of the DuPage River; (**b**) Partial autocorrelation plot of the historical runoff observations of the DuPage River; (**c**) box plot showing the flood events of the DuPage River.

Table 2. The combination of inputs for runoff prediction using Random Forest Regression.

Lag (Days)	The Structure of the Input	Output
5	Discharge of 1 day to the 5-day lag period, Precipitation of 1 day to the 5-day lag period, Sum of 5 days of precipitation (P5 days), Days since last precipitation greater than 0.5 mm. ($p > 0.5$)	One day ahead discharge

The autocorrelation function and the 95% confidence interval are shown in Figure 5a. A strong correlation was found up to 20 lags. The decay of autocorrelation shows the strength of the autoregressive process [29]. Similarly, the partial autocorrelation and 95% confidence interval were calculated. The partial autocorrelation depicted a strong correlation up to a 5-day lag period. Therefore, a lag period of 5 days was selected for the input [29].

2.6. Hydraulic Modeling Using HEC-RAS

Hydraulic modeling using HEC-RAS uses adequate geometry and flow data inputs for an excellent hydraulic model. The 1D HEC-RAS model is commonly employed to analyze flow in mainstream channels and predict the flood extent. Although the 1D model has limited applications, it is cost-effective, durable, and favored when determining flow pathways [85]. When speed is required and flood plain geometry data are scarce, 1D modeling is chosen [86]. HEC-RAS calculates the energy expression using Equation (4), which is based on Saint Venant's equation.

$$Z_2 + Y_2 + \frac{\alpha_2 V_2^2}{2g} = Z_1 + Y_1 + \frac{\alpha_1 V_1^2}{2g} + h_e \tag{4}$$

where

Y_1 and Y_2 = water heights at cross-sections,
Z_1 and Z_2 = elevations of the stream reach,
α_1 and α_2 = velocity weighting coefficients,
V_1 and V_2 = average velocities,
g = acceleration due to gravity, and
h_e = energy head loss.

River Geometry Generation

Hydraulic analysis with HEC-RAS starts with extracting the river section geometry data using the RASMAP, which is available in the HEC-RAS model. The process involved in the hydraulic analysis using HEC-RAS is illustrated in the flowchart in Figure 1. The Lidar 1 m DEM for the hydraulic model was obtained from the USGS website. The DEM data were imported into the RAS Mapper tool in the HEC-RAS model and converted into a Digital Terrain Model. In addition, the georeferenced projection file was assigned in RASMAP for the consistent coordinate system. In the RASMAP, the river centerline, bank lines, flow path lines, and cross-section lines were digitized. The Manning's n value was assigned to each cross-section in the entire reach. After the creation of the river geometry and applying the Manning's n value, the steady discharge was used as input data for the steady flow simulation. The water depth achieved from the simulation was then compared to the water depth at gauging stations downstream of the study reach. The Manning's n values at the main channel and over banks were adjusted for the calibration of a model.

2.7. Statistcal Performance Indicators

The performance of each model should be examined to determine the best models among different model alternatives. The five evaluation metrics (RMSE, RSR, NSE, PBIAS, and R^2) recommended by [87] and the NRMSE were used in this research to assess the performance of the hydrology model. The criteria used to evaluate the proposed model's performance are listed in Table 3.

Table 3. List of statistical indexes used to determine the performance of models.

Indices	Mathematical Expression	Satisfactory Range
Root Mean Square Error (*RMSE*)	$RMSE = \sqrt{\dfrac{\sum_{i=1}^{N}(Q_{s,i}-Q_{o,i})^2}{N}}$	
Nash–Sutcliffe efficiency coefficient (*NSE*)	$NSE = 1 - \left[\dfrac{\sum_{i=1}^{N}(Q_{o,i}-Q_{s,i})^2}{\sum_{i=1}^{N}\left(Q_{o,i}-\overline{Q}_o\right)^2}\right]$	$0.5 < NSE \le 1$
Coefficient of Determination (*R²*)	$R^2 = \dfrac{\left(\sum_{i=1}^{N}(Q_{o,i}-\overline{Q_{0,i}})*\left(Q_{s,i}-\overline{Q_{0,i}}\right)\right)^2}{\sum_{i=1}^{N}\left(Q_{o,i}-\overline{Q_{0,i}}\right)^2 * \sum_{i=1}^{N}\left(Q_{s,i}-\overline{Q_{0,i}}\right)^2}$	>0.5
Standard Deviation Ratio (*RSR*)	$RSR = \dfrac{RMSE}{standard\ Deviation}$	$0 < RSR < 0.7$
Percentage bias (*PBIAS*)	$PBIAS = \dfrac{\sum_{i=1}^{N}(Q_{o,i}-Q_{s,i})*100}{\sum_{i=1}^{N}Q_{o,i}}$	$-25\% < PBIAS < +25\%$
Normalized Root Mean Squared Error (*NRMSE*)	$NRMSE = \dfrac{\frac{1}{N}\sum_{i=1}^{N}(Q_{s,i}-Q_{o,i})^2}{Mean}$	$\le 30\%$

where $Q_{o,i}$ represents the observed data, $Q_{s,i}$ represents the simulated data from the model, $Q_{o,i}$ represents the mean value of the total number of observed data samples, and n represents the total number of data samples.

3. Results and Discussion

This section describes the results of the study, and it covers four main topics. In this section, the results of the precipitation product, hydrology, and hydraulic analyses are presented.

3.1. Precipitation

The rainfall data applied in this research were extracted from satellite-based rainfall products for a time period of 16 years (2006–2021). The daily rainfall data obtained for the studied time period are shown in Figure 6a. The daily precipitation data pattern was consistent with the daily observed discharge data. The result shows that the time of peak rainfall data matched the time of peak discharge data. For example, in this watershed outlet, the highest peak discharge of 33.7 m³/s was observed on 14 September 2008 and, similarly, the extracted precipitation product produced the highest precipitation of 61 mm on the same day. In addition, the validation of the extracted precipitation data was supported by the results of the hydrology analysis, which are presented in the following section.

3.2. HEC-HMS Models

Integration of the Arc-Hydro tool and HEC-GeoHMS successfully generated all the sub-basin parameters needed for the hydrology analysis. HEC-GeoHMS is a sophisticated tool that can be used to delineate natural watersheds and perform automatic basin parameter extraction for the HECHMS model construction. Table 4 lists the basin parameters obtained from HEC-GeoHMS, including sub-basin area, slope, curve number, and basin lag.

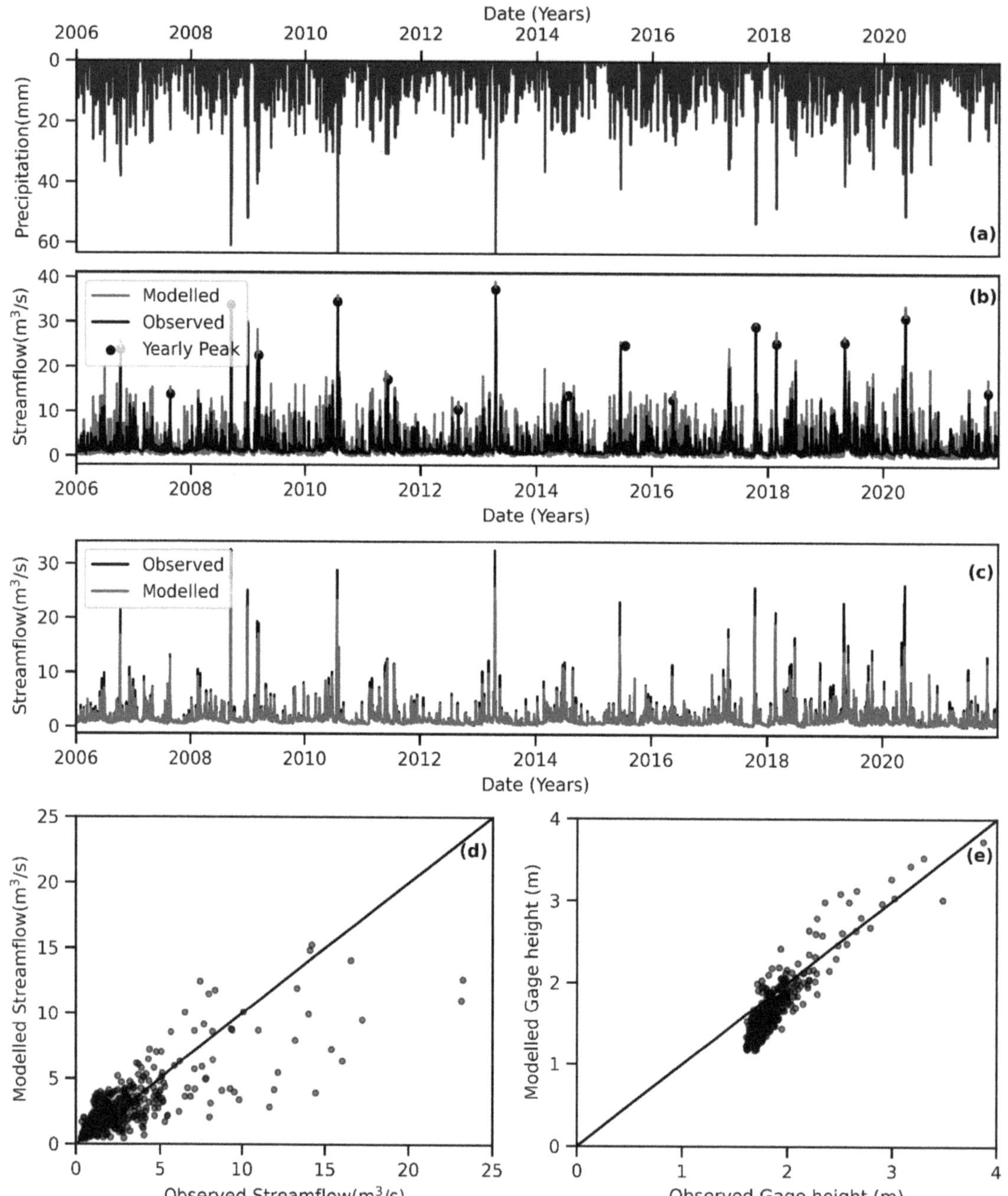

Figure 6. (**a**) Representation of the generated precipitation product; (**b**) training and testing of the HEC-HMS model; (**c**) observed discharge and predicted discharge for Random Forest Regression; (**d**) observed historical and predicted runoff data; (**e**) observed gauge height and simulated gauge height from HEC-RAS.

Table 4. Geographic characteristics of the study watershed.

Sub-Basin	Basin Area (km^2)	Basin Slope (%)	Curve Number (CN)	Basin Lag (min)
W220	4.3	2.6	85.8	150
W210	7.0	2.8	84.7	135
W200	3.6	3.1	83.6	133
W190	6.2	1.9	83.9	84
W180	5.9	3.5	83.2	90
W170	0.3	4.5	86.7	84
W160	3.7	2.6	82.3	81
W150	5.5	3.5	83.7	98
W140	7.4	4.5	83.0	86
W130	5.3	2.2	84.2	20
W120	13.0	3.4	84.0	76

The calibration and validation of the HEC-HMS model in this research were performed by adjusting the Muskingum parameters. The measured discharge from the gauging station was compared to the yearly peak discharge produced from an HEC-HMS simulation. Event 1 January 2006 to 31 December 2018 was considered for the model calibration, and Event 1 January 2019 to 31 December 2019 was used for the model validation. The accuracy of the hydrology model using HEC-HMS was determined using a statistical index. The discharge generated using HEC-HMS for the study period is presented in Figure 6b. The root mean square error is one of the most-used methods for evaluating the validity of predictions. The RMSE value during calibration and validation was 1.45 m^3/s and 2.45 m^3/s, respectively, which is considered a good result. The RSR is calculated by dividing the RMSE by the standard deviation of the measured data, and a value less than 0.7 is considered a good result [88]. The RSR values for the HEC-HMS model were 0.16 and 0.35. The NSE is extensively used in measuring the model performance in hydrology. It ranges from -1 to 1, with 0.5 to 1 being the best values. The NSE method is used to calculate the residual variance in relation to the variance in the measured data. The NSE values were 0.97 and 0.87, respectively, which are close to 1.

The PBIAS shows the average inclination of the calculated data. For a good model, PBIAS values must approach zero or should be between $\pm25\%$ [89]. Positive numbers suggest that the model is underestimated, whereas negative values indicate that the model is overestimated [90]. The HEC-HMS model overestimated the peak discharge by 5.3% and 9.8% during calibration and validation, respectively. The R^2 is used to determine the correlation between calculated and measured flow rates. An R^2 greater than 0.5 indicates satisfactory performance. For the calibration and validation, the R^2 values were 0.99 and 0.96, respectively. The R^2 values close to 1 for the HEC-HMS model validated the accuracy of the model.

3.3. Random Forest Regression Model

Random Forest Regression provided good insights into the prediction of daily discharge data. Figure 6c illustrates the observed discharge data and the Random Forest predicted data during the study period. The scatter plot in Figure 6d demonstrates that the Random Forest prediction data were clustered near the regression line under low- and normal-flow conditions. However, Random Forest Regression slightly underestimated the high discharge value, which can also be termed an extreme event. Table 5 shows the evaluation matrix for Random Forest Regression. The RMSE, RSR, NSE, PBIAS, R^2, and NRMSE values were 0.29 m^3/s, 0.23, 0.94, -0.75%, 0.94, and 0.17 for the training period and 0.47 m^3/s, 0.56, 0.69, $+1.76\%$, 0.72, and 0.260 for testing period, respectively, as shown in Table 5. The statistical index revealed that the Random Forest model's performance was superior during data training. The values of the statistical index dropped sharply during the testing period. The PBIAS values during training and testing were close to 0%, representing the average inclination of the predicted discharge towards the observed discharge. The values of R^2 dropped sharply from 0.94 during training to 0.72 during testing. However, the values of the statistical index were within acceptable ranges during the testing period. Scatter plots were used to analyze the prediction performance of Random Forest Regression with the observed data. In the scatter plot between the observed and predicted values, the more significant deviation was observed for the higher discharge value, which also demonstrates the lower effectiveness of the Random Forest model for peak discharge estimation. The non-peak discharge was more accurately predicted by the machine learning model.

Table 5. Calibration and validation statistics of the HEC-HMS and Random Forest models.

Statistical Index	HEC-HMS Model		Random Forest	
	Calibration	Validation	Training	Testing
RMSE ($\mathrm{m^3/s}$)	1.45	2.45	0.29	0.47
RSR	0.16	0.35	0.23	0.56
NSE	0.97	0.87	0.94	0.69
PBIAS	-5.30%	-9.80%	-0.75%	$+1.76\%$
R^2	0.99	0.96	0.94	0.72
NRMSE	0.06	0.10	0.17	0.26

Random Forest Regression was used for the prediction of discharge for the given input precipitation. The feature selection based on the lag period of precipitation and discharge was used. The validated results of HEC-HMS and Random Forest were compared to determine their ability to predict the discharge for the study period. After the comparison, we observed that the conventional HEC-HMS model needed more parameter optimization than Random Forest Regression. Similarly, the aim of study was also to prove the suitability of the discharge data predicted by Random Forest for hydraulic analysis. The scatter plot shown in Figure 6e shows the observed gauge height in the gauging station versus the simulated gauge height from the HEC-RAS model. During high-flooding events, the water depth predicted by the hydraulic model using the Random Forest generated discharge was slightly underestimated compared with the observed water depth. As the model showed good performance in generating the water depth under non-flooding conditions, the integration of Random Forest and HEC-RAS could be used to derive useful information while planning the water resource infrastructure and flood control measures in the selected study area. As the performance of a watershed model relies on the precision, robustness, and application of the model under other site conditions, the proposed approach could be tested and analyzed for multiple catchment locations, so that the parameters could be fixed to increase the reliability of the result.

3.4. HEC-RAS Model

The hydraulic analysis was carried out for the East DuPage watershed's downstream reach. For calibration purposes, historical discharge data from flood events in 2020 and 2021 were used, and the results are displayed in Table 6. The study reach consists of only one USGS gauging station at the most downstream location of the study reach with gauge height data beginning from 2020. The hydraulic model was calibrated using water depth data from various flooding events in 2021 and 2022. Figure 6e shows the comparison of simulated and observed data at most downstream stations of the study reach. The Manning's n value was adjusted to calibrate the hydraulic model. The water depth produced from the simulation was similar to the observed water depth at the gauging station, as shown in Figure 6e; this result demonstrates the model's consistency and allows it to be used for further investigation. At the upstream cross-section of the reach, daily discharge data from Random Forest were used to calculate the water depth at the downstream reach. The scatter plot in Figure 6e shows that the discharge calculated using Random Forest Regression can be utilized to calculate the flood depth in a river stream. Compared with the observed water depth at the gauging station, the model underestimates the simulated water depth generated from the study.

Table 6. The difference between the observed and simulated water depth.

Event	Discharge (m³/s)	Observed Water Depth (m)	Simulated Water Depth (m)	Difference (m)
11 January 2020	8.78	2.79	2.68	0.11
30 March 2020	3.11	2.09	1.98	0.11
29 March 2020	5.07	2.33	2.58	−0.25
30 April 2020	16.03	3.40	3.02	0.38
18 May 2020	26.42	4.41	3.85	0.56
23 October 2020	6.57	2.45	2.91	−0.46
12 December 2020	8.04	2.52	2.61	−0.09
19 March 2021	2.83	1.96	2.03	−0.07
26 June 2021	10.96	2.90	3.27	−0.37
27 August 2021	2.21	1.81	1.89	−0.08
26 October 2021	8.38	2.50	2.48	0.02

4. Discussion

The results of the hydrology simulation provide strong support for the effectiveness of the satellite precipitation product for the hydrology simulation in an ungauged catchment. Both the HEC-HMS and Random Forest models accurately recreated the discharge characteristics, such as the flood peak and timing, during the study period. These findings are consistent with those of previous studies that showed that PERSIANN-CCS precipitation products could effectively simulate the hydrology in ungauged watersheds [91,92]. The statistical index in Table 5 from the model calibration and validation suggests that Random Forest can be effectively applied for estimating the daily discharge at watershed outlets. The good performance of Random Forest for the hydrology analysis proved its appropriateness for rainfall-runoff simulation in data-scarce regions. The results of Random Forest are in agreement with a previous study's finding of good performance as an alternative prediction method in the hydrology domain [93]. The statistical index in Table 5 proved the suitability of both Random Forest and HEC-HMS for rainfall-runoff simulation. The results illustrate that Random Forest slightly underestimated the peak discharge during the high-flooding events; however, during the non-flooding period, the discharge predicted by Random Forest was better than that predicted by the HEC-HMS model. Figure 6e provided good support for the effectiveness of the Random-Forest-generated discharge for hydraulic simulation. The result indicates that the water depth simulated by HEC-RAS at the most downstream cross-section was slightly underestimated compared with the observed water depth at the gauging station. This result may be due to the use of the slightly underestimated peak discharge obtained from the Random Forest model. The overall result of this research work supports the integration of machine learning and a physical-based model for rainfall-runoff and flood depth prediction in data-scarce regions.

5. Conclusions

This study evaluated the feasibility of HEC-HMS and Random Forest for rainfall-runoff simulation and an integrated approach of machine learning and HEC-RAS for hydraulic analysis. HEC-HMS requires a large number of input variables, which may not always be available in a data-scarce region. In this scenario, the Random Forest model can be used for the prediction of discharge in the watershed. In addition, the Random Forest model is simple to build and takes less time. In this study, a PERSIANN-CCS NetCDF file was used to generate time-series precipitation data. The result supports the usage of PERSIANN-CCS daily precipitation data for rainfall-runoff simulation. Based on the models' reasonably strong performance, the obtained precipitation, LULC, DEM, and SSURGO soil input data are sufficiently dependable for discharge simulation. Because the data sources employed in this study yield reasonably reliable results, they are recommended for hydrology investigations. The continuous simulation of rainfall-runoff processes in the basin using physical and machine learning models yielded good results. Peak flows were underestimated in the Random Forest model and slightly overestimated in the HEC-HMS model. An integrated HEC-RAS and Random Forest Regression model yielded good results in predicting the runoff flood depth downstream of a watershed. Given these findings, it is possible to say that the Random Forest model could aid in rainfall-runoff simulation as a complement to the physical model. This discharge could be used in hydraulic modeling for flood depth and flood extent analysis, which could be helpful to researchers in further research. The model's accuracy in predicting the flow can be increased by removing the outliers; high flood values are considered here in order to compensate for the prediction of the high flood values from Random Forest Regression. In the future, researchers could work in the following areas:

1. In this study, we used the PERSIANN precipitation product, and future work may be more accurate if there is a precipitation gauging station. Furthermore, researchers could also use other precipitation products, such as Next-Generation Weather Data (NEXRAD) and Climate Hazards Group Infrared Precipitation (CHIRPS);
2. In this study, precipitation was only used as an input variable for the Random Forest model; other variables, such as temperature, infiltration, evaporation, and radiation, could be used in future work. In addition, feature selection of input variables could be performed for the most accurate selection;
3. Other machine learning and data-driven models, such as support vector regression (SVR), long short-term memory (LSTM), and artificial neural networks (ANNs), could be used as prediction models. Future research directions could be guided by the selection of the best machine learning model in terms of accuracy, robustness, and reliability;
4. Although the study area is a small watershed in DuPage County, future research could focus on a more dynamic, heterogeneous, and meteorologically unique basin.

Author Contributions: Conceptualization, A.K.; formal analysis, A.B., investigation, A.B., U.P. and S.R.; software, A.B. and U.P.; supervision, A.K; writing—initial draft preparation, A.B., U.P., S.R. and A.K.; writing—review & editing, A.B., U.P. and A.K. All authors have read and agreed to the published version of the manuscript.

Funding: This research received no external funding.

Institutional Review Board Statement: Not applicable.

Informed Consent Statement: Not applicable.

Data Availability Statement: Data available in a publicly accessible repository. The data download information is available in Table 1.

Acknowledgments: The authors would like to thank the reviewers for their valuable suggestions. The authors acknowledge the support of Southern Illinois University and Carbondale's Vice-Chancellor for Research. The research, simulation, and analysis were done with open-source software and datasets.

Conflicts of Interest: The authors declare no conflict of interest.

References

1. Merwade, V.; Olivera, F.; Arabi, M.; Edleman, S. Uncertainty in Flood Inundation Mapping: Current Issues and Future Directions. *J. Hydrol. Eng.* **2008**, *13*, 608–620. [CrossRef]
2. Merz, B.; Kreibich, H.; Schwarze, R.; Thieken, A. Review Article: Assessment of Economic Flood Damage. *Nat. Hazards Earth Syst. Sci.* **2010**, *10*, 1697–1724. [CrossRef]
3. Gaume, E.; Bain, V.; Bernardara, P.; Newinger, O.; Barbuc, M.; Bateman, A.; Blaškovičová, L.; Blöschl, G.; Borga, M.; Dumitrescu, A.; et al. A Compilation of Data on European Flash Floods. *J. Hydrol.* **2009**, *367*, 70–78. [CrossRef]
4. Ghazali, D.; Guericolas, M.; Thys, F.; Sarasin, F.; Arcos González, P.; Casalino, E. Climate Change Impacts on Disaster and Emergency Medicine Focusing on Mitigation Disruptive Effects: An International Perspective. *Int. J. Environ. Res. Public Health* **2018**, *15*, 1379. [CrossRef] [PubMed]
5. Faccini, F.; Luino, F.; Paliaga, G.; Sacchini, A.; Turconi, L.; de Jong, C. Role of Rainfall Intensity and Urban Sprawl in the 2014 Flash Flood in Genoa City, Bisagno Catchment (Liguria, Italy). *Appl. Geogr.* **2018**, *98*, 224–241. [CrossRef]

6. Sapountzis, M.; Kastridis, A.; Kazamias, A.-P.; Karagiannidis, A.; Nikopoulos, P.; Lagouvardos, K. Utilization and Uncertainties of Satellite Precipitation Data in Flash Flood Hydrological Analysis in Ungauged Watersheds. *Glob. NEST J.* **2021**, *23*, 388–399. [CrossRef]

7. Pathak, P.; Kalra, A.; Ahmad, S. Temperature and Precipitation Changes in the Midwestern United States: Implications for Water Management. *Int. J. Water Resour. Dev.* **2017**, *33*, 1003–1019. [CrossRef]

8. Jenkins, K.; Surminski, S.; Hall, J.; Crick, F. Assessing Surface Water Flood Risk and Management Strategies under Future Climate Change: Insights from an Agent-Based Model. *Sci. Total Environ.* **2017**, *595*, 159–168. [CrossRef]

9. Kundzewicz, Z.W.; Kanae, S.; Seneviratne, S.I.; Handmer, J.; Nicholls, N.; Peduzzi, P.; Mechler, R.; Bouwer, L.M.; Arnell, N.; Mach, K.; et al. Flood Risk and Climate Change: Global and Regional Perspectives. *Hydrol. Sci. J.* **2014**, *59*, 1–28. [CrossRef]

10. Guerreiro, S.B.; Dawson, R.J.; Kilsby, C.; Lewis, E.; Ford, A. Future Heat-Waves, Droughts and Floods in 571 European Cities. *Environ. Res. Lett.* **2018**, *13*, 034009. [CrossRef]

11. Min, S.-K.; Zhang, X.; Zwiers, F.W.; Hegerl, G.C. Human Contribution to More-Intense Precipitation Extremes. *Nature* **2011**, *470*, 378–381. [CrossRef] [PubMed]

12. Vörösmarty, C.J.; de Guenni, L.B.; Wollheim, W.M.; Pellerin, B.; Bjerklie, D.; Cardoso, M.; D'Almeida, C.; Green, P.; Colon, L. Extreme Rainfall, Vulnerability and Risk: A Continental-Scale Assessment for South America. *Philos. Trans. R. Soc. A* **2013**, *371*, 20120408. [CrossRef] [PubMed]

13. Woznicki, S.A.; Baynes, J.; Panlasigui, S.; Mehaffey, M.; Neale, A. Development of a Spatially Complete Floodplain Map of the Conterminous United States Using Random Forest. *Sci. Total Environ.* **2019**, *647*, 942–953. [CrossRef] [PubMed]

14. Archer, D.R.; Fowler, H.J. Characterising Flash Flood Response to Intense Rainfall and Impacts Using Historical Information and Gauged Data in Britain: Flash Flood Response to Intense Rainfall in Britain. *J. Flood Risk Manag.* **2018**, *11*, S121–S133. [CrossRef]

15. Kastridis, A.; Stathis, D. The Effect of Rainfall Intensity on the Flood Generation of Mountainous Watersheds (Chalkidiki Prefecture, North Greece). In *Perspectives on Atmospheric Sciences*; Karacostas, T., Bais, A., Nastos, P.T., Eds.; Springer Atmospheric Sciences; Springer International Publishing: Cham, Switzerland, 2017; pp. 341–347. ISBN 978-3-319-35094-3.

16. Schoppa, L.; Disse, M.; Bachmair, S. Evaluating the Performance of Random Forest for Large-Scale Flood Discharge Simulation. *J. Hydrol.* **2020**, *590*, 125531. [CrossRef]

17. Talei, A.; Chua, L.H.C.; Quek, C. A Novel Application of a Neuro-Fuzzy Computational Technique in Event-Based Rainfall–Runoff Modeling. *Expert Syst. Appl.* **2010**, *37*, 7456–7468. [CrossRef]

18. Singh, V.P.; Frevert, D.K. *Watershed Models*; Taylor and Francis: Abingdon, UK, 2005.

19. Halwatura, D.; Najim, M.M.M. Application of the HEC-HMS Model for Runoff Simulation in a Tropical Catchment. *Environ. Model. Softw.* **2013**, *46*, 155–162. [CrossRef]

20. US Army Corps of Engineers. *Hydrologic Modeling System (HEC-HMS) Application Guide Version 3.1.0*; Institute for Water Resources: Davis, CA, USA, 2008.

21. Bajwa, H.S.; Tim, U.S. *Toward Immersive Virtual Environments for GIS-Based Floodplain Modeling and Visualization*; ESRI: Redlands, CA, USA, 2002.

22. Senthil Kumar, A.; Sudheer, K.; Jain, S.; Agarwal, P. Rainfall-runoff modelling using artificial neural networks: Comparison of network types. *Hydrol. Process.* **2005**, *19*, 1277–1291. [CrossRef]

23. Rezaeianzadeh, M.; Stein, A.; Tabari, H.; Abghari, H.; Jalalkamali, N.; Hosseinipour, E.Z.; Singh, V.P. Assessment of a Conceptual Hydrological Model and Artificial Neural Networks for Daily Outflows Forecasting. *Int. J. Environ. Sci. Technol.* **2013**, *10*, 1181–1192. [CrossRef]

24. Kim, B.; Sanders, B.F.; Famiglietti, J.S.; Guinot, V. Urban Flood Modeling with Porous Shallow-Water Equations: A Case Study of Model Errors in the Presence of Anisotropic Porosity. *J. Hydrol.* **2015**, *523*, 680–692. [CrossRef]

25. Sahoo, S.; Russo, T.A.; Elliott, J.; Foster, I. Machine Learning Algorithms for Modeling Groundwater Level Changes in Agricultural Regions of the U.S. *Water Resour. Res.* **2017**, *53*, 3878–3895. [CrossRef]

26. Rajaee, T.; Khani, S.; Ravansalar, M. Artificial Intelligence-Based Single and Hybrid Models for Prediction of Water Quality in Rivers: A Review. *Chemom. Intell. Lab. Syst.* **2020**, *200*, 103978. [CrossRef]

27. Zounemat-Kermani, M.; Batelaan, O.; Fadaee, M.; Hinkelmann, R. Ensemble Machine Learning Paradigms in Hydrology: A Review. *J. Hydrol.* **2021**, *598*, 126266. [CrossRef]

28. Jordan, M.I.; Mitchell, T.M. Machine Learning: Trends, Perspectives, and Prospects. *Science* **2015**, *349*, 255–260. [CrossRef] [PubMed]

29. Ghimire, S.; Yaseen, Z.M.; Farooque, A.A.; Deo, R.C.; Zhang, J.; Tao, X. OPEN Streamflow Prediction Using. *Sci. Rep.* **2021**, *11*, 17497. [CrossRef]

30. Mewes, B.; Oppel, H.; Marx, V.; Hartmann, A. Information-Based Machine Learning for Tracer Signature Prediction in Karstic Environments. *Water Resour. Res.* **2020**, *56*, e2018WR024558. [CrossRef]

31. Parisouj, P.; Mohebzadeh, H.; Lee, T. Employing Machine Learning Algorithms for Streamflow Prediction: A Case Study of Four River Basins with Different Climatic Zones in the United States. *Water Resour. Manag.* **2020**, *34*, 4113–4131. [CrossRef]

32. Adnan, R.M.; Petroselli, A.; Heddam, S.; Santos, C.A.G.; Kisi, O. Short Term Rainfall-Runoff Modelling Using Several Machine Learning Methods and a Conceptual Event-Based Model. *Stoch. Environ. Res. Risk Assess.* **2021**, *35*, 597–616. [CrossRef]

33. Shamshirband, S.; Hashemi, S.; Salimi, H.; Samadianfard, S.; Asadi, E.; Shadkani, S.; Kargar, K.; Mosavi, A.; Nabipour, N.; Chau, K.-W. Predicting Standardized Streamflow Index for Hydrological Drought Using Machine Learning Models. *Eng. Appl. Comput. Fluid Mech.* **2020**, *14*, 339–350. [CrossRef]

34. Nguyen, D.T.; Chen, S.-T. Real-Time Probabilistic Flood Forecasting Using Multiple Machine Learning Methods. *Water* **2020**, *12*, 787. [CrossRef]

35. Zhou, Y.; Cui, Z.; Lin, K.; Sheng, S.; Chen, H.; Guo, S.; Xu, C.-Y. Short-Term Flood Probability Density Forecasting Using a Conceptual Hydrological Model with Machine Learning Techniques. *J. Hydrol.* **2022**, *604*, 127255. [CrossRef]

36. Kalra, A.; Ahmad, S.; Nayak, A. Increasing Streamflow Forecast Lead Time for Snowmelt-Driven Catchment Based on Large-Scale Climate Patterns. *Adv. Water Resour.* **2013**, *53*, 150–162. [CrossRef]

37. Rezaei, K.; Pradhan, B.; Vadiati, M.; Nadiri, A.A. Suspended Sediment Load Prediction Using Artificial Intelligence Techniques: Comparison between Four State-of-the-Art Artificial Neural Network Techniques. *Arab. J. Geosci.* **2021**, *14*, 215. [CrossRef]

38. Choubin, B.; Darabi, H.; Rahmati, O.; Sajedi-Hosseini, F.; Kløve, B. River Suspended Sediment Modelling Using the CART Model: A Comparative Study of Machine Learning Techniques. *Sci. Total Environ.* **2018**, *615*, 272–281. [CrossRef] [PubMed]

39. Rezaei, K.; Vadiati, M. A Comparative Study of Artificial Intelligence Models for Predicting Monthly River Suspended Sediment Load. *J. Water Land Dev.* **2020**, *45*, 107–118. [CrossRef]

40. Wang, S.; Peng, H.; Liang, S. Prediction of Estuarine Water Quality Using Interpretable Machine Learning Approach. *J. Hydrol.* **2022**, *605*, 127320. [CrossRef]

41. Deng, T.; Chau, K.-W.; Duan, H.-F. Machine Learning Based Marine Water Quality Prediction for Coastal Hydro-Environment Management. *J. Environ. Manag.* **2021**, *284*, 112051. [CrossRef]

42. Melesse, A.M.; Khosravi, K.; Tiefenbacher, J.P.; Heddam, S.; Kim, S.; Mosavi, A.; Pham, B.T. River Water Salinity Prediction Using Hybrid Machine Learning Models. *Water* **2020**, *12*, 2951. [CrossRef]

43. Asadollah, S.B.H.S.; Sharafati, A.; Motta, D.; Yaseen, Z.M. River Water Quality Index Prediction and Uncertainty Analysis: A Comparative Study of Machine Learning Models. *J. Environ. Chem. Eng.* **2021**, *9*, 104599. [CrossRef]

44. Hussein, E.A.; Thron, C.; Ghaziasgar, M.; Bagula, A.; Vaccari, M. Groundwater Prediction Using Machine-Learning Tools. *Algorithms* **2020**, *13*, 300. [CrossRef]

45. Khedri, A.; Kalantari, N.; Vadiati, M. Comparison Study of Artificial Intelligence Method for Short Term Groundwater Level Prediction in the Northeast Gachsaran Unconfined Aquifer. *Water Supply* **2020**, *20*, 909–921. [CrossRef]

46. Zhu, S.; Piotrowski, A.P. River/Stream Water Temperature Forecasting Using Artificial Intelligence Models: A Systematic Review. *Acta Geophys.* **2020**, *68*, 1433–1442. [CrossRef]

47. Chang, H.; Psaris, M. Local Landscape Predictors of Maximum Stream Temperature and Thermal Sensitivity in the Columbia River Basin, USA. *Sci. Total Environ.* **2013**, *461–462*, 587–600. [CrossRef] [PubMed]

48. Weierbach, H.; Lima, A.R.; Willard, J.D.; Hendrix, V.C.; Christianson, D.S.; Lubich, M.; Varadharajan, C. Stream Temperature Predictions for River Basin Management in the Pacific Northwest and Mid-Atlantic Regions Using Machine Learning. *Water* **2022**, *14*, 1032. [CrossRef]

49. Feigl, M.; Lebiedzinski, K.; Herrnegger, M.; Schulz, K. Machine-learning methods for stream water temperature prediction. *Hydrol. Earth Syst. Sci.* **2021**, *25*, 2951–2977. [CrossRef]

50. Zhang, J.; Xu, J.; Dai, X.; Ruan, H.; Liu, X.; Jing, W. Multi-Source Precipitation Data Merging for Heavy Rainfall Events Based on Cokriging and Machine Learning Methods. *Remote Sens.* **2022**, *14*, 1750. [CrossRef]

51. Radhakrishnan, C.; Chandrasekar, V.; Reising, S.C.; Berg, W. Rainfall Estimation from TEMPEST-D CubeSat Observations: A Machine-Learning Approach. *IEEE J. Sel. Top. Appl. Earth Obs. Remote Sens.* **2022**, *15*, 3626–3636. [CrossRef]

52. Guo, W.-D.; Chen, W.-B.; Yeh, S.-H.; Chang, C.-H.; Chen, H. Prediction of River Stage Using Multistep-Ahead Machine Learning Techniques for a Tidal River of Taiwan. *Water* **2021**, *13*, 920. [CrossRef]

53. Chiang, S.; Chang, C.-H.; Chen, W.-B. Comparison of Rainfall-Runoff Simulation between Support Vector Regression and HEC-HMS for a Rural Watershed in Taiwan. *Water* **2022**, *14*, 191. [CrossRef]

54. Ni, L.; Wang, D.; Singh, V.P.; Wu, J.; Wang, Y.; Tao, Y.; Zhang, J. Streamflow and Rainfall Forecasting by Two Long Short-Term Memory-Based Models. *J. Hydrol.* **2020**, *583*, 124296. [CrossRef]

55. Yin, H.; Wang, F.; Zhang, X.; Zhang, Y.; Chen, J.; Xia, R.; Jin, J. Rainfall-Runoff Modeling Using Long Short-Term Memory Based Step-Sequence Framework. *J. Hydrol.* **2022**, *610*, 127901. [CrossRef]

56. Tikhamarine, Y.; Souag-Gamane, D.; Ahmed, A.N.; Sammen, S.S.; Kisi, O.; Huang, Y.F.; El-Shafie, A. Rainfall-Runoff Modelling Using Improved Machine Learning Methods: Harris Hawks Optimizer vs. Particle Swarm Optimization. *J. Hydrol.* **2020**, *589*, 125133. [CrossRef]

57. Tamiru, H.; Dinka, M.O. Application of ANN and HEC-RAS Model for Flood Inundation Mapping in Lower Baro Akobo River Basin, Ethiopia. *J. Hydrol. Reg. Stud.* **2021**, *36*, 100855. [CrossRef]

58. Samantaray, S.; Das, S.S.; Sahoo, A.; Satapathy, D.P. Monthly Runoff Prediction at Baitarani River Basin by Support Vector Machine Based on Salp Swarm Algorithm. *Ain Shams Eng. J.* **2022**, *13*, 101732. [CrossRef]

59. Adnan, R.M.; Liang, Z.; Heddam, S.; Zounemat-Kermani, M.; Kisi, O.; Li, B. Least Square Support Vector Machine and Multivariate Adaptive Regression Splines for Streamflow Prediction in Mountainous Basin Using Hydro-Meteorological Data as Inputs. *J. Hydrol.* **2020**, *586*, 124371. [CrossRef]

60. Worland, S.C.; Farmer, W.H.; Kiang, J.E. Improving Predictions of Hydrological Low-Flow Indices in Ungaged Basins Using Machine Learning. *Environ. Model. Softw.* **2018**, *101*, 169–182. [CrossRef]
61. Breiman, L. Random Forests. *Mach. Learn.* **2001**, *45*, 5–32. [CrossRef]
62. Zhou, P.; Li, Z.; Snowling, S.; Baetz, B.W.; Na, D.; Boyd, G. A Random Forest Model for Inflow Prediction at Wastewater Treatment Plants. *Stoch. Environ. Res. Risk Assess.* **2019**, *33*, 1781–1792. [CrossRef]
63. Meng, Y.; Yang, M.; Liu, S.; Mou, Y.; Peng, C.; Zhou, X. Quantitative Assessment of the Importance of Bio-Physical Drivers of Land Cover Change Based on a Random Forest Method. *Ecol. Inform.* **2021**, *61*, 101204. [CrossRef]
64. Li, B.; Yang, G.; Wan, R.; Dai, X.; Zhang, Y. Comparison of Random Forests and Other Statistical Methods for the Prediction of Lake Water Level: A Case Study of the Poyang Lake in China. *Hydrol. Res.* **2016**, *47*, 69–83. [CrossRef]
65. Bachmair, S.; Svensson, C.; Prosdocimi, I.; Hannaford, J.; Stahl, K. Developing Drought Impact Functions for Drought Risk Management. *Nat. Hazards Earth Syst. Sci.* **2017**, *17*, 1947–1960. [CrossRef]
66. Erdal, H.I.; Karakurt, O. Advancing Monthly Streamflow Prediction Accuracy of CART Models Using Ensemble Learning Paradigms. *J. Hydrol.* **2013**, *477*, 119–128. [CrossRef]
67. Muñoz, P.; Orellana-Alvear, J.; Willems, P.; Célleri, R. Flash-Flood Forecasting in an Andean Mountain Catchment—Development of a Step-Wise Methodology Based on the Random Forest Algorithm. *Water* **2018**, *10*, 1519. [CrossRef]
68. Tyralis, H.; Papacharalampous, G.; Langousis, A. A Brief Review of Random Forests for Water Scientists and Practitioners and Their Recent History in Water Resources. *Water* **2019**, *11*, 910. [CrossRef]
69. Wang, Z.; Lai, C.; Chen, X.; Yang, B.; Zhao, S.; Bai, X. Flood Hazard Risk Assessment Model Based on Random Forest. *J. Hydrol.* **2015**, *527*, 1130–1141. [CrossRef]
70. Feng, Q.; Liu, J.; Gong, J. Urban Flood Mapping Based on Unmanned Aerial Vehicle Remote Sensing and Random Forest Classifier—A Case of Yuyao, China. *Water* **2015**, *7*, 1437–1455. [CrossRef]
71. Quirogaa, V.M.; Kurea, S.; Udoa, K.; Manoa, A. Application of 2D Numerical Simulation for the Analysis of the February 2014 Bolivian Amazonia Flood: Application of the New HEC-RAS Version 5. *Ribagua* **2016**, *3*, 25–33. [CrossRef]
72. Brunner, G. *HEC-RAS, River Analysis System Hydraulic Reference Manual*; U.S. Army Corps of Engineers: Davis, CA, USA, 2016.
73. İcaga, Y.; Tas, E.; Kilit, M. Flood Inundation Mapping by GIS and a Hydraulic Model (Hec Ras): A Case Study of Akarcay Bolvadin Subbasin, in Turkey. *Acta Geobalcanica* **2016**, *2*, 111–118. [CrossRef]
74. Abaya, S.W.; Mandere, N.; Ewald, G. Floods and Health in Gambella Region, Ethiopia: A Qualitative Assessment of the Strengths and Weaknesses of Coping Mechanisms. *Glob. Health Action* **2009**, *2*, 2019. [CrossRef]
75. US Army Corps of Engineers. *Dupage River, Illinois Feasibility Report and Integrated Environmental Assessment*; US Army Corps of Engineers: Chicago, IL, USA, 2019.
76. StreamStats. Available online: https://Streamstats.Usgs.Gov/Ss/ (accessed on 15 June 2022).
77. Nguyen, P.; Shearer, E.J.; Tran, H.; Ombadi, M.; Hayatbini, N.; Palacios, T.; Huynh, P.; Braithwaite, D.; Updegraff, G.; Hsu, K.; et al. The CHRS Data Portal, an Easily Accessible Public Repository for PERSIANN Global Satellite Precipitation Data. *Sci. Data* **2019**, *6*, 180296. [CrossRef]
78. Mockus, V. *National Engineering Handbook Section 4 HydrologY*; US Soil Conservation Service: Washington, DC, USA, 1972; p. 127.
79. Saadi, M.; Oudin, L.; Ribstein, P. Random Forest Ability in Regionalizing Hourly Hydrological Model Parameters. *Water* **2019**, *11*, 40. [CrossRef]
80. Müller, A.; Guido, S. *Introduction to Machine Learning with Python: A Guide for Data Scientists*, 1st ed.; O'Reilly: Farnham, UK, 2016.
81. Park, H.; Kim, K.; Lee, D.K. Prediction of Severe Drought Area Based on Random Forest: Using Satellite Image and Topography Data. *Water* **2019**, *11*, 705. [CrossRef]
82. Biau, G.; Scornet, E. A Random Forest Guided Tour. *Test* **2016**, *25*, 197–227. [CrossRef]
83. Gregorutti, B.; Michel, B.; Saint-Pierre, P. Correlation and Variable Importance in Random Forests. *Stat. Comput.* **2017**, *27*, 659–678. [CrossRef]
84. Hussain, D.; Khan, A.A. Machine Learning Techniques for Monthly River Flow Forecasting of Hunza River, Pakistan. *Earth Sci. Inf.* **2020**, *13*, 939–949. [CrossRef]
85. Gharbi, M.; Soualmia, A.; Dartus, D.; Masbernat, L. Comparison of 1D and 2D Hydraulic Models for Floods Simulation on the Medjerda Riverin Tunisia. *J. Mater. Environ. Sci.* **2016**, *7*, 3017–3026.
86. Pathan, A.I.; Agnihotri, P.G. Application of New HEC-RAS Version 5 for 1D Hydrodynamic Flood Modeling with Special Reference through Geospatial Techniques: A Case of River Purna at Navsari, Gujarat, India. *Model. Earth Syst. Environ.* **2021**, *7*, 1133–1144. [CrossRef]
87. Hydrologic and Water Quality Models: Performance Measures and Evaluation Criteria. *Trans. ASABE* **2015**, *58*, 1763–1785. [CrossRef]
88. Kumar, N.; Singh, S.K.; Srivastava, P.K.; Narsimlu, B. SWAT Model Calibration and Uncertainty Analysis for Streamflow Prediction of the Tons River Basin, India, Using Sequential Uncertainty Fitting (SUFI-2) Algorithm. *Model. Earth Syst. Environ.* **2017**, *3*, 30. [CrossRef]
89. Abbaspour, K.C.; Rouholahnejad, E.; Vaghefi, S.; Srinivasan, R.; Yang, H.; Kløve, B. A Continental-Scale Hydrology and Water Quality Model for Europe: Calibration and Uncertainty of a High-Resolution Large-Scale SWAT Model. *J. Hydrol.* **2015**, *524*, 733–752. [CrossRef]

90. Gupta, H.V.; Sorooshian, S.; Yapo, P.O. Status of Automatic Calibration for Hydrologic Models: Comparison with Multilevel Expert Calibration. *J. Hydrol. Eng.* **1999**, *4*, 135–143. [CrossRef]
91. Hong, Y.; Gochis, D.; Cheng, J.; Hsu, K.; Sorooshian, S. Evaluation of PERSIANN-CCS Rainfall Measurement Using the NAME Event Rain Gauge Network. *J. Hydrometeorol.* **2007**, *8*, 469–482. [CrossRef]
92. Joshi, N.; Bista, A.; Pokhrel, I.; Kalra, A.; Ahmad, S. Rainfall-Runoff Simulation in Cache River Basin, Illinois, Using HEC-HMS. In *World Environmental and Water Resources Congress 2019*; American Society of Civil Engineers: Pittsburgh, PA, USA, 2019; pp. 348–360.
93. Desai, S.; Ouarda, T.B.M.J. Regional Hydrological Frequency Analysis at Ungauged Sites with Random Forest Regression. *J. Hydrol.* **2021**, *594*, 125861. [CrossRef]

Article

Assessment of Deep Convective Systems in the Colombian Andean Region

Nicolás Velásquez

IIHR Hydroscience & Engineering, Iowa Flood Center, The University of Iowa, Iowa City, IA 52240, USA; nicolas-giron@uiowa.edu; Tel.: +319-512-4194

Abstract: In tropical regions, deep convective systems are associated with extreme rainfall storms that usually detonate flash floods and landslides in the Andean Colombian region. Several studies have used satellite data to address the structure and formation of tropical convective storms. However, there is a local gap in the characterization, which is essential for a better understanding of flash floods and preparedness, filling a gap in a region with scarce information regarding extreme events. In this work, we assess the deep convective storms in a mountainous region of Colombia using meteorological radar observations between 2014 and 2017. We start by identifying convective and stratiform formations. We refine the convective identification by classifying convective systems into enveloped (contained in a stratiform system) and unenveloped (not contained). Then, we analyze the systems' temporal and spatial distributions and contrast them with the watersheds' features. According to our results, unenveloped convective systems have higher reflectivity and hence higher rainfall intensities. Moreover, they also have a well-defined spatial and temporal distribution and are likely to occur in watersheds with elevation gradients of around 2000 m and an aspect contrary to the wind direction. Our assessment of the convective storms is of significant value for the hydrologic community working on flash floods. Moreover, the spatiotemporal description is highly relevant for stakeholders and future local analysis.

Keywords: deep convective systems; extreme rainfall; flash floods

Citation: Velásquez, N. Assessment of Deep Convective Systems in the Colombian Andean Region. *Hydrology* **2022**, 9, 119. https://doi.org/10.3390/hydrology9070119

Academic Editors: Davide Luciano De Luca and Andrea Petroselli

Received: 12 May 2022
Accepted: 20 June 2022
Published: 28 June 2022

Publisher's Note: MDPI stays neutral with regard to jurisdictional claims in published maps and institutional affiliations.

1. Introduction

Convective systems usually turn into intense storms that could develop flooding events. Moreover, in regions with a steep terrain, convective storms are linked to the occurrence of flash floods [1–4], which are likely to produce human and infrastructure losses [5,6]. At the mesoscale, convective systems cover areas of around 600 km^2 [7], but it can be smaller at a local scale. Moreover, convective-detonated flash floods are usually limited to small catchments (less than 1000 km^2) [8–10] and their effects have been described in different world regions [9,11,12]. Nevertheless, there is a lack of analysis in the tropical Andean region, where topography-driven convective storms detonate shallow landslides and flash floods [13,14].

Several authors have described how the topography is intertwined with the occurrence of convective storms [15–18], with evidence of a strong connection [19–22]. Additionally, the described topographical influence increases with the elevation [23,24] and the aspect of the hillslopes relative to the preferential wind direction [24–27]. However, most of the work has been conducted in the Himalayas [7,23,28–30], the US [31,32], and in the southern Andean region [21,33,34]. Therefore, additional work is needed to understand the connection of the topography-convective system in the tropical Andes, specifically in the Colombian Andean region.

The Colombian Andean is made up of three mountainous rages, with its weather being dominated by the Intertropical Convergence Zone (ITCZ), the El Niño Southern Oscillation (ENSO), and the Pacific and Atlantic oceans' oscillations [35–37]. The ITCZ develops two wet periods with higher rainfall accumulation (March to May and September to November) [38]. In addition, the La Niña phase of the ENSO usually increases the occurrence of convective storms [35] with significant socioeconomic impacts [39]. Nevertheless, there is a gap in the literature exploring convective systems and their connections with the topography. Most of the work conducted around the topic has been performed at a mesoscale using TRMM data [40,41], which has a coarse resolution in contrast with the local processes. However, local radar data improve the analysis resolution [31,42,43]. In the case of the Colombian Andean region, meteorological radar information became available in 2012 in the surroundings of the city of Medellin (see Figure 1). However, the continuity and quality of the data improved near the end of 2013.

Figure 1. Region of analysis, radar localization is presented at the center of the image, and gray circles correspond to radar radii of 30 and 90 km, respectively. Yellow to dark blue colors represents the stream network Horton orders from 3 to 6.

This work uses local meteorological radar data to assess convective systems in the central Colombian Andean region between 2014 and 2017. Some of these convective systems detonated flash floods and landslides during this period [13,14]. First, we identify the stratiform and convective systems observed. Then, we classify the convective systems into enveloped and unenveloped, allowing us to describe the most intense ones better. We analyze the convective systems' size, reflectivity, and spatial and temporal occurrences using radar data acquired by the Sistema de Alertas Tempranas Ambientales (SIATA). Additionally, we compare the localization of the identified convective systems with 18 watersheds. In the comparison, we explore how the aspect and elevation gradient of the watersheds are intertwined with the convective systems' formation. Our main goal is to develop a comprehensive regional analysis of the storms linked to the occurrence of rainfall-related catastrophes in the Colombian Andean region. Additionally, we seek to fill a knowledge gap in a relevant topic that could worsen in the coming years due to climate change and population growth.

We begin this paper by describing the radar and topographic data in Section 2. Then, in the methodology Section 3, we describe the algorithms to identify convective systems, extract their features, and the methodology followed to contrast them against the selected watersheds. In Section 4, we present the results of the convective systems' identification and the comparison with the watersheds. Finally, Section 5 offers our conclusions and remarks for future work.

2. Data and Information

We used radar reflectivity, digital elevation data, and wind data. Using radar reflectivity, we identified and analyzed the convective systems. Moreover, we delineated the watersheds and extracted their properties using a digital elevation model. Subsequently, we describe in detail the data used.

2.1. Radar Data

The polarimetric 350 Kw Doppler C-band radar manufactured by Enterprise Electronics Corporation is in the occidental central hill of the Aburra Valley watershed (Rio Medellin in Figure 1). The radar operation includes four plan position indicator sweeps (PPIs) at 0.5, 1.0, 2.0, and 4.0°. We used radar reflectivity every 5 min at the PPI of 1.0° with a Cartesian projected resolution of 128 m. The radar beam has a beam width of 1°ʼ and, depending on the PPI, it reaches a distance between 50 and 200 km. It has a wavelength of 5.3 cm and an antenna gain of 45 dBZ. However, to avoid issues due to the bright band interception, we limited the information to a radius of 90 km (second gray ring in Figure 1).

Radar data have been available since 2012. However, it has continuous quality data since 2014. The radar image quality control uses the co-polar correlation coefficient (CC). CC tends towards one when water particles have a uniform distribution. On the other hand, CC decreases when there is a heterogeneous distribution of the particles' shape and orientation. Moreover, CC values above one represents errors due to noise. Between 2014 and 2017, around 93% of the radar data has a good quality.

2.2. Topography and Wind Data

We used an ALOS-PALSAR digital elevation model (DEM) to describe the region's topography with a resolution of 12.7 m. Considering the area of the study domain, we resampled the DEM to a resolution of 40 m (background in Figure 1). We applied the A^T algorithm [44] over the resampled DEM to estimate the direction map (DIR). Using the DEM and DIR maps, we extracted the watersheds shown in Figure 1 and their respective boundary sub-watersheds of orders 3, 4, and 5. Additionally, for each watershed, we estimated the total area, the elevation difference, and the predominant aspect. We used the watershed modeling framework (WMF) (https://github.com/nicolas998/WMF accessed on 12 May 2022) to perform the watershed delineation and analysis.

Additionally, we used ERA-5 observations at 750 hPa to determine the preferential wind direction. The selected pressure level was around 2500 m above sea level, a value close to the mean elevation of the region.

3. Methodology

The region's mountainous terrain induced extra challenges in our convective system assessment. Hillslopes with high elevation gradients promoted the formation of deep convective systems and, at the time, added noise to the radar images [45,46], increasing the uncertainty in the classification. Therefore, we divided the convective systems into two categories: enveloped and unenveloped. Enveloped systems are embedded into a stratiform formation, while unenveloped ones are not. With the convective discretization, we performed a more comprehensive assessment.

Additionally, we explored the relationship between convective systems' spatial distribution and the topography. To this end, we delineated 18 watersheds (see Figure 1) and their boundary sub-watersheds with Horton orders 3, 4, and 5. Boundary sub-watersheds share a boundary with their containing watersheds. We measured the overlap between the convective systems and the watersheds and contrasted it with their area, aspect, and elevation difference. In this analysis, we also included the preferential direction of the wind near an elevation of 2500 m.

Subsequently, we describe in detail each one of the methodology steps.

3.1. Systems' Classification and Analysis

We first identified the convective systems present in each radar image between 2014 and 2017. To perform the identification, we implemented in Fortran90 and Python the algorithm proposed by [47], and then modified by [48]. We also included functions to measure the features of the convective systems and classified them into enveloped and unenveloped. The code used can be found at the following GitHub repository: https://github.com/nicolas998/Radar (accessed on 12 May 2022).

3.1.1. Convective and Stratiform Systems' Classifications

Convective storms exhibit high reflectivity and are prone to develop extreme rainfall events. On the other hand, stratiform formations have lower reflectivity and are associated with low-intensity rainfall. Considering the described differences, we started by separating convective and stratiform systems using the algorithm proposed by [47], and then modified by [48]. In our implementation, we used a background radius of 11 km and a threshold for peaks of 40 dBZ. We validated the algorithm's accuracy by comparing the classified systems with vertical profiles obtained by the radar [13]. Nevertheless, we identified noise due to convective systems embedded in stratiform formations. Therefore, we created the enveloped and unenveloped convective categories.

3.1.2. Enveloped and Unenveloped Convective Systems

Depending on the meteorological conditions and the temporal evolution of the storm, convective systems may occur as individual systems or as part of a cluster [49]. Single storms usually cover more area and have higher reflectivity than clustered systems. Moreover, during the storm, clustered systems are generally enveloped by a stratiform formation [50,51]. Therefore, we categorized convective systems into enveloped and unenveloped. Enveloped systems are embedded into stratiform formations, while unenveloped ones are not. We implemented the following procedures to identify both:

1. First, we separated convective and stratiform objects into two binary images (Bin_c and Bin_s, respectively). Bin_s is equal to one where there are stratiform formations and zero elsewhere. Bin_c is equal to two in regions with convective systems and zero elsewhere. Figure 2a presents a schematic of Bin_s (left) and Bin_c (right).
2. Then, we eroded Bin_s using a 3×3 kernel. In the erosion, each Bin_s element with a value of one and at least one neighbor equal to zero in the kernel became zero. From the erosion, we obtained $BinE_s$ (light blue image in the left panel in Figure 2b).
3. After the erosion, we filled the holes left in $BinE_s$. From this procedure, we obtained the eroded and then filled stratiform binary $BinEF_s$ (Figure 2b, right).
4. Then, we computed the superposition between Bin_c and $BinEF_s$ as $SupBin = Bin_c + BinEF_s$ (Figure 2c). In $SupBin$, values equal to 1 correspond to stratiform formations, 2 to unenveloped convective systems, and 3 to enveloped convective systems.
5. Finally, we classified each convective system as enveloped or unenveloped using their corresponding modal values in $SupBin$. For example, if 90 pixels of a convective storm were 2 and 10 pixels were 3, the system was considered unenveloped (or 2).

The described methodology was applied to each radar image. In Figure 3, we present an example showing the results obtained. The process starts with the reflectivity information (Figure 3a), from where we performed the convective and stratiform classifications (Figure 3b). Then, we identified each convective object (Figure 3c). An object is formed by all the adjacent pixels previously identified as convective. Finally, we obtained the enveloped and unenveloped convective systems' identification (Figure 3d).

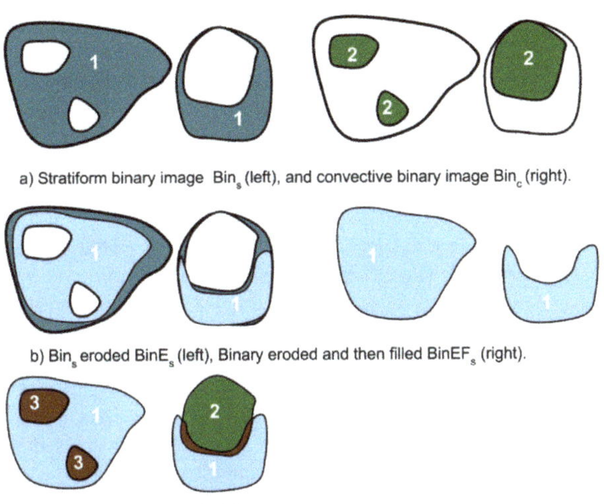

a) Stratiform binary image Bin_s (left), and convective binary image Bin_c (right).

b) Bin_s eroded $BinE_s$ (left), Binary eroded and then filled $BinEF_s$ (right).

c) Superposition binary SubBin

Figure 2. A schematic representation of the proposed methodology to separate enveloped and unenveloped systems. (**a**) Binaries of convective (Bin_c) and stratiform (Bin_s) systems are separated, Bin_c takes values of 0 and 2, Bin_s takes values of 0 and 1. (**b**) Bin_s is eroded and $BinE_s$ is obtained, then $BinEF_s$ is obtained by filling holes in the binary. (**c**) $SubBin$ is computed as the sum of Bin_c and $BinEF_s$. $SubBin$ values equal to 2 correspond to unenveloped convective systems, and values of 3 correspond to enveloped systems.

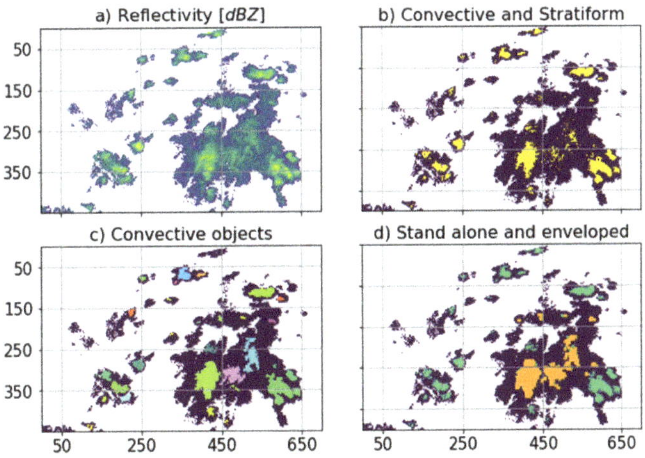

Figure 3. Example of a classified radar image. (**a**) Original reflectivity image in dBZ, (**b**) identification of convective (yellow) and stratiform systems (purple), (**c**) ID assignation to each convective system, and (**d**) identification of enveloped (orange) and unenveloped (green) systems.

3.1.3. Reflectivity Statistics and Morphometrics

Using the described methodology, we processed the radar images between 2014 and 2017, obtaining a classified image every 5 min. We obtained a collection of convective objects (systems) from the classification. Then, we extracted the reflectivity statistics and morphometric features of each object. We used the equivalent reflectivity factor Z to compute the mean, μ_{ref}, and deviation, σ_{ref}, of each system, and transformed them to reflectivity by using their logarithm. Additionally, we computed the area, $A \left[\text{km}^2 \right]$, the centroid coordinates, C $[lat, lng]$, and the time of occurrence of each convective system. The area corresponded to the pixel count in a system multiplied by 16,684 m^2 (the square of 128 [m], the radar resolution), and the centroid was the system center of mass in X and Y. Moreover, we marked each object as enveloped (3) or unenveloped (2).

The described information allowed us to obtain an extensive collection of data and perform a comprehensive assessment of the convective systems in the region.

3.2. Watersheds Analysis

According to several studies, convective system formation is intertwined with topography [20,32]. The connection has also been reported in tropical regions [17,22] with links between rainfall rates and elevation [18,40]. However, most of the work used mesoscale information and did not consider the characteristics of the watersheds. Therefore, we compared the convective system localization with features of the watersheds in the region.

We analyzed the 18 watersheds delimited by the black divisor lines in Figure 1, covering most of the domain. The areas of the watersheds oscillated between 230 and 2000 km^2, and the Horton stream orders at their outlets oscillated between 5 and 7. Additionally, we included the boundary sub-watersheds of the 18 watersheds with orders 3, 4, and 5 (see Figure 4a–c). A boundary sub-watershed has no upstream tributaries; in most cases, it shares its divisor line with its containing watershed. Then, we computed the following features of the watersheds: boundary elevation, maximum elevation gradient, predominant aspect (direction), and the upstream area [km^2].

We also measured the intersection between the convective systems and the watersheds using the overlap index, O_i, as follows:

$$O_i = \frac{A_o}{A_s} \tag{1}$$

where A_o $\left[\text{km}^2\right]$ is the intersected area between the convective system and the watershed and A_s $\left[\text{km}^2\right]$ is the system area. Figure 4d presents three different cases of overlap: There is no overlap when the convective system falls outside of a watershed and A_o becomes equal to zero. There is a weak overlap when A_o is less than 15% of the total area, and it is strong when A_o is greater than 15%. Nevertheless, the described categories are subjective and will change depending on the watershed area and the resolution of the images used. Therefore, the given values are a reference that allows us to contrast the overlap between the convective systems and the watersheds.

Finally, we compared O_i with the described features of the watersheds and the direction of the wind at 750 hPa (ERA-5 data). Using the described procedure, we explored how the topographical features of the watersheds were intertwined with the occurrence of deep convective systems at different scales.

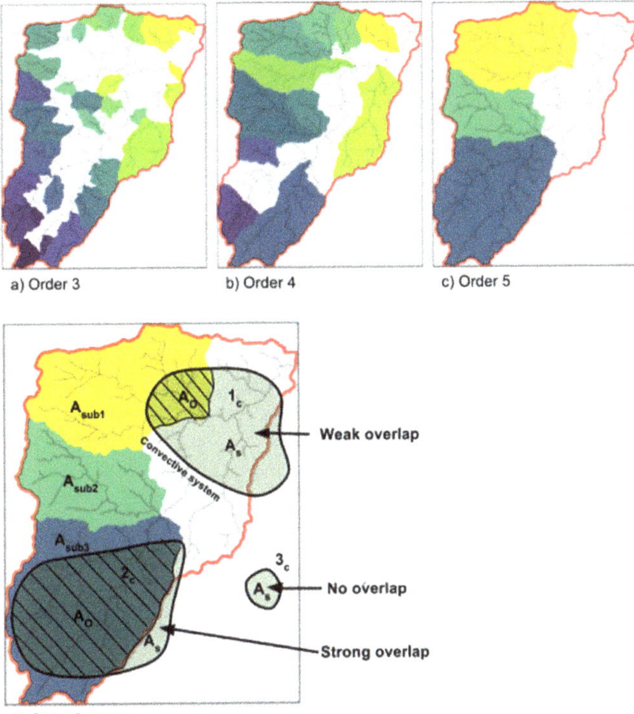

Figure 4. Example of boundary sub-watersheds definitions (**a–c**) and overlap index estimations (**d**). The three top panels describe the boundary sub-watersheds of orders 3, 4, and 5 for the Samana watershed.

4. Results and Discussion

The main goal of our work was to present a characterization of the tropical convective systems observed by a meteorological radar. We also contrasted the convective systems with the region's orientation, elevation, and aspects of several watersheds. We presented and discussed the obtained results after applying the methods described in Section 3.

4.1. Convective Systems Analyses

We started by identifying the observed convective and stratiform systems following the method proposed by Steiner (1995). Figure 5 presents the area, A $\left[\text{km}^2\right]$ (X axis), mean reflectivity, μ_{ref} [dBZ] (Y axis), and reflectivity deviation, σ_{ref} [dBZ] (colors), of both kind of systems for a sample of 10,000 elements. According to it, the mean reflectivity distribution was different for the convective and stratiform systems. Convective μ_{ref} distribution (green line in the vertical pdf panel) was less skewed and centered around 27 dBZ. On the other hand, the stratiform μ_{ref} pdf (purple line) was skewed and centered around 12 dBZ. We also observed some differences in the distributions of their areas (top pdf panel). While stratiform areas oscillated around 0.01 and 13,000 km^2 with 80% below 40 km^2, convective systems oscillated between 0.01 and 80 km^2 with 80% below 9 km^2.

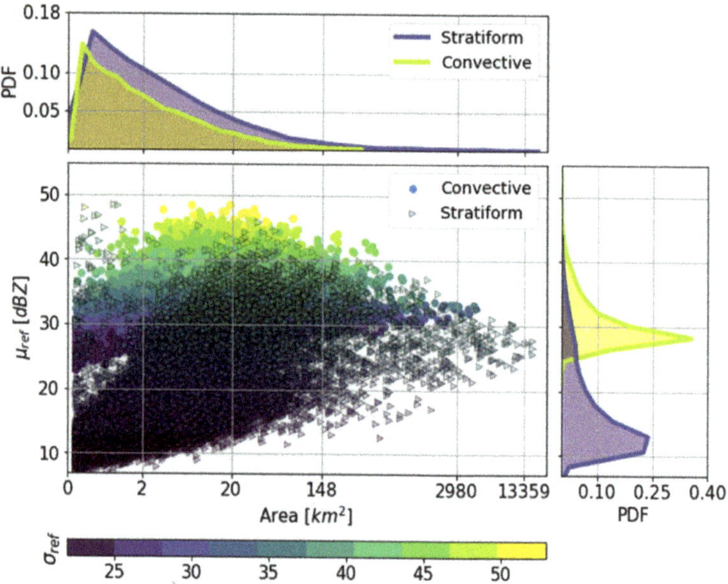

Figure 5. Reflectivity and area for 10,000 randomly selected systems. Circles correspond to convective systems, triangles to stratiform systems, and the color is associated with the variability of the reflectivity inside the system. Vertical histograms present the PDF for the mean reflectivity for both types of systems, and the horizontal histograms correspond to the PDF of the area.

Convective and stratiform systems' reflectivity statistics changed with the area of the system. For large areas, convective μ_{ref} tended towards 30 dBZ while the stratiform μ_{ref} towards 25 dBZ. Additionally, for areas around 20 km^2, the convective μ_{ref} and σ_{ref} reached their maximum values of 48 dBZ and 50 dBZ, respectively. Moreover, the stratiform μ_{ref} maximized for areas around 50 km^2. In both cases, the mean reflectivity decreased for systems with areas below or above the ones mentioned. The described result may be related to the size of the watersheds where flash floods usually occur (areas below 100 km^2). According to the convective μ_{ref} vs. area scatter in Figure 5, extensive convective system usually had a lower μ_{ref}.

By comparing the area and μ_{ref} distributions, we observed many relatively small convective systems with reduced reflectivity. The described pattern is also similar in the case of the stratiform systems. Both cases are likely attributed to misclassification caused by noise in the radar images. Therefore, we improved our classification by dividing the convective systems into two categories.

Using the methodology described in sub-Section 3.1.2, we classified 80% of the systems as unenveloped and 20% as enveloped. Figure 6 compares the area and μ_{ref} of the enveloped (triangles) and unenveloped (circles) systems. Compared with Figure 5, the enveloped and unenveloped systems' area and reflectivity are more similar. The mean reflectivity distributions of the enveloped and unenveloped categories were 28 and 32 dBZ, and the areas had mean values of 15 and 30 km^2, respectively. Moreover, the unenveloped convective system's σ_{ref} was higher and peaked along with μ_{ref} for systems with areas between 20 and 40 km^2. Furthermore, the expected σ_{ref} value of the unenveloped was considerably higher than the enveloped system (see Figure 6).

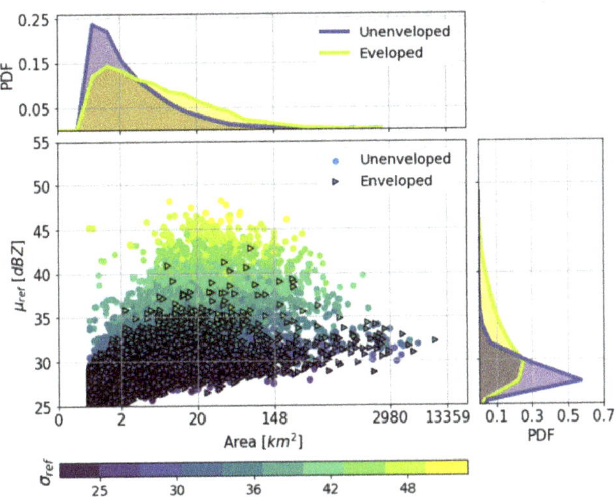

Figure 6. Reflectivity and area comparison for enveloped (triangles) and unenveloped (circles) connective systems. Colors represent the standard deviation. Vertical and horizontal histograms present the PDFs for the reflectivity and areas, respectively.

Table 1 summarizes the statistics of the stratiform, convective, enveloped, and unenveloped systems. According to the results, stratiform systems have an area with greater mean and deviation values. However, they have a lower mean reflectivity. On the other hand, convective systems have an area that oscillates around 24 km^2. Enveloped systems are smaller and have a reflectivity lower than unenveloped systems. The described results mark the differences between both types of convective systems, highlighting unenveloped ones as good descriptors of deep convective systems. Moreover, our results also match the given description of convective clusters enveloped in stratiform formations [52].

Table 1. Summary values of the systems identified.

Feature	Stratiform	Convective	Unenveloped	Enveloped
Maximum area $\left[\text{km}^2\right]$	43,000	14,000	14,000	13,200
Mean area $\left[\text{km}^2\right]$	47	24	30	15
Area deviation $\left[\text{km}^2\right]$	445	160	180	131
Area 10th percentile $\left[\text{km}^2\right]$	1.50	1.43	1.52	1.34
Area 90th percentile $\left[\text{km}^2\right]$	60	33	43	16
Maximum μ_{ref} [dBZ]	40	60	60	46
Minimum μ_{ref} [dBZ]	0.26	22	22	22
Expected μ_{ref} [dBZ]	16	30	32	28
Expected σ_{ref} [dBZ]	6.6	3.7	4.1	1.6
μ_{ref} 10th percentile [dBZ]	10	27	27	27
μ_{ref} 90th percentile [dBZ]	27	36	38	30

The defined convective categories help us explore the region's storm systems. Moreover, the studied features of both classes have relevant differences. Considering the described differences between the enveloped and unenveloped systems, we also explored their spatial and temporal differences.

4.2. Spatiotemporal Behavior

We first analyzed the convective systems' diurnal and spatial distributions. We counted the hour of the day and the spatial localization of the convective, enveloped, and unenveloped systems. Then, we divided each count by the total of observed systems obtaining their probability of occurrence. Figure 7 presents the daily distribution observed in the region, starting at midnight and ending at 23:00. According to the figure, convective and undeveloped systems develop around noon, reaching the peak probability between 15:00 and midnight. On the other hand, enveloped systems are more likely to occur late at night and early morning. We attributed the observed temporal oscillations to surface heating and the eventual deep convection. Moreover, the increase in enveloped systems during the morning might be explained by the late stage of the deep convective systems.

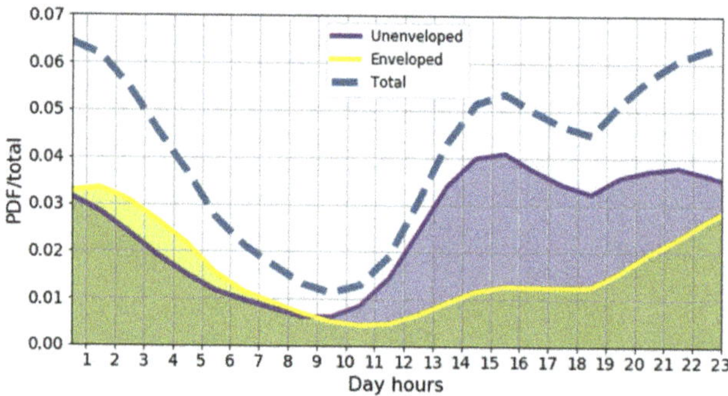

Figure 7. Hourly distributions of the occurrence of total convective (dashed blue), unenveloped (purple), and enveloped (yellow) systems.

In Figure 7, we closely observe the described temporal evolution of the convective system. In it, we colored the spatial domain using the total count of systems falling in pixels of 1 km^2. Considering the significant count differences, we used different ranges to color each column of the figure. Moreover, we divided the results using a time step of six hours starting at 1:00 A.M. Since the total count of cases for the columns was significantly different, we kept independent color bars for each.

Column **a** in Figure 7 presents the spatiotemporal results of all the convective systems and expands the description presented in Figure 6. According to Figure 7, the morning hours have some convective activity (1st row). This activity reaches its minimum values between 7 A.M. and noon (2nd row). Then, during the afternoon, convective systems occur over the West and center regions of the domain (3rd row). Finally, during the night, the storms occur over the East (4th row).

Columns **b** and **c** in Figure describe the results for the unenveloped and enveloped system cases. Unenveloped systems (column **b**) exhibit a predominant formation over the eastern region between the afternoon and midnight (third and fourth rows). Moreover, they decrease during the morning (first row) and mostly disappear during the day (second row). Enveloped systems dominate the East, with some occurrences over the West.

Moreover, the lag described in Figure 6 between the enveloped and unenveloped systems is also present. During the afternoon, unenveloped systems intensify while some envelopes occur (third row). During the night (fourth row), both categories increase. Then, in the morning, enveloped systems become dominant. According to our results, enveloped and unenveloped systems categories' temporal distributions coincide with the described evolution of the convective system [40]. Moreover, their occurrence coincides with the region's descriptions of orographic rainfall formation [53,54].

According to Figure 7, the convective systems have well-defined spatiotemporal distributions. Most convective systems occurred in the East between noon and midnight. Their occurrence coincided with reported flash floods and shallow landslide events [13,14].

They also coincided with descriptions of the Colombian Andean climatology [54,55]. The results presented here are likely intertwined with several climatological variables. However, exploring that connection is out of our scope. Future work may explore that link and how to use the results presented in the vulnerability analysis and risk assessment.

Figure 8 summarizes the spatial distribution regardless of the daytime. According to Figure 8a, most convective systems occur in the East, specifically in Rio Playas, Rio Calderas, and Rio Verde (see Figure 1). There are also minor convective accumulations in the West near Rio Penderisco and Rio San Juan watersheds, and in the north near the outlet of Rio Grande. Unenveloped systems follow a similar localization (Figure 8b), with a significant occurrence increase near the boundaries of the watersheds. On the other hand, enveloped systems are more sparse over the region, being dominant in the West and in the East in the watersheds of Rio Verde and Samana.

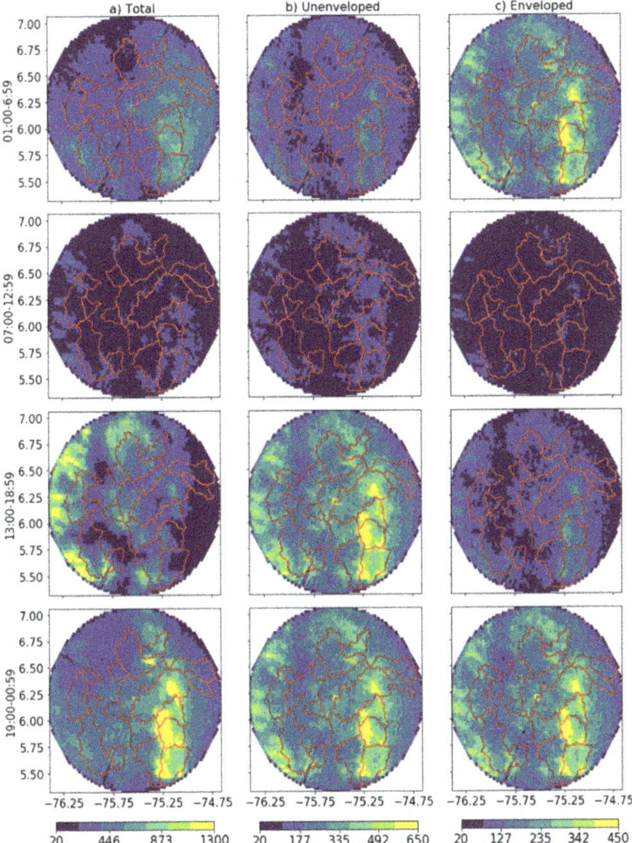

Figure 8. Two-dimensional histogram of convective centroid localization inside the radar region. Column (**a**) corresponds to total convective localization, (**b**) unenveloped systems, and (**c**) enveloped ones. The colors represent the count of elements. Color bars correspond to each of the columns.

The localization of the convective systems coincides with the region's topography. Unenveloped systems mainly occur over the watersheds' boundary lines, usually the steepest areas. Moreover, they occur more often in watersheds with a significant elevation difference, in this case, the Rio Caldera, Rio Verde, and Samana watersheds with elevation gradients of around 2000 m. On the other hand, enveloped systems behave like a remanent of unenveloped ones being distributed downstream of the described regions.

The given description is of significant value, and the localization of the unenveloped systems can help determine vulnerable regions.

4.3. Connections with the Topography

According to our results, several convective systems tend to occur near the divisor lines of watersheds with high elevation gradients (see Figure 9a,b). However, some watersheds had a high elevation gradient, but the count of convective systems was low. In this section, we explored further the connection between the convective systems' occurrence and the characteristics of the watersheds. Here, we compared the localization of the convective systems with the area, aspect, and elevation gradient of the region's watersheds.

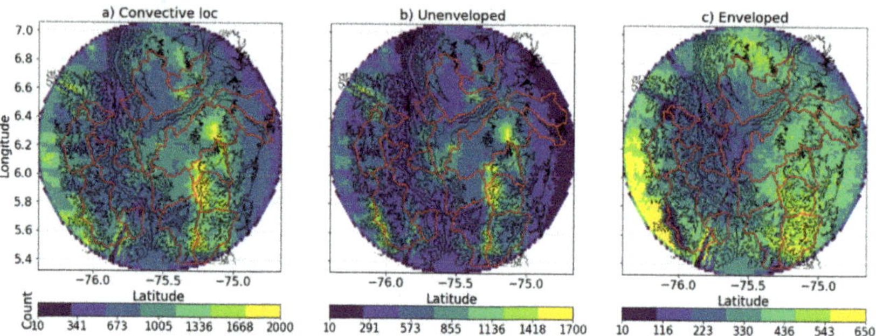

Figure 9. Two-dimensional histogram of convective centroid localization inside the radar region. (**a**) Total convective systems. Figures (**b**,**c**) correspond to unenveloped and enveloped systems, respectively. The black lines are topographical contour lines representing increments of 1000 m.

Figure 10 presents the 18 watersheds, their aspects, the topography, and the prevailing direction of the wind at 750 hPa (around 2500 m above sea level). Moreover, in the figure, we colored the aspects of the watersheds with the overlapping index, O_i. The index measures the area shared between a convective system and a watershed. In our case, O_i oscillated between 0.12 and 0.22.

According to Figure 10, the Samana, Rio Verde, Guadalupe (East), and Rio San Juan (West) watersheds have the highest O_i values. In the four cases, the aspect is opposite to the wind, and the elevation difference is above 2000 m, except for Guadalupe where it is 1843 m. Nevertheless, this analysis has been conducted using the preferential wind direction, and, in many cases, the results presented here (3) may not be fulfilled. Moreover, it seems that additional features are involved, since some watersheds exhibit similar topographical features, but lower O_i values.

Additionally, we compared O_i with the upstream area, aspect, and elevation gradient of the watersheds (Figure 11). In the figure, the position of the arrow coincides with the area and the overlap index, O_i, computed for each watershed. The direction represents the aspect, and the color the elevation gradient. Moreover, the circles represent the index for the enveloped (green) and unenveloped (purple) systems.

According to Figure 11, the unenveloped and enveloped O_i values change among watersheds. In thirteen out of the eighteen watersheds, unenveloped O_i is dominant. Moreover, enveloped O_i is foremost in watersheds with upstream areas of around 1000 km^2 or less. The difference between the enveloped and unenveloped O_i decreases towards 0.16 for watersheds with areas around 500 km^2. On the other hand, with some exceptions, watersheds with high elevation differences tend to exhibit high O_i values, a condition also influenced by their aspect. Watersheds, such as Samana, Rio Verde, and Rio San Juan, have a high O_i, areas greater than 500 km^2, and elevation differences above 2,000. The given description presents more information regarding the required local conditions for the occurrence of convective systems. However, the analyzed watersheds are large compared to the mean area of the convective systems (24 km^2, see Table 1). Therefore, we expanded our analysis using the boundary sub-watersheds of orders 3, 4, and 5.

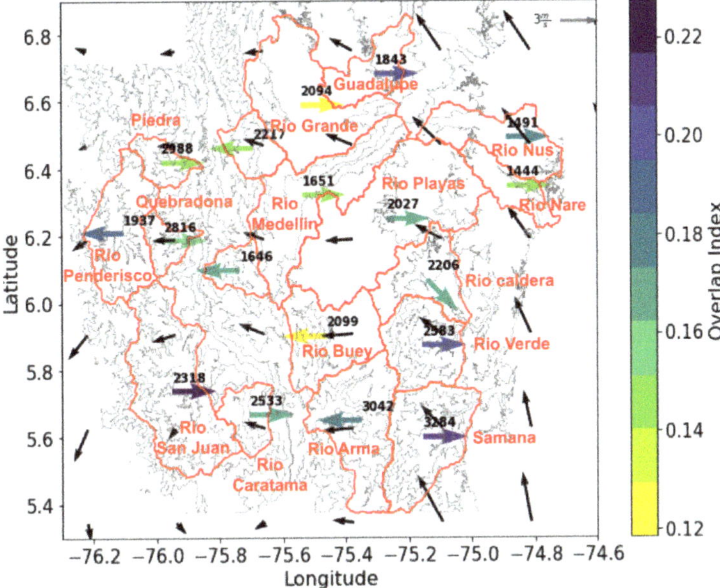

Figure 10. Spatial distribution of the overlap index. The bold arrows represent the aspect of the watersheds, while their color corresponds to the overlap index. Black arrows represent the ERA-5 wind speed and direction. The red and black lines are the watershed boundaries and topography (500 m). The numbers represent the elevation gradient.

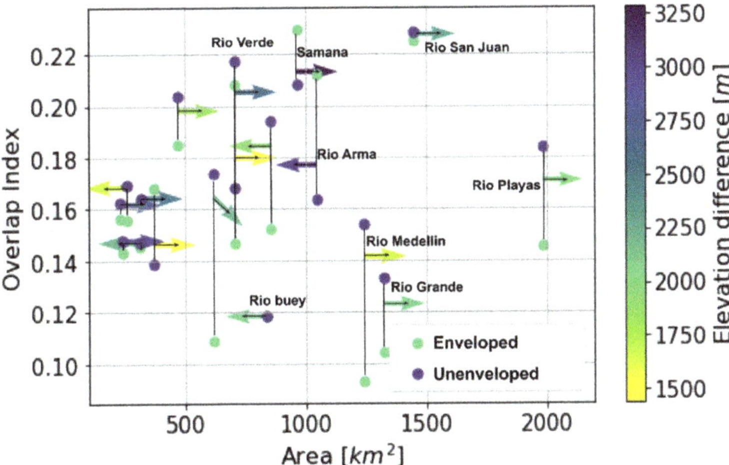

Figure 11. Overlap index in function of the area, aspect, and elevation gradient of the watersheds. The arrows correspond to the analyzed watershed denoting its main aspect. The arrows represent the watershed aspect, and the color corresponds to the elevation gradient ($H_{max}-H_{min}$). The purple and green circles represent the overlap index, I_{oc}, for the enveloped and unenveloped systems.

The boundary sub-watersheds had no upstream tributaries and, in most cases, shared the divisor line with its containing watershed. We used boundary sub-watersheds of orders

3, 4, and 5. In the case of order 5, we obtained between two and three sub-watersheds per watershed (blue borders in Figure 12a). Moreover, each sub-watershed had its preferential aspect. According to Figure 12a, watersheds with an aspect opposite to the wind direction tended to have a larger O_i. In addition, the relationship between the upstream area, elevation difference, and O_i became more evident (Figure 12b). Larger areas usually had increased elevation gradients and O_i values.

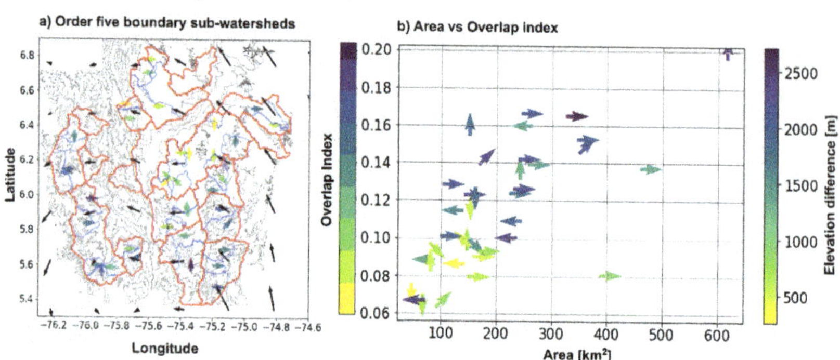

Figure 12. Spatial distribution of the overlap index for the boundary sub-watersheds or orders 5. In panel (**a**), the blue lines represent the sub-watersheds, the colored arrows the aspect (direction) and the overlap index (color), and the black arrows represent the ERA-5 wind at 750 hPa. In panel (**b**), the arrows represent the aspect, and their color is the elevation difference.

Boundary sub-watersheds of orders 3 and 4 (Figure 13a,b) had a pattern where the overlap index increased with the area and the elevation difference. However, this trend had noise. In order 3 sub-watersheds (Figure 13a), areas between 10 and 40 km^2 had similar maximum O_i values. Nevertheless, most of the cases with a high index were sub-watersheds facing East with elevation differences above 1500 m. In the case of order 4 sub-watersheds (Figure 13b), there was an increase in the O_i values and the trend with the area was evident. Moreover, the aspect and the elevation difference keep playing a significant role.

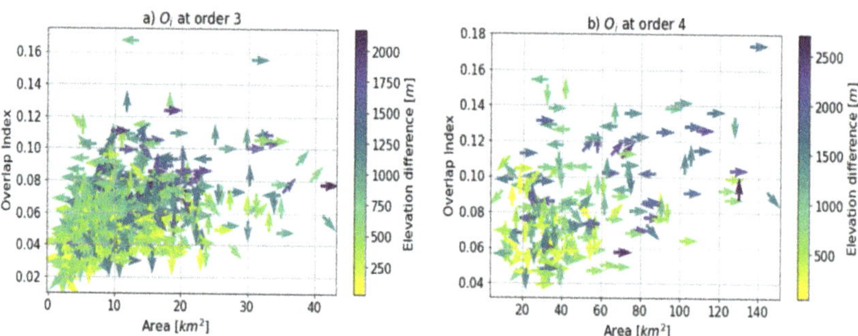

Figure 13. Overlap index for boundary sub-watersheds of orders 3 (**a**) and 4 (**b**). The arrows represent the aspect and colors the elevation difference.

The occurrence of convective systems was linked to the elevation difference and aspect of the watersheds at several scales. We analyzed this relationship by defining an overlap index between watersheds and the identified convective systems. According to our results, watersheds with a high elevation gradient and an aspect opposite to the mean wind direction are more likely to develop deep convective systems. Nevertheless, further work is required to establish a link with the topography.

5. Conclusions

This work assessed the convective systems observed in the Colombian Andean region using local meteorological radar data between 2014 and 2017. Furthermore, we expanded our identification by defining two categories: enveloped and unenveloped convective systems. In addition, we compared the area, reflectivity, and localization of the systems. Finally, we proposed a novel method to explore possible interactions between the region's topography and the occurrence of convective systems. In the topography interaction analysis, we used 18 watersheds and their boundary sub-watersheds of orders 3, 4, and 5. We identified some features of the convective storms of the region along with the areas and times where they usually occurred.

The characteristics of the deep convective systems depended on the stage of their evolution and the mechanisms behind them. A convective storm usually starts as an individual or as a collection of small systems that form a deep system and finally evolves into a stratiform formation with small convective systems [56]. The proper identification of the described evolution requires a tracking algorithm, such as TITAN [57]. Moreover, additional identification algorithms may present more insights into their properties [58]. However, implementing such algorithms in the region is beyond the work presented here. Instead, we introduced a classification of enveloped and unenveloped convective systems.

We described the differences between the region's deep convective (unenveloped) and clustered (enveloped) systems. Unenveloped systems exhibited higher reflectivity, larger areas, and a higher connection with the topography. On the other hand, enveloped systems behave similar to a late stage of deep convection. This conclusion is also supported by the occurrence time of both types of systems. While unenveloped ones developed between the afternoon and midnight, enveloped systems occurred during the night and morning. The described oscillations may be linked to the energy and humidity availability during the day and the eventual cooling during the night. Nevertheless, more work is required in this direction to develop a more robust conclusion. Furthermore, our approach is a step forward in understanding the mechanisms behind the occurrence of intense storms in the region.

Our analysis also found that deep convective storms occur more often over topographies with specific characteristics. The elevation gradient of the terrain and its aspect relative to the local wind direction were two critical features. We explored the relationship with the topography by analyzing the overlap between the convective systems, the ERA-5 wind at 750 hPa, and hundreds of watersheds. According to our results, convective systems occur more often in areas with high topographical gradients, where the aspect is opposite to the wind direction. This result coincides with other studies performed in the region around rainfall and topography [55,59]. Moreover, similar results linking topography and convective rainfall have been reported in other regions [60–64]. However, our analysis does not fully describe the convective storms' spatial distributions. Future work may include additional meteorological variables and use different ways to assess the connection with topography.

Our results are significant for the hydrologic community and the region's stakeholders. The characterization presented here identified the areas prone to deep convective storms. Furthermore, joint with landscape information, it could help identify areas in which flash floods and landslides may occur. Moreover, the analysis presented here could be easily replicated and improved, helping to identify vulnerable areas in other regions.

Funding: This work was completed with support from the Iowa Flood Center and SIATA. MID-America Transportation Center (Grant number: 69A3551747107).

Informed Consent Statement: Not applicable.

Data Availability Statement: The data that support the findings of this study are available from the corresponding author upon reasonable request.

Acknowledgments: The authors wish to thank the Sistema de Alertas Tempranas Ambientales (SIATA).

Conflicts of Interest: The authors declare no conflict of interest.

References

1. Agyakwah, W.; Lin, Y.-L. Generation and enhancement mechanisms for extreme orographic rainfall associated with Typhoon Morakot (2009) over the Central Mountain Range of Taiwan. *Atmospheric Res.* **2020**, *247*, 105160. [CrossRef]
2. Anders, A.M.; Roe, G.H.; Hallet, B.; Montgomery, D.R.; Finnegan, N.J.; Putkonen, J. Spatial Patterns of Precipitation and Topography in the Himalaya. *Geol. Soc. Am.* **2006**, *2398*, 39–53.
3. Barrett, B.S.; Garreaud, R.; Falvey, M. Effect of the Andes Cordillera on Precipitation from a Midlatitude Cold Front. *Mon. Weather Rev.* **2009**, *137*, 3092–3109. [CrossRef]
4. Barros, A.P.; Kim, G.; Frajka-Williams, E.; Nesbitt, S. Probing Orographic Controls in the Himalayas during the Monsoon Using Satellite Imagery. *Nat. Hazards Earth Syst. Sci.* **2004**, *4*, 29–51. [CrossRef]
5. Barros, A.P.; Joshi, M.; Putkonen, J.; Burbank, D.W. A Study of the 1999 Monsoon Rainfall in a Mountainous Region in Central Nepal Using TRMM Products and Rain Gauge Observations. *Geophys. Res. Lett.* **2000**, *15*, 3683–3686. [CrossRef]
6. Barros, A.P.; Lang, T.J. Monitoring the Monsoon in the Himalayas: Observations in Central Nepal, June 2001. *Mon. Weather. Rev.* **2003**, *131*, 1408–1427. [CrossRef]
7. Bedoya-Soto, J.M.; Aristizábal, E.; Carmona, A.M.; Poveda, G. Seasonal Shift of the Diurnal Cycle of Rainfall Over Medellin's Valley, Central Andes of Colombia (1998–2005). *Front. Earth Sci.* **2019**, *7*, 92. [CrossRef]
8. Bowman, K.P.; Fowler, M.D. The Diurnal Cycle of Precipitation in Tropical Cyclones. *J. Clim.* **2015**, *28*, 5325–5334. [CrossRef]
9. Braud, I.; Ayral, P.-A.; Bouvier, C.; Branger, F.; Delrieu, G.; Le Coz, J.; Nord, G.; Vandervaere, J.-P.; Anquetin, S.; Adamovic, M.; et al. Multi-Scale Hydrometeorological Observation and Modelling for Flash Flood Understanding. *Hydrol. Earth Syst. Sci.* **2014**, *18*, 3733–3761. [CrossRef]
10. Camarasa-Belmonte, A.M. Flash Floods in Mediterranean Ephemeral Streams in Valencia Region (Spain). *J. Hydrol.* **2016**, *541*, 99–115. [CrossRef]
11. Dixon, M.; Wiener, G. TITAN: Thunderstorm Identification, Tracking, Analysis, and Nowcasting-A Radar-Based Methodology. *J. Atmos. Ocean. Technol.* **1993**, *10*, 785–797. [CrossRef]
12. Doswell, C.A.; Harold, E.B.; Robert, A.M. Flash Flood Forecasting: An Ingredients-Based Methodology. *Weather. Forecast.* **1996**, *11*, 560–581. [CrossRef]
13. Dowell, D.C. *Severe Convective Storms*; American Meteorological Society: Boston, MA, USA, 2001.
14. Ehlschlaeger, C. Using the A^T Search Algorithm to Develop Hydrologic Models from Digital Elevation Data. In Proceedings of the International Geographic Information Systems (IGIS) Symposium'89, Baltimore, MD, USA, March 1989; Volume 56, pp. 275–281.
15. Froidevaux, P.; Martius, O. Exceptional integrated vapour transport toward orography: An important precursor to severe floods in Switzerland. *Q. J. R. Meteorol. Soc.* **2016**, *142*, 1997–2012. [CrossRef]
16. Futyan, J.M.; Del Genio, A.D. Deep Convective System Evolution over Africa and the Tropical Atlantic. *J. Clim.* **2007**, *20*, 5041–5060. [CrossRef]
17. Garreaud, R.; Rutllant, J. Coastal Lows in North-Central Chile: Numerical Simulation of a Typical Case. *Mon. Weather. Rev.* **2003**, *131*, 891–908. [CrossRef]
18. Garreaud, R.; Fuenzalida, H.A. The Influence of the Andes on Cutoff Lows: A Modeling Study*. *Mon. Weather. Rev.* **2007**, *135*, 1596–1613. [CrossRef]
19. Garrote, J.; Alvarenga, F.; Diez-Herrero, A. Quantification of Flash Flood Economic Risk Using Ultra-Detailed Stage–Damage Functions and 2-D Hydraulic Models. *J. Hydrol.* **2016**, *541*, 611–625. [CrossRef]
20. Gultepe, I. 56 Advances in Geophysics. In *Mountain Weather: Observation and Modeling*; Elsevier Ltd.: Amsterdam, The Netherlands, 2015. [CrossRef]
21. Hilgendorf, E.R.; Johnson, R.H. A Study of the Evolution of Mesoscale Convective Systems Using WSR-88D Data. *Weather. Forecast.* **1998**, *13*, 437–452. [CrossRef]
22. Houze, R.A., Jr. Mesoscale Convective Systems. *Rev. Geophys.* **2004**, *42*, 1–43. [CrossRef]
23. Houze, R.A. Orographic Control of Precipitation: What Are We Learning from MAP. *MAP Newsl.* **2001**, *14*, 3–5.
24. Orographic Effects on Precipitating Clouds. *Rev. Geophys.* **2012**, *50*, 1–47.
25. Houze, R.A., Jr.; Rasmussen, K.L.; Zuluaga, M.D.; Brodzik, S.R. The variable nature of convection in the tropics and subtropics: A legacy of 16 years of the Tropical Rainfall Measuring Mission satellite. *Rev. Geophys.* **2015**, *53*, 994–1021. [CrossRef] [PubMed]
26. Hoyos, C.D.; Ceballos, L.I.; Pérez-Carrasquilla, J.S.; Sepúlveda, J.; López-Zapata, S.M.; Zuluaga, M.D.; Velásquez, N.; Herrera-Mejía, L.; Hernández, O.; Guzmán-Echavarría, G.; et al. Meteorological conditions leading to the 2015 Salgar flash flood: Lessons for vulnerable regions in tropical complex terrain. *Nat. Hazards Earth Syst. Sci.* **2019**, *19*, 2635–2665. [CrossRef]
27. Jaramillo, L.; Poveda, G.; Mejía, J.F. Mesoscale Convective Systems and Other Precipitation Features over the Tropical Americas and Surrounding Seas as Seen by TRMM. *Int. J. Climatol.* **2017**, *37*, 380–397. [CrossRef]
28. Kingsmill, D.E.; Persson, P.O.G.; Haimov, S.; Shupe, M.D. Mountain Waves and Orographic Precipitation in a Northern Colorado Winter Storm. *Q. J. R. Meteorol. Soc.* **2015**, *142*, 836–853. [CrossRef]
29. Li, Y.; Zhang, G.; Doviak, R.J.; Lei, L.; Cao, Q. A New Approach to Detect Ground Clutter Mixed with Weather Signals. *IEEE Trans. Geosci. Remote Sens.* **2012**, *51*, 2373–2387. [CrossRef]
30. Lilly, D.K. A Severe Downslope Windstorm and Aircraft Turbulence Event Induced by a Mountain Wave. *J. Atmos. Sci.* **1978**, *35*, 59–77. [CrossRef]

31. Lin, Y.-L.; Chiao, S.; Wang, T.-A.; Kaplan, M.L.; Weglarz, R.P. Some Common Ingredients for Heavy Orographic Rainfall. *Weather Forecast.* **2001**, *16*, 633–660. [CrossRef]
32. Llasat, M.C.; Marcos, R.; Turco, M.; Gilabert, J.; Llasat-Botija, M. Trends in Flash Flood Events versus Convective Precipitation in the Mediterranean Region: The Case of Catalonia. *J. Hydrol.* **2016**, *541*, 24–37. [CrossRef]
33. Maddox, R.A.; Hoxit, L.R.; Chappell, C.F.; Caracena, F. Comparison of Meteorological Aspects of the Big Thompson and Rapid City Flash Floods. *Mon. Weather Rev.* **1978**, *106*, 375–389. [CrossRef]
34. Mohr, K.I.; Slayback, D.; Yager, K. Characteristics of Precipitation Features and Annual Rainfall during the TRMM Era in the Central Andes. *J. Clim.* **2014**, *27*, 3982–4001. [CrossRef]
35. Posada-Marín, J.A.; Rendón, A.M.; Salazar, J.F.; Mejía, J.F.; Villegas, J.C. WRF Downscaling Improves ERA-Interim Representation of Precipitation around a Tropical Andean Valley during El Niño: Implications for GCM-Scale Simulation of Precipitation over Complex Terrain. *Clim. Dyn.* **2019**, *52*, 3609–3629. [CrossRef]
36. Poveda, G.; Jaramillo, A.; Gil, M.M.; Quiceno, N.; Mantilla, R. Seasonally in ENSO-Related Precipitation, River Discharges, Soil Moisture, and Vegetation Index in Colombia. *Water Resour. Res.* **2001**, *37*, 2169–2178. [CrossRef]
37. Poveda, G.; Alvaro, J.; Marta, M.G.; Natalia, Q.; Ricardo, M. Water Resources Research—2001—Poveda—Seasonally in ENSO-related Precipitation River Discharges Soil Moisture and.Pdf. *Water Resour. Res.* **2001**, *37*, 20169–20178.
38. Germán, P.; Oscar, M.; Paula, A.; Juan, Á.; Paola, A.; Hernán, M.; Luis, S.; Vladimir, T.; Sara, V. Influencia Del ENSO, Oscilación Madden-Julian, Ondas Del Este, Huracanes y Fases de La Luna En El Ciclo Diurno de Precipitación En Los Andes Tropicales de Colombia. *Meteorol. Colomb.* **2002**, *5*, 3–12.
39. La Hidroclimatología De Colombia: Una Síntesis Desde La Escala Inter-Decadal Hasta La Escala Diurna. *Rev. Acad. Colomb. Cienc.* **2004**, *XXVIII*, 201–222.
40. Purnell, D.J.; Kirshbaum, D.J. Synoptic Control over Orographic Precipitation Distributions during the Olympics Mountains Experiment (OLYMPEX). *Mon. Weather Rev.* **2018**, *146*, 1023–1044. [CrossRef]
41. Rasmussen, K.L.; Chaplin, M.M.; Zuluaga, M.; Houze, R.A., Jr. Contribution of Extreme Convective Storms to Rainfall in South America. *J. Hydrometeorol.* **2015**, *17*, 353–367. [CrossRef]
42. Rasmussen, K.L.; Houze, R.A. Convective Initiation near the Andes in Subtropical South America. *Mon. Weather Rev.* **2016**, *144*, 2351–2374. [CrossRef]
43. Rasmussen, K.L.; Robert, A.H. A Flash-Flooding Storm at the Steep Edge of High Terrain. *Bull. Am. Meteorol. Soc.* **2012**, *93*, 1713–1724. [CrossRef]
44. Rasmussen, K.L.; Zuluaga, M.D.; Houze, R.A., Jr. Severe Convection and Lightning in Subtropical South America. *Geophys. Res. Lett.* **2014**, *41*, 7359–7366. [CrossRef]
45. Roe, G.H. Orographic Precipitation. *Annu. Rev. Earth Planet. Sci.* **2005**, *33*, 645–671. [CrossRef]
46. Romatschke, U.; Houze, R.A. Characteristics of Precipitating Convective Systems in the South Asian Monsoon. *J. Hydrometeorol.* **2001**, *12*, 3–26. [CrossRef]
47. Romatschke, U.; Medina, S.; Houze, R.A. Regional, Seasonal, and Diurnal Variations of Extreme Convection in the South Asian Region. *J. Clim.* **2010**, *23*, 419–439. [CrossRef]
48. Roy Bhowmik, S.K.; Roy, S.S.; Kundu, P.K. Analysis of Large-Scale Conditions Associated with Convection over the Indian Monson Region. *Int. J. Climatol.* **2008**, *4*, 797–821. [CrossRef]
49. Rozalis, S.; Morin, E.; Yair, Y.; Price, C. Flash Flood Prediction Using an Uncalibrated Hydrological Model and Radar Rainfall Data in a Mediterranean Watershed under Changing Hydrological Conditions. *J. Hydrol.* **2010**, *394*, 245–255. [CrossRef]
50. Ruiz-Villanueva, V.; Borga, M.; Zoccatelli, D.; Marchi, L.; Gaume, E.; Ehret, U. Extreme Flood Response to Short-Duration Convective Rainfall in South-West Germany. *Hydrol. Earth Syst. Sci.* **2012**, *16*, 1543–1559. [CrossRef]
51. Characterisation of Flash Floods in Small Ungauged Mountain Basins of Central Spain Using an Integrated Approach. *Catena* **2013**, *110*, 32–43. [CrossRef]
52. Schumacher, R.; Johnson, R.H. Organization and Environmental Properties of Extreme-Rain-Producing Mesoscale Convective Systems. *Mon. Weather Rev.* **2005**, *133*, 961–976. [CrossRef]
53. Chumchean, S.; Seed, A.; Sharma, A. An Operational Approach for Classifying Storms in Real-Time Radar Rainfall Estimation. *J. Hydrol.* **2008**, *363*, 1–17. [CrossRef]
54. Smith, R.B.; Barstad, I.; Bonneau, L. Orographic Precipitation and Oregon's Climate Transition. *J. Atmos. Sci.* **2005**, *62*, 177–191. [CrossRef]
55. Steiner, M.; Houze, R.A.; Yuter, S. Climatological Characterization of Three-Dimensional Storm Structure from Operational Radar and Rain Gauge Data. *J. Appl. Meteorol.* **1995**, *34*, 1978–2007. [CrossRef]
56. Urrea, V.; Andrés, O.; Oscar, M. Seasonality of Rainfall in Colombia. *Water Resour. Res.* **2019**, *55*, 4149–4162. [CrossRef]
57. Valenzuela, R.A.; David, E.K. Terrain-Trapped Airflows and Orographic Rainfall along the Coast of Northern California. Part II: Horizontal and Vertical Structures Observed by a Scanning Doppler Radar. *Mon. Weather Rev.* **2018**, *146*, 2381–2402. [CrossRef]
58. Vargas, G.; Hernández, Y.; Pabón, J.D. La Niña Event 2010–2011: Hydroclimatic Effects and Socioeconomic Impacts in Colombia. In *Climate Change, Extreme Events and Disaster Risk Reduction: Towards Sustainable Development Goals*; Mal, S., Singh, R.B., Huggel, C., Eds.; Springer International Publishing: Cham, Switerland, 2018; pp. 217–232. [CrossRef]
59. Velásquez, N.; Hoyos, C.D.; Vélez, J.I.; Zapata, E. Reconstructing the 2015 Salgar Flash Flood Using Radar Retrievals and a Conceptual Modeling Framework in an Ungauged Basin. *Hydrol. Earth Syst. Sci.* **2020**, *24*, 1367–1392. [CrossRef]

60. Vivekanandan, J.; Yates, D.N.; Brandes, E.A. The Influence of Terrain on Rainfall Estimates from Radar Reflectivity and Specific Propagation Phase Observations. *J. Atmos. Ocean. Technol.* **1999**, *16*, 837–845. [CrossRef]
61. Wang, Y.; Tang, L.; Chang, P.-L.; Tang, Y.-S. Separation of Convective and Stratiform Precipitation Using Polarimetric Radar Data with a Support Vector Machine Method. *Atmos. Meas. Tech.* **2021**, *14*, 185–197. [CrossRef]
62. Xie, S.-P.; Xu, H.; Saji, N.H.; Wang, Y.; Liu, W.T. Role of Narrow Mountains in Large-Scale Organization of Asian Monsoon Convection. *J. Clim.* **2006**, *19*, 3420–3429. [CrossRef]
63. Yu, C.-K.; Smull, B.F. Airborne Doppler Observations of a Landfalling Cold Front Upstream of Steep Coastal Orography. *Mon. Weather Rev.* **2000**, *128*, 1577–1603. [CrossRef]
64. Zuluaga, M.; Houze, R.A. Extreme Convection of the Near-Equatorial Americas, Africa, and Adjoining Oceans as Seen by TRMM. *Mon. Weather Rev.* **2015**, *143*, 298–316. [CrossRef]

MDPI
St. Alban-Anlage 66
4052 Basel
Switzerland
Tel. +41 61 683 77 34
Fax +41 61 302 89 18
www.mdpi.com

Hydrology Editorial Office
E-mail: hydrology@mdpi.com
www.mdpi.com/journal/hydrology